The Urban Book Series

Editorial Board

Fatemeh Farnaz Arefian, University of Newcastle, Singapore, Singapore; Silk Cities & Bartlett Development Planning Unit, UCL, London, UK
Michael Batty, Centre for Advanced Spatial Analysis, UCL, London, UK
Simin Davoudi, Planning & Landscape Department GURU, Newcastle University, Newcastle, UK
Geoffrey DeVerteuil, School of Planning and Geography, Cardiff University, Cardiff, UK
Andrew Kirby, New College, Arizona State University, Phoenix, AZ, USA
Karl Kropf, Department of Planning, Headington Campus, Oxford Brookes University, Oxford, UK
Karen Lucas, Institute for Transport Studies, University of Leeds, Leeds, UK
Marco Maretto, DICATeA, Department of Civil and Environmental Engineering, University of Parma, Parma, Italy
Fabian Neuhaus, Faculty of Environmental Design, University of Calgary, Calgary, AB, Canada
Vitor Manuel Aráujo de Oliveira, Porto University, Porto, Portugal
Christopher Silver, College of Design, University of Florida, Gainesville, FL, USA
Giuseppe Strappa, Facoltà di Architettura, Sapienza University of Rome, Rome, Roma, Italy
Igor Vojnovic, Department of Geography, Michigan State University, East Lansing, MI, USA
Jeremy W. R. Whitehand, Earth & Environmental Sciences, University of Birmingham, Birmingham, UK

The Urban Book Series is a resource for urban studies and geography research worldwide. It provides a unique and innovative resource for the latest developments in the field, nurturing a comprehensive and encompassing publication venue for urban studies, urban geography, planning and regional development.

The series publishes peer-reviewed volumes related to urbanization, sustainability, urban environments, sustainable urbanism, governance, globalization, urban and sustainable development, spatial and area studies, urban management, transport systems, urban infrastructure, urban dynamics, green cities and urban landscapes. It also invites research which documents urbanization processes and urban dynamics on a national, regional and local level, welcoming case studies, as well as comparative and applied research.

The series will appeal to urbanists, geographers, planners, engineers, architects, policy makers, and to all of those interested in a wide-ranging overview of contemporary urban studies and innovations in the field. It accepts monographs, edited volumes and textbooks.

Now Indexed by Scopus!

More information about this series at http://www.springer.com/series/14773

Andrzej Kowalczyk · Marta Derek
Editors

Gastronomy and Urban Space

Changes and Challenges in Geographical Perspective

Springer

Editors
Andrzej Kowalczyk
Department of Tourism Geography
and Recreation, Faculty of Geography
and Regional Studies
University of Warsaw
Warsaw, Poland

Marta Derek
Department of Tourism Geography
and Recreation, Faculty of Geography
and Regional Studies
University of Warsaw
Warsaw, Poland

ISSN 2365-757X   ISSN 2365-7588 (electronic)
The Urban Book Series
ISBN 978-3-030-34494-8   ISBN 978-3-030-34492-4 (eBook)
https://doi.org/10.1007/978-3-030-34492-4

© Springer Nature Switzerland AG 2020
This work is subject to copyright. All rights are reserved by the Publisher, whether the whole or part of the material is concerned, specifically the rights of translation, reprinting, reuse of illustrations, recitation, broadcasting, reproduction on microfilms or in any other physical way, and transmission or information storage and retrieval, electronic adaptation, computer software, or by similar or dissimilar methodology now known or hereafter developed.
The use of general descriptive names, registered names, trademarks, service marks, etc. in this publication does not imply, even in the absence of a specific statement, that such names are exempt from the relevant protective laws and regulations and therefore free for general use.
The publisher, the authors and the editors are safe to assume that the advice and information in this book are believed to be true and accurate at the date of publication. Neither the publisher nor the authors or the editors give a warranty, expressed or implied, with respect to the material contained herein or for any errors or omissions that may have been made. The publisher remains neutral with regard to jurisdictional claims in published maps and institutional affiliations.

This Springer imprint is published by the registered company Springer Nature Switzerland AG
The registered company address is: Gewerbestrasse 11, 6330 Cham, Switzerland

Foreword

This volume focuses on the gastronomy–urban space nexus within a positivist framework. It adopts a holistic, geographical perspective to highlight the changes that shape this relationship and the challenges it faces. In emphasising the nature, role, locations and dynamism of restaurants, bars, street food stalls, food trucks and cafés, the theoretical perspectives of the work are complemented by and comprehensively illustrated with case study exemplification from around the world. The excellent range and quality of maps, diagrams and photographs amply support this holistic approach to articulating relationships between gastronomy and urban space in a wide range of contexts. As such, in the final part of Preface of the work, highlighting 'what the book is not about' rather underplays the actual breadth and significance of the text.

Although the empirical bias of the book is towards Central and Eastern Europe, drawing considerable exemplification from the 'region', this volume holds much wider pertinence in the conceptual frameworks, issues, problems, analyses and arguments it addresses and employs. Further, its case studies emphasise the ways in which globalisation meets and interacts with local neighbourhoods, traditions, habits and people through gastronomy produced, supplied and consumed within the spaces of towns and cities.

As such, this volume provides a very useful addition to the existing literature on both urban gastronomy and urban structure. Although the micro-geography of Central and Eastern Europe is now much better researched (at least as expressed in English-language publications),[1] there is still clearly much to be gained from focussing on the relationships between detailed local case studies and wider analytical/theoretical perspectives, as this volume well illustrates.

As the book highlights, the supply of gastronomy in urban places can be shaped by a wide range of factors, both geographical and non-geographical. An understanding of the historical evolution of such provision and its political context can contribute

---

[1] In no way is this meant to underestimate the enormous amount of high-quality research previously (and currently) undertaken that, unfortunately, has not reached English-language audiences.

important analytical insights. For example, human migration, which may have been economically or politically driven, can both help to establish 'ethnic' culinary traditions, cuisines and tastes in particular urban places as a consequence of the settled residence of such migrants. The development and provision of particular cuisines may also come about as a response to the diverse gastronomic demands of those *temporary* migrants—international tourists and business travellers—whose mobilities may generate demand for both 'authentic' and inauthentic cuisines. Further, when such travellers return home, they may seek to reproduce their 'foreign' gastronomic experiences by visiting and/or creating demand for restaurants and other food providers that serve the cuisine of the country or region they have visited. In this way patterns of tourism and travel may influence the gastronomy–urban structure nexus at both destination and source areas of such travellers.

This volume is, however, driven by a focus on supply rather than demand, and it highlights the increasing functional importance of the supply of gastronomy for many towns, cities and regions, as expressed, for example, in the growing use of gastronomy in place marketing, branding, image projection, visitor and investor promotion.

The spatial and analytical perspectives of urban geography form a central strand of this book, both in terms of understanding the internal structure of urban places and of critically evaluating the range of explicit and implicit linkages that may obtain between towns and cities, regionally, nationally and internationally. Thus, for example, in highlighting the 'challenges' wrought by change, this volume draws attention to changing trends in the location of restaurants and bars within urban space, the growth of 'street food' and its associated consumption in open spaces, and the dynamics of telecommunications and physical transport to produce and deliver food to the home or workplace. In the latter case, India provides a clear example of how enormous numbers of people can be involved in such a (currently) labour-intensive service provision, and how crucially important the logistics of multiple spatial interactions within urban areas can be for the production and supply of everyday gastronomic needs.

The structure and location of gastronomic supply is also required to respond to changing patterns of demand that can be driven by spatial, social and cultural factors, as with the slow food 'movement' and the 'healthy lifestyle' reaction to the epidemic of obesity that appears to bedevil societies that over-indulge in fast food. Vegetarianism and veganism, present in many traditional societies for both cultural and practical reasons, are further important dynamic factors exerting influences on the production and supply of gastronomy in 'Western' urban societies.

Overall, therefore, the editors of this volume have produced a welcome addition to the literature, which should appeal to a wide range of readers. In the ways in which it raises many questions addressing the gastronomy–urban structure nexus, the text provides a useful springboard from which further studies and research can, and no doubt will, be developed.

Maidens, Scotland, UK Derek Hall
derekhall@seabankscotland.co.uk

Preface: The Essence of Gastronomy in Urban Context

According to the title of Flint's (2014) recent article, 'Restaurants really can determine the fate of cities and neighbourhoods'. This seems like a good place to start: while the statement in the title may be making a bold—some would say *exaggerated*—claim, it highlights the increasing importance of gastronomy for many cities. This is supported by the growing use of gastronomy in territorial marketing and tourist promotion.

This book, *Gastronomy and urban space. Changes and challenges in a geographical perspective*, discusses two fundamental issues: *gastronomy* and *urban spaces* (mostly cities for our purposes), and puts them into a synthesising geographical perspective. This means that the spatial perspective of urban geography, together with its 'real analytical value' (Pacione 2005, 24), will be central in this book.

Gastronomy and Urban Space

Gastronomy

Gastronomy is complex, ambiguous and difficult to define. The word itself made its official debut in France in 1801, when Jacques Berchoux titled his playful poem '*La Gastronomie, ou l'homme des champs à table*' (Gastronomy, or the peasant at the table). This was about 40 years after the appearance of the first modern restaurants in Paris, the then gastronomic capital of the world (Rambourg 2013). In 1825 another breakthrough French book, '*La physiologie du gout*' (The physiology of taste), was published. Its author, Jean-Anthelme Brillat-Savarin, saw gastronomy as a science, which is a part of natural history, physics, chemistry, cookery, business and political economy (Brillat-Savarin 2009). He states that 'the subject matter of gastronomy is whatever can be eaten; its direct end is the conservation of individuals; and its means of execution are the culture which produces, the commerce which exchanges, the industry which prepares, and the experience which invents means to dispose of everything to the best advantage' (Brillat-Savarin 2009). Ten years later, in 1835, 'gastronomy' was included in the dictionary of the French Academy as "the art of good eating' (Scarpato 2002)—the meaning which still can be found in many dictionaries (e.g. The Merriam-Webster Dictionary, The Oxford Dictionary, Collins Dictionary).

The original definition of gastronomy has extremely broadened in recent years. Originally for the nobility, gastronomy is now perceived as a widespread, common socio-cultural practise, or—as Richards (2002) suggests—a global industry. The broad spectrum of its definitions can be reduced into two main, very general categories. First, gastronomy as enjoyment of food and drink; second, gastronomy as a far-reaching discipline that encompasses everything into which food enters (Scarpato 2002). Although many authors underline its cultural meaning ('gastronomy is a social practise of a table; a culture'; Cohen and Csergo 2012), gastronomy, like any other social behaviour, is both culturally and spatially differentiated. Its immaterial character is reflected in, among others, material places like restaurants (Csergo 2008). More than this, though, the concept of gastronomy nowadays also involves several forms of economic activity, which makes it an important part of the services sector. In this book, we will, therefore, use the term *gastronomy* to refer to '…any business activity located in the geographical space consisting of offering customers food and drink in exchange for money'.

Urban Space

Unlike gastronomy, *urban space* is not defined in any dictionary, including The Dictionary of Human Geography (Gregory et al. 2009). This can lead to some difficulties, because the implication is that the term is understood implicitly. The term appears, however, in many books and scientific journals. Examples include articles on urban space issues in prestigious periodicals (e.g., Harms 2016; He et al. 2017; Mallick 2018; Mele 2017) and academic books (e.g. Eckardt 2008). The term can also be found on websites used by professionals (New mobility and urban space: how can cities adapt? 2018). In all of these publications, *urban space* is understood both in a strictly spatial sense (three dimensional) and in a multidimensional sense. The first perspective covers issues such as spatial distribution and urban planning, and is typically used in traditional geography and urbanism (this problem will be described in Chap. 2.2.3). The second perspective, however, extends the concept to encompass non-material aspects, such as ideas, and lifestyle (this perspective will be described in more detail in Chap. 2.4).

Geographical Perspective

"If history is about time, geography is about space (…). If time is the dimension (…) of succession, then space is the dimension of (…) simultaneity [and] (…) multiplicity. (…) Whereas historians concentrate on the temporal dimension, how things change over time; what geographers concentrate on is the way in which things are arranged—we would often say geographycally (...), 'over space'" (Massey 2013, pp. 2–3).

This citation from the interview given to Social Science Bites by Doreen Massey, one of the leading British geographers, is essential to the understanding of a 'geographical perspective'. Although the academic subject of geography is changing as the planet changes (Dorling and Lee 2016), the distinctive character of the discipline remains the same. It is explained by the 'spatial lens' through which geographers look at, and analyse, different phenomena which they observe around them. In one of her works, when she states that cities, both individually and in the relations between them, are spatial phenomena, Massey (2000) underlines that she means something more than just that cities are spaces and they exist in space. She argues that "by thinking spatially, and through taking account of changes in the way in which space is imagined and organised, we may contribute something specific, and useful, to an understanding of what Serageldin called in his article 'the coming urban century'" (Massey 2000, p. 155).

Urban geography, which we refer to in this book, maybe defined as 'the geographical study of urban spaces and urban ways of being' (Gregory et al. 2009, p. 784). Pacione (2005) notes that one of the principal characteristics of geographical analysis of the city is the centrality of a spatial perspective, which distinguishes urban geography from cognate areas of urban study such as urban economics, urban sociology or urban politics. He differentiates two basic approaches to urban geography. The first refers to the spatial distribution of towns and cities and the linkages between them. The second refers to the internal structure of urban places (Pacione 2005).

By adopting the geographical perspective, this book contributes to the literature on gastronomy and food with an emphasis on its spatial aspects. We mainly use Pacione's second approach to urban geography, i.e. the internal structure of urban places, but some references to the first one (spatial distribution and linkages between towns and cities) are also made. Following Massey's (2013, p. 3) understanding of space as a 'cut through the myriad of stories in which we are all living at any one moment', we will attempt to show gastronomy not only as a cultural, economical, tourist or sociological phenomenon, but, above all, as a spatial (i.e. geographical) phenomenon.

Changes and Challenges

Changes

From the very beginning of civilisation, gastronomy has been an important part of human activity. In traditional hunter-gatherer communities and pastoral and agrarian civilisations, however, it was limited to the eating of food. While food was sometimes eaten communally (e.g. during tribal and religious ceremonies), most of people's food needs were catered for at home. Only the development of trade and increases in spatial activity made eating outside the home a more common activity. This meant that gastronomy—still a socio-cultural phenomenon—began to become an integral part of the economy.

In the oldest civilisations (Indus Valley, Mesopotamian, Egyptian as well as Persian or Chinese), gastronomy may well have been one of the services provided in cities and towns. Research indicates that it was only in ancient Rome, however, that it became an important element of economic activity. Several authors have pointed out that eating out of the home was not only typical for merchants, sailors, pilgrims and other travellers, but was also becoming part of the lifestyle of Rome's residents.

The Roman conquests made the largest cities including (and especially) Rome, ethnically and culturally diverse. Rome itself became divided into various quarters that were inhabited by a range of ethnic groups including Greeks, Jews, Libyans and Germans (Montanari 2013, p. 166). Each brought their own culinary traditions as well as their own places to eat out. A similar situation occurred later in Byzantium (later renamed Constantinople, then Istanbul), where there were separate quarters inhabited by Genoese, Venetians and Slavs.

In the Middle Ages, this diversification became more widespread as a result of internal (rural to urban) migration and international migration. In countries that were expanding their territory, some of the districts in the larger cities were inhabited by ethnic and religious minorities. For instance, in the cities on the Iberian Peninsula they were Jewish and Moorish, in Central European cities they were Jewish, and in the eastern part of the former Kingdom of Poland, they were Jewish, Armenian and Ruthenian. After the end of the Middle Ages, this diversification through migration continued to spread further afield, especially during the Reformation (sixteenth century), after the end of the Thirty Years War and during the Counter-Reformation (seventeenth century) in Germanic countries, Austria, the Netherlands and France.

Then, in the seventeenth and eighteenth centuries, in Western Europe a new factor appeared that significantly influenced the development of gastronomy in cities. Colonial conquests and the almost continuous military conflicts between Austria, Venice and Poland on the one hand and the Ottoman Empire on the other led to the appearance of new food products and new ways of processing and serving them. Cafés and tea rooms began to appear in many European cities—including London, Paris, Venice and Madrid in Western Europe and Vienna, Berlin and Warsaw in Central Europe—fulfilling the same roles as they did in their homelands of the Middle East and China.

In the nineteenth and twentieth centuries, the importance of gastronomy continued to grow in European and North American cities. At the same time, they were exporting ways of preparing, and serving meals to the countries of Asia, South America and Africa. The prime example here is the spread of fast food offered by international corporations such as McDonald's, Pizza Hut, Burger King and KFC. At the same time restaurants and bars in European and North American cites were also offering food from other continents (mainly Asia and South America). The net effect was that at the start of the twenty-first century the gastronomic offerings in cities and towns across the world, on the one hand, reflected their own culinary culture and traditions, but on the other became an increasingly standardised combination of different cultures and traditions, as a result of globalisation. Some exceptions of these include, for example, countries of Central and Eastern Europe, where a market economy started to develop only in 1990. Before that an expansion of many international corporations was not possible, as well as a phenomenon of eating out, as such, was very limited. In some countries (Poland, for example) the catering sector was almost non-existent before the 1980s. Under communism, the vast majority of Poles ate at home. Therefore the political shift, consisting of the transition to a market economy, brought tremendous changes to gastronomy in cities of this part of the world. Changes in the context of urban space can be observed not only in the longer historical term, but in the short term (e.g. years, or even months). This is due to the fact that the restaurant industry is very flexible and dynamic. There are several reasons for this, including the seasonality of food products (particularly fruit and vegetables) served in restaurants or as street food, food prices, lifestyle choices, current fashions and advertising. The most important reason, though, is the high probability of failure in the restaurant business. Research in the United States has shown that 57.22% of franchise restaurants and 61.36% of independent ones closed down within 3 years of opening (Parsa et al. 2005, p. 309). This is particularly true of those cities where there is strong competition.

Challenges

As a part of the economy and as a lifestyle issue, gastronomy is subject to constant change. Some create major challenges for the people and institutions involved in gastronomy.

The first major challenge is the result of changing trends in the location of restaurants and bars in the urban space. In cities of all sizes, a significant proportion of eating establishments are situated in shopping centres (e.g. food courts). When shopping centres are on the outskirts of the city, the threat from these eating establishments to existing restaurants in the city centre is relatively small. Shopping malls are also being built in the inner city, however, which is likely to lead to existing restaurants and bars losing customers as a result of competition. Until recently, gastronomy in shopping centres was associated with fast-food bars, but more upmarket restaurants are appearing in the newer generation of malls. Also, fans of ethnic cuisines, who previously had to travel to areas frequented by immigrants (e.g. Chinatown, Little Italy) can now visit a restaurant with Thai, Indonesian or Brazilian cuisine in a shopping gallery.

Another part of this trend is the increasing number of meals that are consumed in open spaces. In large cities and towns, culinary events are increasingly being organised in parks, at rivers, in post-industrial areas, etc. This poses a threat to the current model of restaurant business operation. The third strand of this trend is the renaissance of street food in European and, to a lesser extent, North American cities, which is a serious challenge to traditional restaurants and bars that are situated in a fixed place in the urban space. Entrepreneurs offering street food or operating from food trucks do not have to pay property rental charges, which means that their costs are lower. Although the street food often scrolls in this book, a much more detailed picture of this phenomenon can be found in the works of Tinker (1997), who presents examples of seven cities in developing countries, and in Parham (2015).

The second major challenge for the traditional restaurant industry is changes to the way that dining services are provided. The move to the information society, plus the increasing use of different forms of transportation mean that the consumer no longer has to visit a restaurant in order to eat a meal. People in the larger cities are increasingly ordering restaurant meals online, or by phone. These are prepared in the restaurant's kitchen and then delivered to homes (or offices) by a specialised company. Although this trend may not threaten restaurants and bars themselves, it does force them to change the way they operate in order to remain competitive. For example, by reducing spending on waiters and increasing expenditure on employing someone to take orders online and by telephone.

Urban Gastronomy in the Context of Urban Space

The central ideas presented in the book are encapsulated in the model shown in Fig. 1. Urban gastronomy lies at its centre. This is immediately surrounded by the *physical* urban space and *socio-economic* urban space, based on our understanding of urban space as multidimensional, rather than just three dimensional. Urban space is in turn associated with the next layer, global space. This means that what is happening in a given city is subject to external influences that are regional, national and international.

What happens in the urban space is influenced by a range of geographical and non-geographical factors. They consist of numerous sub-agents that share feedback. These relations occur both within and between geographical and non-geographic factors. The spatial development of a city, for example, can be partly attributed to the dissemination of new technologies in the field of logistics (including transportation), while the city's functions are influenced by economic, socio-cultural and historical factors.

The most important geographical factors are the city's location; size; rank in the wider settlement system; functions; physical expansion; internal structure; external relations, etc.

Here, when we talk about location, we mean location in the climatic zone, its topography, its position relative to the sea, river or lake, etc. Size is (normally) related to its rank in the wider settlement structure. In addition, both factors influence the city's functions. A large city can dominate the state's settlement structure, and other large (but slightly smaller) cities. This is clearly seen in the United States, Canada, Germany, Italy or Spain. In the case of the United States, the eating function in urban space is equally important in Chicago and Los Angeles, the same is true for Hamburg and Munich in Germany. Another factor that is very

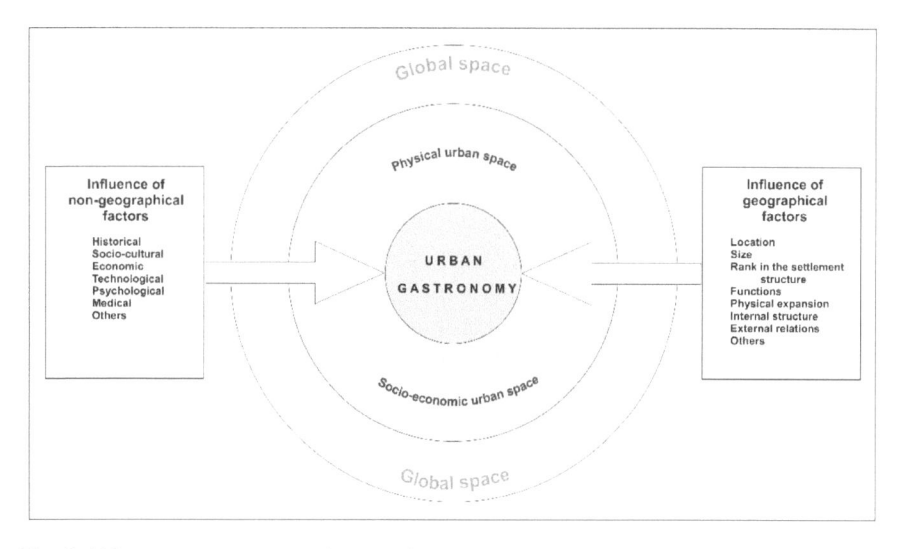

Fig. 1 Urban gastronomy as an element of urban space

important for gastronomy in the context of the urban space is the relationship between the size of the city and its internal structure. In big cities, where the distance from the downtown area to the outskirts is several kilometres, eating establishments are generally more dispersed. In medium-sized cities and towns, however, gastronomy is more concentrated in the city centre. Another important factor is the city's external relations: within the region; within the country; and internationally. In this case, the most important element is tourism, but other aspects include the importance of the city as a place for organising international fairs or its function as a capital city.

There are also several non-geographical factors that are important. In addition to historical factors, the most important are socio-cultural (e.g. the presence of immigrants) and economic factors. Together, these two aspects determine the quality of life of a city's residents: the better the quality of life, the greater the demand for gastronomy. Technological factors include both the presence of an efficient public transport system (which facilitates the spatial accessibility of restaurants and bars), as well as innovations in the field of technology (web based and otherwise). Psychological factors can be important too. For example, the development of a Slow Food philosophy leads to an increase in demand for restaurants offering local and regional cuisine. Psychological factors sometimes coincide with medical factors. Criticism of fast-food gastronomy in recent years has led to a fall in the number of outlets in the vicinity of schools, for example, but led to a rise in the number of vegetarian and vegan restaurants. Other reasons are indirectly related to the factors that have already been mentioned. As an example, halal restaurants are increasingly apparent in European cities in recent years. This expansion results from the inflow of refugees from the Middle East and North

Africa and immigration from some parts of South Asia, as well as doctrinal changes taking place in the Muslim religion. It could be said that the development of halal gastronomy in Europe is caused by both political and cultural factors.

Note that while economics, cultural anthropology, sociology, medical sciences and psychology are all important factors that are linked to gastronomy in urban space, they are largely beyond the scope of this book. Here, we focus our attention on the most important geographical factors. In other words, the central issue we are addressing is the provision of dining services in a spatial context, i.e., the location of eating establishments and their functions. When we use the term *eating establishments* (and synonyms such as dining facilities) we are referring to places in the public space in which people can satisfy their food and drink needs. In other words, restaurants, bars, cafés, canteens, tea houses, ice-cream shops, food trucks, street food stalls, etc.

The most important eating establishments are restaurants. There are many different taxonomies of restaurant types. For instance, Muller and Woods (1994, p. 28) proposed five categories (a) quick service, (b) mid-scale, (c) moderate up-scale, (d) business dining and (e) multi-unit. In contrast, Walker (2014, quoted by Longart 2015, p. 3) only has three: (a) fine dining, (b) casual dining and (c) quick-service restaurants. The first category, sometimes referred to as gourmet restaurants, have sophisticated service, an elegant ambience and spectacular views or location (Mehta and Maniam 2002). These are divided into sub-categories: themed restaurants; steak houses; ethnic; and celebrity owned (Longart 2015, p. 3). The casual dining category of restaurants is less formal and targeted more towards family dining. They include some restaurant chains such as TGI Friday's and Hard Rock Café. The quick-service (or fast food) category offers a limited menu, moderate prices and fast service. Many belong to international restaurant chains as McDonald's, Burger King, KFC and Subway. Yen et al. (2015, p. 7) use similar criteria, but have five categories: (a) fast-food restaurants (orders are made at the counter with fast service food partly prepared off-site), (b) fast-casual (the cost is higher than the fast-food type, and food is mostly prepared onsite), (c) casual dining (with table service and possibly bar service, cost is moderate), (d) family-style (table service, cost moderate, but no bar service) and (e) fine dining (full service provided by waiters, fine décor and bar services).

One of the factors that is mentioned in each of the taxonomies identified above is the price range. This can be used as the primary criterion for defining three categories of restaurants: (a) budget; (b) mid-price; and (c) upmarket.

The budget segment is often referred to as fast-food restaurants, but the category also includes quick-service restaurants. The term *fast food* emphasises not only the technology, but also the speed of service. Operations in this segment of the market range from small-scale street vendors with food carts or food trucks to global companies such as Pizza Hut, Taco Bell, KFC, Burger King and McDonald's. In fast-food restaurants, food is ordered from a front counter or, in some cases, using an electronic terminal. Customers usually carry their own food from the counter to a table of their choosing and afterwards dispose of any waste from their trays. Some fast- food restaurants offer drive-through and takeaway services. Fast-food restaurants were initially situated along the main streets in the outer zone of a city, at road junctions, near railway stations, etc. In motels, they were oriented towards clients who were travelling between cities. They eventually appeared in almost all parts of the city, including on main streets in city centres such as *Avenue* des *Champs-Élysées in Paris, Fifth Avenue in New York City and Oxford Street in London, and in historical areas such as Piazza di Spagna in Rome and* Rambla dels Caputxins *in Barcelona.*

The mid-price segment is made up of casual dining restaurants. These serve moderately priced food in a casual atmosphere and usually provide table service. They often have a full bar with separate bar staff, and a larger selection of beers than wines. Generally, they are independently owned and operated, but sometimes they are a part of a restaurant chain. Casual dining restaurants include what are sometimes described as family restaurants. The food is often brought to the table on platters and consumers then serve themselves. This type of restaurant is close to fast-casual restaurants. They are mainly part of a chain, but unlike typical fast-food gastronomy, food is prepared at the restaurant. Fast-casual restaurants generally do not offer full table service, but many offer non-disposable plates and cutlery. The quality of food and prices are higher than those of a conventional fast-food restaurant but lower than casual dining. The example of chain with this type of restaurant in Poland is Sphinx Restauracje. These restaurants, popular among the young, families with children and tourists, offer mainly Oriental-style dishes (shoarma or shawarma, kofta, etc.), but some dishes are close to traditional Polish cuisine. Since many of these chain restaurants are located in central parts of larger cities, they are also often visited by foreign tourists. Casual dining restaurants are located in different parts of the cities and towns. Some are in the central zone, e.g. in historic areas of Italian cities and towns, but others are situated in the outer city. They may also be situated in new-build housing estates, in commercial centres, along main streets or at crossroads, at sport and recreation zones, etc. Casual dining restaurants are used by residents as well as visitors, particularly tourists travelling with children.

The upmarket segment covers luxury (or fine dining) restaurants. Luxury restaurants offer a full service (table service, bar service, etc.) and often have established rules of dining and behaviour which visitors are expected to follow, such as a particular dress code. The décor, furniture and tableware are all high quality. Luxury restaurants, similar to five-star hotels, boutique shops and banks, are willing to pay higher rents to secure properties in the most desirable locations (in the inner city). The core zone of the city is considered very valuable because it is traditionally the most accessible location for both city residents and visitors. As a result, they are willing to pay very high property rental prices. They maximise the potential of their site by charging high prices, but still need a large number of customers to generate enough revenue to cover their (high) costs.

These different types of eating establishments are central to the discussions in this book, and we will refer to them throughout. Other terms related to gastronomy in urban space are defined as, and when necessary.

What This Book Is not About and What It Is

First of all, 'Gastronomy and urban space: changes and challenges in a geographical perspective' is not about gastronomic tourism, nor about tourism *per se*. There is an extensive literature devoted to tourism and gastronomy, food tourism, etc., so those who want to look at gastronomy only as a tourism phenomenon should go no further and look elsewhere (e.g. Hjalager and Richards 2002; Hall et al. 2003; Cohen and Avieli 2004; Mak et al. 2012, Molz 2007), including newly published 'The Routledge Handbook on Gastronomic Tourism' (Dixit 2019). The starting point for this book is not the demand side (tourism or tourists), but the supply side, i.e. gastronomical facilities in urban space. In fact, although gastronomy is an important part of tourist studies, and catering facilities are among the most frequently used tourism services (Ashworth and Tunbridge 1990; Jansen-Verbeke 1986; Law 1993; Page and Hall 2003), many such facilities are used primarily by local residents. In cities, where it is difficult to distinguish between residents and tourists, many establishments reflect the needs of the local community and tourism only complements the existing usage pattern (Page 1995). So although tourists are an important group of consumers of urban eateries, they are not the only, and very often not even dominating, group. In this book, we treat them in the same way as any other group of urban gastronomy users.

Second, this book is not a book about the importance of food for urban population. In other words—the issue of catering for city residents is not the purpose of this book, unless it is about meals prepared in restaurants or bars and purchased by consumers. The question of food in the context of urban space was recently described in detail in Susan Parham's book 'Food and urbanism: the convivial city and sustainable future' (Parham 2015). For this reason, problems related to the methods of obtaining food by households, which are then processed and consumed at home, are omitted here. Although the issue of food market (market halls, farmers' markets, etc.) is described in some parts of the book, food marketplaces are understood only in the context of places to eat (due to the concentration of restaurants and bars and other dining facilities). The question of purchasing agricultural products (fresh or pre-processed) to further processing is not the issue of this book. One exception is the so-called 'Breakfast Market' phenomenon in Warsaw (Poland) (see Chap. 16.4), but the main role of these events is to provide eating services as such (offering ready meals and drinks).

Third, this book is not a book about gastronomy understood as a food in the context of culture. The authors are familiar with the concept of habitus by Pierre Bourdieu (which is often cited in the works on the culinary behaviour of city residents), but in our opinion sociologists and anthropologists use this theory much better than geographers. So, the book is not about gastronomy understood as culinary preferences and behaviours of people living within the cities. But this problem, which is obviously very important, appear in various parts of the book (e.g. in Chap. 12 on Polish students' tourist experiences and their food preferences for ethnic cuisines). However, the authors believe that the problem of food culture and culinary behaviour is within the competences of anthropologists and sociologists, not geographers.

Moreover, this book has no aspirations to deal with the importance of restaurants, pubs or cafés in social life (interpersonal relations) and their impact on the development of people's personality. There is no deeper reflection on the concept of conviviality and 'convivial cities', which are thoroughly analysed in the works of Parham (2008, pp. 16–19, 2012, pp. 10–15). We assume that this is a very important problem but it is a subject of research of humanities and social sciences (especially by anthropology, sociology and psychology), not geography. It is also not a book about the influence of restaurants and cafés on art and philosophy. The authors are aware of the importance of Café de Guerbois for painting (impressionism) at the end of the nineteenth century, and Les Deux Magots and Café de Flore (existentialism) in Paris for philosophy and literature in the twentieth century. They are aware of the role of Viennese cafés for the development of Art Nouveau style (Wiener secession or Jugendstil), bars in Buenos Aires for the popularity of tango dance, etc. But these are very important issues for historians, literary scholars, etc., not for socio-economic geographers.

Fourth, the book does not address issues related to the production of food in urban areas. Also, this problem is described in detail in 'Food and urbanism: the convivial city and sustainable future'. However, omitting the problem of food production in the city does not mean that it is not identified by the authors of this book. Especially in the context of urban space, agriculture within the city urban gardening is a very important problem, especially since it occurs in most of the metropolitan areas in Asia, Africa and Latin America, as well as in European and North American urban agglomerations. This results both from the history of urban development (territorial expansion and absorption of rural areas), as well as new trends in urban politics and planning. The concept of a sustainable city, or the idea of a slow city, makes urban gardening visible in more and more cities. The authors of this book do not ignore this very important question, but believe that it is the domain of urban studies and planning, landscape architecture and agricultural sciences, and, to a lesser extent, socio-economic geography.

In reference to the above, it has to be clearly stated that this book does not deal with the impact of food consumed by city residents on their health.[2] It is not about 'obesogenic environment' (Parham 2008, pp. 11 and 53–54), as the phenomenon of obesity, so characteristic for the population of contemporary cities (and not only), is sometimes called. Although some actions undertaken in order to face this problem have spatial dimensions (e.g. location of fast-food bars) this problem is mainly approached by sciences such as nutrition, medicine and education.

Fifth, this is not a book about the functioning of the restaurant sector. Issues related to restaurant operating costs (e.g. labour costs), turnover, profits, work organisation, labour productivity, as well as restaurant business management, etc.,

---

[2]According to Schram-Bijkerk et al. (2015, p. 20) urban gardening has a positive impact for people because they eat more fresh fruits and vegetables than before.

are not taken into account here. These topics are in the sphere of interest of economics and management sciences, not geography.

Sixth, it is not about the design of restaurants and bars, their interior design, equipment in the kitchen, furniture, etc., too. In the authors' opinion, these problems are subject to research in architecture, design and technical sciences.

Finally, the history of gastronomy plays a secondary role here. Although many fragments of this book contain historical themes, they are only a background for describing the current situation in which gastronomy is located in the urban space.

The same applies to the links between gastronomy and politics, although the authors are familiar with the case of a meeting at the Bürgerbräukeller in Munich in November 1923 with his repercussions in the form of the so-called Beer Hall Putsch.

This is not to say that all these issues are not important or of interest, or that they are not mentioned in the book. In fact, we do refer to many of them in the context of gastronomy as a geographical phenomenon. They simply are not the purpose or the main focus of this book.

So what is the book 'Gastronomy and urban space: changes and challenges in a geographical perspective' about? The answer to this question will be short. **This book is primarily a presentation of problems related to the location (geographical location) of restaurants, bars, cafés, food trucks, stalls with ready-made food, etc., in urban areas**. We understand gastronomy as eating places in urban public space, and try to analyse it accordingly. As mentioned above, the spatial aspects of eating facilities are in the centre of our approach.

Finally, the authors must present an important declaration. In contemporary human geography, there are various research approaches that largely result from different philosophical approaches. The most important are positivism, humanism, Marxism, poststructuralism, realism, feminism (Aitken and Valentine 2006, p. 4, fig. 1.2) and behavioural approach (Golledge 2006, p. 75). Because this book is dedicated primarily to students and people interested in the problem of gastronomy in a city, not for academicians, its authors adopted mainly the positivist and realistic approaches.

Book Structure

This book consists of three main parts. The first deals with theoretical issues, while the second and third cover the changes and challenges referred to in the title, respectively.

The first, *Theoretical aspects of the studies on gastronomy in urban space* comprises three sections. The first, introductory chapter, *Relations between gastronomy and the city* provides a brief overview of the history of gastronomy in cities and towns. The second chapter, *Theories and concepts related to the gastronomy in urban space* covers concepts related to spatial issues and social science. The final section, *Dimensions of gastronomy in contemporary cities* analyses the restaurant industry in the context of economic, socio-demographic and socio-cultural processes. It considers the place of gastronomy in urban spatial policy (urban planning and revitalisation and gentrification) and in national marketing, paying particular attention to the city's identity and promotional activities.

The second part, *Changes*, comprises 10 chapters. It starts by Introduction, and then looks at the culinary attraction of cities, both old and new, based on existing ranking systems and explores the subjective nature of rankings and their impact on culinary tourism. The gastronomic attraction of cities also depends on the local cuisine, as described in *Traditional and regional cuisine in urban space*, which looks at the cases of Warsaw and Kraków (Poland), Madrid (Spain) and Prague (Czech Republic). The next sections focus on the distribution of eating establishments across the various zones of a city. The impact of suburbanisation on the distribution and function of dining outlets outside the core area of Warsaw is explored in some detail. This is followed by a look at the particular types of gastronomy in medium-sized cities, which are increasingly attracting more culinary tourists (foodies). The growth in ethnic cuisine and its importance is explored next, using examples from Poland, Spain, the Czech Republic and the Netherlands. This phenomenon results partly from the inflow of immigrants, but partly from the growing demand, from culinary tourists, for exotic dishes. This issue is also touched upon in the next section from the point of view of Polish students in Warsaw. This part of the book concludes with an exploration of the restaurant industry's supply chain in Warsaw, from the viewpoints of both a company that supplies the restaurants and bars, and the owners of one of Warsaw's local restaurant chains.

The final part, *Challenges*, is made up of five chapters. It looks at how changes in gastronomy are creating new challenges for the industry. After the Introduction, it begins by looking at new trends in gastronomy in urban space, before focusing on the growing importance of eating establishments situated in shopping galleries, often in food courts. There has also been a rise in gastronomy in green (e.g. parks and recreation areas) and blue spaces (e.g. rivers and lakes) in cities. Consumers are increasingly adopting a greener (ecologically aware or pro-ecological) outlook and local authorities are looking at how they can revitalise and reactivate areas that have not previously provided gastronomic functions. The recent renaissance in street food is also discussed, along with the rise of the food truck, and the particular challenges they present to the traditional restaurant industry. The book closes by looking at food home delivery services. This is an area of gastronomy that has old roots (in pizza delivery, for example) but with advances in technology and changes in the available delivery options is becoming increasingly important and is beginning to pose another threat to the older bricks and mortar restaurants.

Although this book draws upon many different examples to illustrate both changes and challenges in the gastronomy–urban space nexus, most are drawn from Central and Eastern Europe. This is not only a relatively under-researched region in terms of the changes that shape urban space, it is also a region where these changes are particularly clear because, in many cases, they are very recent and dynamic. Fast and dynamic geopolitical changes after 1990 are reflected in fast and dynamic changes in the urban space. This makes Central and Eastern Europe an especially interesting laboratory for exploring the development of gastronomy in urban space.

Concluding Remarks

All of the book's contributors have worked in urban geography or the geography of tourism for several years. Some have also published texts on gastronomy in the Polish literature and beyond. *Gastronomy and urban space. Changes and challenges in a geographical perspective*, presents the results of our research supplemented by examples from the literature. Although we are all geographers, we have referenced the economics, sociology and cultural anthropology literature where it helps to support our arguments and discussion. Naturally, though, the geographical point of view is predominant in what is, after all, a book about the spatial aspects of urban gastronomy.

Warsaw, Poland
Andrzej Kowalczyk
akowalczyk@uw.edu.pl

Marta Derek
m.derek@uw.edu.pl

References

Ashworth GJ, Tunbridge JE (1990) The tourist-historic city. Belhaven, London
Brillat-Savarin JA (2009) The physiology of taste, or mediations on transcendental gastronomy (trans: MFK Fisher). Everyman's Library, New York, London, Toronto
Cohen E, Avieli N (2004) Food in tourism: attraction and impediment. Ann Tourism Res 31(4):755–778
Cohen É, Csergo J (2012) L'artification du culinaire. Sociétés & Représentations 2(34):7–11
Csergo J (2008) Patrimoine et pot-au-feu, Liberation, 10 October, https://next.liberation.fr/vous/2008/10/10/patrimoine-et-pot-au-feu_114153. Accessed on 20 Apr 2019
Dixit SK (ed) (2019) The routledge handbook of gastronomic tourism. Routledge, Oxon, New York
Dorling D, Lee C (2016) Ideas in profile: geography. series: Profile books Ltd., London
Eckardt F (ed) (2008) Media and urban space: understanding, investigating and approaching mediacity. Frank & Timme GmbH Verlag, Berlin
Flint A (2014) Restaurants really can determine the fate of cities and neighborhoods. CityLab, July 22, 2014, http://www.citylab.com/politics/2014/07/how-food-drives-cities-resurgences/374806. Accessed 20 Jan 2017
Gastronomy, https://www.collinsdictionary.com/dictionary/english/gastronomy. Accessed on 15 Apr 2018
Gastronomy, https://www.merriam-webster.com/dictionary/gastronomy. Accessed on 15 Apr 2018
Gastronomy, https://en.oxforddictionaries.com/definition/gastronomy. Accessed on 15 Apr 2018

Gregory D, Johnston R, Pratt G, Watts MJ, Whatmore S (eds) (2009) The dictionary of Human Geography, 5th edn. Blackwell Publishing Ltd, Chichester

Hall CM, Sharples L, Mitchell R, Macionis N Cambourne B (eds) (2003) Food tourism around the world: development, management and markets. Butterworth-Heinemann, Oxford–Burlington

Harms E (2016) Urban space and exclusion in Asia. Ann Rev Antrophology 45 (October), Supplement, pp 45–61

He S, Kong L, Lin GCS (2017) Interpreting china's new urban spaces: state, market, and society in action. Urban Geogr 38(5):635–642

Hjalager AM, Richards G (eds) (2002) Tourism and gastronomy. Routledge, London, UK

Jansen-Verbeke, M (1986) Inner-city tourism: Resources, tourists and promoters. Ann Tourism Res 13(1):79–100

Law CM (1993) Urban tourism: attracting visitors to large cities. Mansell, London, New York

Longart P (2015) Consumer decision making in restaurant selection (volume I), April 2015. Faculty of Design, Media and Management, Buckinghamshire New University, Coventry University. Ph.D. dissertation. http://collections.crest.ac.uk/9388/1/1._Volume_I_Pedros_final_thesis_170715.pdf. Accessed on 01 June 2017

Mak AHN, Lumbers M, Eves A (2012) Globalisation and food consumption in tourism. Ann Tourism Res 39(1):171–196

Molz JG (2007) Eating difference: the cosmopolitan mobilities of culinary tourism. Space Cult 1 (10):77–93

Mallick A (2018) Urban space and (the limits of) middle class hegemony in Pakistan. Urban Geogr 39(7):1113–1120. https://www.academia.edu/35956553/Urban_space_and_the_limits_of_middle_class_hegemony_in_Pakistan. Accessed on 10 June 2018

Massey D (2013) *Doreen Massey on Space*, a podcast by Social Science Bites. February 1, 2013. https://www.socialsciencespace.com/2013/02/podcastdoreen-massey-on-space/. Accessed on 29 Apr 2019

Massey D. (2000) On space and the city. In: Massey D, Allen J, Pile S (eds) City worlds. Routledge, London and New York, pp 151–170

Mehta SS, Maniam B (2002) Marketing determinants of customers' attitudes towards selecting a restaurant. Acad Mark Stud J 6(1) (January):27–44 https://www.thefreelibrary.com/Marketing+determinants+of+customers%27+attitudes+towards+selecting+a…-a0166751793. Accessed on 01 June 2017

Mele Ch (2017) Spatial order through the family: the regulation of urban space in Singapore. Urban Geogr 38(7):1084–1108

Montanari M (2013) Romans, Barbarians, Christians: the dawn of European food culture. In: Flandrin JL, Montanari M (eds) Food: a culinary history from antiquity to the present. (trans: Sonnenfeld A). Columbia University Press, New York, Chichester, pp 165–167

Muller CC, Woods RH (1994) An expected restaurant typology. Cornell Hotel Restaurant Adm Q 35(3) (June):27–37

New mobility and urban space: how can cities adapt? Workshop 24/05/2018. UITP Advancing Public Transport, London, UK, http://www.uitp.org/events/new-mobility-and-urban-space-how-can-cities-adapt. Accessed on 30 May 2018

Pacione M (2005) Urban geography: a global perspective. Routledge, London and New York

Page S (1995) Urban tourism. Routledge, London

Page S, Hall CM (2003) Managing urban tourism. Pearson Education, Essex

Parham S (2008) Exploring London's food quarters: urban design and social process in three food-centred spaces: a thesis submitted to the department of sociology of the london school of economics. London, December 2008

Parham S (2012) Market place: food quarters, design and urban renewal in london. Cambridge Scholars Publishing, Newcastle upon Tyne

Parham S (2015) Food and urbanism: the convivial city and sustainable future. Bloomsbury Publishing, London, New Delhi, New York, Sydney

Parsa HG, Self JT, Njite D, King T (2005) Why restaurants fail. Cornell Hotel Restaurant Adm Q 46(3) (August):304–322
Rambourg P (2013) Des métiers de bouche à la naissance du restaurant: l'affirmation de paris comme capitale gastronomique (XVIe-XVIIIe siècle). In: Belleguic, T, Turcot, L (eds) Les Histoires de Paris (XVIe-XVIIIe siècle). Editions Hermann, Paris, pp 185–197
Richards G (2002) Gastronomy: an essential ingredient in tourism production and consumption? In: Hjalager AM, Richards G (eds) Tourism and gastronomy. Routledge, London, UK, pp 3–20
Scarpato R (2002) Gastronomy as a tourist product: the perspective of gastronomy studies. In: Hjalager AM, Richards G (eds) Tourism and gastronomy. Routledge, London, UK, pp 51–70
Royal geographical society (2019). https://www.rgs.org/geography/what-is-geography/. Accessed on 26 Apr 2019
Tinker I (1997) Street foods: urban food and employment in developing countries. Oxford University Press, New York, Oxford
Yen B, Burke M, Tseng C, Ghafoor MMT, Mulley C, Moutou C (2015) Do restaurant precincts need more parking? differences in business perceptions and customer travel behaviour in Brisbane, Queensland, Australia. Paper presented at the 37th Australasian Transport Research Forum (ATRF), Sydney, Australia, 29th September, 2nd October, 2015. http://atrf.info/papers/2015/files/ATRF2015_Resubmission_84.pdf. Accessed on 01 June 2017

Contents

Part I Theoretical Aspects of Studies on Gastronomy in Urban Space

1 **Relations Between Gastronomy and the City** 3
Andrzej Kowalczyk and Marta Derek

2 **Theories and Concepts Related to Gastronomy in Urban Space** ... 53
Andrzej Kowalczyk

3 **Dimensions of Gastronomy in Contemporary Cities** 91
Andrzej Kowalczyk

Part II Changes

4 **Changes in Gastronomy and Urban Space—Introduction to Part II** 121
Marta Derek

5 **Culinary Attractiveness of a City—Old and New Destinations** 125
Andrzej Kowalczyk

6 **Traditional and Regional Cuisine in Urban Space** 135
Andrzej Kowalczyk and Magdalena Kubal-Czerwińska

7 **Changes in the Distribution of Gastronomic Services in the City Centre** 159
Marta Derek, Andrzej Kowalczyk, Konstantin A. Kholodilin,
Leonid Limonov, Magdalena Kubal-Czerwińska and Dana Fialová

8 **Restaurants and Bars in the Outer City** 183
Andrzej Kowalczyk and Aleksandra Korpysz

9 **Suburbanisation and Gastronomic Services on the Outskirts of Warsaw (Poland): Piaseczno** 199
Andrzej Kowalczyk and Sylwia Seremak

10	**Eating Establishments in Smaller Cities and Towns in Poland (on Selected Examples)** .... Małgorzata Durydiwka	209
11	**Ethnic Cuisine in Urban Space** .... Marta Derek	225
12	**Tourist Experience and Change in Culinary Tastes. An Example of Polish Students in Warsaw** .... Katarzyna Gwiazdowska and Andrzej Kowalczyk	239
13	**The Food Supply Chain in the Restaurant Industry: A Case Study from Warsaw, Poland** .... Andrzej Kowalczyk	247

Part III Challenges

14	**New Trends in Gastronomy in the Context of the Urban Space— Introduction to Part III** .... Andrzej Kowalczyk	263
15	**New Gastronomic Hotspots in the Urban Space. Food Courts in Poland** .... Andrzej Kowalczyk, Magdalena Kubal-Czerwińska, Katarzyna Duda-Gromada and Aleksandra Korpysz	273
16	**Challenges to Urban Gastronomy: Green and Blue Spaces** .... Sylwia Kulczyk, Monika Kordowska and Katarzyna Duda-Gromada	295
17	**Street Food and Food Trucks: Old and New Trends in Urban Gastronomy** .... Andrzej Kowalczyk and Magdalena Kubal-Czerwińska	309
18	**Home Delivery Services** .... Andrzej Kowalczyk and Piotr Kociszewski	329

Conclusion 343
Marta Derek

List of Figures

Fig. 1.1	The distribution of taverns, bars, etc. in Pompeii (Roman Empire). Source: Andrzej Kowalczyk after Ellis (2008, Fig. 42)	5
Fig. 1.2	Schematic map of a historical city or town showing the concentration of gastronomy facilities: A—surrounding the marketplace (Aa) and main street (Ab); B—merchant port; C—religious sanctuary; D—city gate	7
Fig. 1.3	Indoor markets (market halls) in European cities. Sukiennice market hall on the main market square in Kraków (Poland, 1556–1559, renovated in 1875–1879) (**a**). The entrance to the Mercat de Sant Josep de la Boqueria on Las Ramblas in Barcelona (Spain, 1840–1853) (**b**). Source: Anna Kowalczyk 2006, 2015	7
Fig. 1.4	The Naghshe Jahan Traditional Restaurant on Naghshe Jahan Square in the old part of Isfahan (Iran). The entrance to the former *caravanserai* (**a**) and courtyard with restaurant on the 2nd floor (**b**). For the wealthy residents of Isfahan, this restaurant is a "place" in the meaning of Yi-Fu Tuan's (1975, 1979) and Edward Relph's (1976) concepts (more on these concepts in Sect. 2.2.4).	8
Fig. 1.5	King Abbasi's *caravanserai* at Neyshabur (Iran). Courtyard and entrance (**a**). Interior of the teahouse (**b**). Traditional stone containers for cooking and serving the Iranian stew called *dizi*, *abgoosht* or *ābgusht* (**c**)	9
Fig. 1.6	**a** Restaurante Sobrino de Botin in Madrid (Spain). This restaurant, which opened in 1725, is among the oldest restaurants in Europe. **b** Antico Caffè Greco in Rome (Italy). This coffee house was established on Via dei Condotti 86 in 1760 and is one of the oldest coffee houses in Europe. It is a one of the must-see tourist sites in Rome. Its many guests have included historic figures such as the poets Goethe,	

	Byron and Keats, composers Liszt, Wagner and Mendelssohn, writers Stendhal and Andersen, the playwright Ibsen and last, but not least, the adventurer Casanova. In October 2016 a cup of espresso cost about €7. These restaurants are other examples of "places" (Tuan 1975). Source: Anna Kowalczyk 2015....................	11
Fig. 1.7	Market halls in the central part of Barcelona (Spain). **a** Mercat de Santa Caterina (built in 1848) in the area of Santa Caterina. **b** Mercat del Born (built in 1878) in the area of La Ribera. **c** Mercat d'Hostafrancs (built in 1888) in the area of Hostafrancs. The revitalisation of the Mercat de Santa Caterina was completed in 2005 and it now hosts many restaurants and bars. The Mercat del Born was restored in 2013. Source: Anna Kowalczyk 2015.................	12
Fig. 1.8	The main roofed market halls (*mercat* in Catalan language) in Barcelona (Spain)................................	13
Fig. 1.9	Market halls in Berlin (Germany), Lisbon (Portugal) and Wrocław (Poland). **a** The Arminiushalle in Berlin-Moabit (built in 1891). **b** Mercado de Santa Clara (built in 1877) in the area of São Vicente (on the eastern edge of Alfama). Since 2011, the building has been home to the Centre of Culinary Arts (Centro das Artes Culinárias). **c** The Hala Targowa at Piaskowa Street in Wrocław (formerly Markthalle Nummer 1) was built in 1906–1908. It was established in order to organise street trading in the city centre. In addition to being a typical market place, it is now a hotspot for foodies. Source: Anna Kowalczyk 2014, 2017.............	16
Fig. 1.10	Distribution of registered activity in the gastronomy sector in Madrid (Spain) in 2017. **a** Per 100 inhabitants and **b** per hectare showing differences between downtown, inner city and outer city areas................................	19
Fig. 1.11	Registered activity in the gastronomy sector in Prague (Czech Republic) in 2016 (by district) per 100 inhabitants....	20
Fig. 1.12	Registered activity in the gastronomy sector in Prague (Czech Republic) in 2016 (by district) per hectare..........	21
Fig. 1.13	Main concentrations of gastronomy in the inner city........	22
Fig. 1.14	**a** Eating establishments in the historical centre of Prague (Czech Republic). On the right-hand side is the restaurant U Zlaté studně, which is located in the Dům U Zlaté studně (built in seventeenth century in Renaissance style) at the junction of Seminářská and Karlová Streets. In the house to the right there is a fast-food bar, the café Clementin (green building) and the Hotel at Black Star restaurant. On the left side is one of the terraces of the café Clementin. **b** The	

	Restaurante Lope de Vega in the old part of Madrid. This restaurant is situated on the Calle de Lope de Vega 37, not far from the museum Casa Museo Lope de Vega (Calle de Cervantes 11), where the famous Spanish playwright, poet and novelist, Félix Lope de Vega y Carpio (1562–1635) lived. Both cases refer to the concept of "place". Source: Anna Kowalczyk 2009, 2015 24
Fig. 1.15	**a** Nerudova Street in the Malá Straná quarter of the old centre of Prague (Czech Republic). This street has the biggest concentration of restaurants and bars on The Royal Way (Královská cesta). **b** Roof terrace restaurant in one of the medieval houses on Nerudova Street (viewed from Hradčany castle). Source: Anna Kowalczyk 2006................... 24
Fig. 1.16	Eating establishments in the historical parts of Italian cities. **a** *Taverne* with terraces in the old part of Sienna. **b** *Ristoranti* on the banks of the Canal Grande in Venice. In both cases, the dominant groups of visitors are overseas tourists. Source: Anna Kowalczyk 2004 25
Fig. 1.17	Eating establishments in the Taikoo Place business zone in eastern Victoria (Hong Kong Special Administrative Region of the People's Republic of China) in 2017 25
Fig. 1.18	**a** Restaurants and bar offering different cuisines on the Calle Isabel la Católica in central Madrid. On the right is Gran Via, the main street in the commercial centre. **b** The Hard Rock Cafe Rome restaurant on Via Vittorio Veneto, one of the busiest streets in central Rome. The Hard Rock café is a chain of themed restaurants that was founded in London in 1971. By June 2017 it had 181 bars worldwide. **c** Coffee bar operated by Green Cafè Nero, part of the Polish–British joint-venture chain in the Gdański Business Centre, a new office park in the inner city of Warsaw. Buildings in this business park mainly house the headquarters of transnational corporations from Poland and the Central Eastern European region. Source: Anna Kowalczyk 2015................... 26
Fig. 1.19	**a** The main entrance to *Obecní dům* (Municipal House) in Náměstí Republiky square, in the centre of Prague. The building, which opened in 1912, is in the Art Nouveau (*Jugendstil* in German, *Secessiosstil* in Austrian; *Modernismo* in Spanish) architectural style. The ceremonial halls on the first floor house the works of Alphonse Mucha (1860–1939), the painter and decorative artist. The building also includes five restaurants and cafés, and is used as a concert hall (with 1200 seats) and ballroom. **b** Restaurant tables in front of the *Stadttheater* in Baden bei Wien (Austria). **c** Eating

	establishments close to the theatre in Wrocław (Poland). On the right is Café Kulisy in the building of the Teatr Polski theatre. In the background are the bars Mekong and Hot Seven, which are situated under the railway viaduct. Source: Anna Kowalczyk 2009, 2013, 2017	27
Fig. 1.20	**a** Restaurant Parc Güell near the main entrance to Parc Güell, one of the main sightseeing spots in Barcelona. **b** Café Lennon on Plaça de John Lennon in the Gràcia quarter in Barcelona. This café is located some 600–700 m from La Sagrada Família and Parc Güell. Source: Anna Kowalczyk 2015	28
Fig. 1.21	**a** A Cartouxina restaurant at the Rua das Farinhas 7 in the Mouraria quarter of the old part of Lisbon. This small, family owned restaurant offers exotic African meals from the Republic of São Tomé and Príncipe. Although this is a typical example of an ethnic restaurant, it is not located in a part of the city where ethnic minorities live. **b** Mustafa's Gemuese Kebab in the Kreuzberg district of Berlin. This street food bar has gained cult status (note the long queue) among residents and visitors to Berlin. It is situated in an area that has a high percentage of immigrants (in 2010, 10.9% of residents were of Turkish origin). Source: Anna Kowalczyk 2014	29
Fig. 1.22	**a** Restaurant in the building of the former textile factory at Żyrardów (Poland). An example of an eating establishment in a redeveloped industrial area. **b** Food trucks in the revitalised SOHO Factory in Warsaw (Poland). In this post-industrial zone, a few casual dining restaurants have opened, which sometimes host culinary events (e.g. food truck meetings). Source: Anna Kowalczyk 2015	29
Fig. 1.23	**a** Café Kavárna Arco opposite the Masarykovo nádraží (Masaryk's Railway Station) in Prague (Czech Republic). This coffee house was one of the places that was often visited by the author Franz Kafka. **b** Café bar inside Atocha Railway Station (Estación de Atocha) in Madrid. Source: Anna Kowalczyk 2009, 2015	30
Fig. 1.24	**a** Biergart'l im Stadtpark in Vienna (Austria). This 170-seat restaurant and café is located in Stadtpark, the first public park in Vienna, which opened in 1862. At weekends, visitors can attend open-air concerts. **b** Restaurante Eleven in the Jardim Amália Rodrigues public park in the centre of Lisbon (Portugal). **c** The dining facility for visitors to Wrocław's (Poland) zoological garden. On the right is a signpost showing directions to attractions. In the background is the bar. Source: Anna Kowalczyk 2013, 2014, 2017	31

Fig. 1.25	**a** Bars along the Donaukanal in the centre of Vienna (Austria). **b** Restaurants on the bank of the Vltava River in the old part of Prague (Czech Republic). The main difference between these two examples is that bars and cafés on the bank of the Donaukanal are visited mostly by residents of Vienna, whereas in Prague high-priced restaurants are more tourist-oriented. Source: Anna Kowalczyk 2013, 2009	31
Fig. 1.26	The main gastronomy zones in the outer city	32
Fig. 1.27	**a** Fast-food bars at the urban rapid transit station (U-Bahn Linie U 6) Alterlaa in south-west Vienna (Austria). **b** The fast-food bar Konnopke's Imbiß in the district of Berlin-Prenzlauer Berg (Berlin, Schönhauser Allee 44B). The Konnopke's Imbiß was established in 1930 close to the Eberswalder Straße metro station on Linie-U 2. It is one of the most iconic places to eat for *Currywurst* lovers in Berlin. Source: Anna Kowalczyk 2013, 2014	33
Fig. 1.28	Distribution of coffee chain shops in the Służewiec area (Mokotów borough) of Warsaw (Poland) July 2017	34
Fig. 1.29	**a** The view from the car park of the Kentucky Fried Chicken fast-food restaurant between The Grove Mall and the MegaCentre in Windhoek (Namibia). This restaurant lies in the mall area in the Klein Kuppe district in the southern part of Windhoek. This district is inhabited mainly by the middle classes. **b** The entrance to another fast-food restaurant in The Grove Mall. This restaurant belongs to Wimpy SA (created in 1934 in the United States as Wimpy Grills), a multinational chain of fast-food restaurants headquartered in Johannesburg (South Africa) since 2007. Source: Anna Kowalczyk 2016 . . .	35
Fig. 1.30	Gastronomy in suburban areas	36
Fig. 1.31	**a** Entrance to the Gordon Ramsay Plane Food restaurant in Terminal 5 at Heathrow Airport in London (United Kingdom). **b** The food court, My City Helsinki, at Helsinki-Vantaa International Airport. The difference between these two dining facilities is clearly visible—the first is isolated from its surroundings, whereas the second is in the public space. According to Marc Augé (1995), both cases can be considered as "non-places" (read more in the Sect. 2.2.4). Source: Anna Kowalczyk 2011, 2016.	36
Fig. 1.32	**a** Gaststätte Neu Venedig restaurant in the Köpenick district of Berlin (Germany). This is an example of a restaurant located in a former suburb of the city. It is closed on Tuesdays and can be entered either from the street, or directly from a boat. **b** The Grinzing-Heuriger Rudolfshof restaurant	

	in Vienna (Austria). This is an example of a *Heuriger*: a tavern that is typical of the outer zone of Vienna and its surroundings. Source: Anna Kowalczyk 2014, 2013.	38
Fig. 1.33	The main zones of concentration of gastronomy at seaside tourist cities	41
Fig. 1.34	Examples of eating establishments in historical buildings situated in tourist cities. **a** Funchal, Madeira (Portugal). **b** Arrecife, Lanzarote (Spain). **c** Playa del Carmen, Lanzarote (Spain). **d** Lindos, Rhodes (Greece). Where a restaurant lies inside an historic building, this becomes an additional attraction for visitors. Source: Anna Kowalczyk 2009, 2010, 2011, 2016	42
Fig. 1.35	In coastal tourist cities and towns eating outlets are an integral part of fishing ports and marinas. **a** Restaurant at the quay in Lošinj, Lošinj Island (Croatia). **b** Restaurants are mainly occupied tourists visiting Alanya (Turkey) harbour in the evenings. Source: Anna Kowalczyk 2016.	43
Fig. 1.36	Tourist-oriented amenities on the seafront promenade in Agadir (Morocco) showing global fast-food restaurants. **a** Pizza Hut and McDonald's restaurants. **b** KFC. The location of these restaurants is an example of a gastronomic cluster. Source: Anna Kowalczyk 2017.	43
Fig. 1.37	Eating establishments in typical seaside tourist resorts. **a** Hotel restaurant terrace at Varadero (Cuba). **b** Cafés near the beach at Swakopmund (Namibia). In both **a** and **b**, the beautiful sea view is an additional attraction for visitors. **c** Restaurant with views of the sea and tourist boats in Alanya (Turkey). **d** Exterior of the restaurant at Bitez (Turkey) with views of the beach and sea (on the right) and the walkway that leads along the coast (on the left). Source: Katarzyna Kowalczyk 2015 (a), Anna Kowalczyk 2016 (b), 2010 (c and d)	44
Fig. 1.38	**a** The esplanade at Morro Jable, Fuerteventura (Spain). **b** The English pub at Costa Teguise, Lanzarote (Spain). The sign for The Cutty Sark and the (English) flag beneath it are intended to attract tourists from the United Kingdom (particularly England). Source: Anna Kowalczyk 2010.	44
Fig. 1.39	Alanya (Turkey)—an example of restaurant location along the river. Such localization is another factor in attracting guests. Source: Anna Kowalczyk 2015.	45
Fig. 1.40	Gastronomy facilities do not have to be dining places. **a** Winery in the outskirts of Corfu town, Corfu Island (Greece). In addition to learning about wine technology and tasting, tourists can also buy wine. **b** Food court in Puerto de	

	la Cruz, Tenerife (Spain), which is considered to be a good place for walking and taking photographs. Source: Anna Kowalczyk 2010, 2014...............................	46
Fig. 1.41	**a** Restaurants and bars on the ground floor of the 4-star Grand Hotel Palace Bellevue in Opatija, Croatia. **b** The Hole in One pub in Funchal, Madeira. This Irish-style pub is located on Estrada Monumental opposite the 5-star CS Madeira Atlantic Resort & Sea SPA (which became the VidaMar Resort Hotel Madeira in July 2017). This was the world's first Crowne Plaza Resort (an American brand established in 1983 by the Holiday Inn Corporation) when it opened in 1999. Source: Anna Kowalczyk 2012, 2016...........................	46
Fig. 1.42	Location of eating establishments in Bitez, Muğla Province (Turkey) in 2013. The linear arrangement along the beach, which is clearly visible, is typical of coastal tourist towns....	47
Fig. 1.43	Examples of the touristscape linked to gastronomy. **a**, **b** Maspalomas, Gran Canaria (Spain). **c** Lignano Sabbiadoro, Friuli-Venezia Giulia (Italy). In both cases photographs were taken during the winter season when restaurants were closed and tourists were absent. Source: Anna Kowalczyk 2015 (Maspalomas), 2004 (Lignano Sabbiadoro)...............	48
Fig. 2.1	Catchment areas of eating establishments in a hypothetical city according to central-place theory (Berry 1963).........	56
Fig. 2.2	Distribution of restaurants, bars, etc. in the urban space according to bid-rent theory...........................	57
Fig. 2.3	Application of models of the city using Burgess' concentric zone theory (**a**), Hoyt's sector theory (**b**) and Harris and Ullman's multiple nuclei theory (**c**) to explain the concentration of eating establishments. Source: Adapted from 5 urban models. https://www.slideshare.net/ecumene/5-urban-models. Accessed 10 June 2017...................	58
Fig. 2.4	Vance's urban realms model. This model shows that contemporary cities have several functional centres: the CBD (sometimes in crisis), the new downtown, suburban downtowns and commercial centres. There are eating establishments in all of these locations. Source: Adapted from Chen (2015).......................................	60
Fig. 2.5	Kearsley's model of the city. This model is much more detailed than that of Vance. It includes districts with ethnic minorities, gentrification zones (where expensive and fashionable restaurants are located) and post-industrial areas. It also shows the development of gastronomy in suburban areas, notably new towns and inter-urban commercial sites. Source: Adapted from 5 urban models.	

xxxii   List of Figures

	https://www.slideshare.net/ecumene/5-urban-models. Accessed 10 June 2017..........................	61
Fig. 2.6	Eating establishments are an integral part of revitalisation programmes. **a** Kowloon (Hong Kong Special Administrative Region of the People's Republic of China)—Avenue of Stars in the post-industrial zone created from former railway yards. **b** *Barcelona Centro Commercial Maremagnum* at Port Vell in Barcelona (Spain)—formerly a commercial sea port. Source: Anna Kowalczyk 2009, 2013...................	62
Fig. 2.7	Gastronomy at Victoria and Alfred Waterfront in Cape Town (South Africa). **a** Casual restaurants on Quay Four; **b** V&A Food market in the old power station (1882).............	64
Fig. 2.8	Eating establishments (open to the public) at Ölüdeniz-Belceğiz in Muğla Province (Turkey) in 2011. Although Ölüdeniz-Belceğiz is not a town (it is a part of the Hisarönü—Ovacık—Ölüdeniz-Belceğiz tourist zone) it is a good example of the typical distribution of restaurants, bars, etc. in tourist cities and towns. Dining facilities are concentrated along the seaside and main streets (usually transformed into pedestrian zones) leading to the beach......	65
Fig. 2.9	Gastronomy from the perspectives of the theory of cultural diffusion and the theory of globalization. **a** The food court at Rehoboth Shopping Mall at Rehoboth (Namibia). The restaurant names suggest South African cuisine (biltong—on the left), as well as worldwide culinary trends (coffee shop and Steers—in the centre; pizza outlet on the right). The Steers restaurant chain is South Africa's leading burger brand. **b** McDonald's restaurant in the Defence Housing Authority district of Lahore (Punjab, Pakistan)...................	76
Fig. 2.10	In Europe, perhaps the best example of the theory of the 'city as an entertainment machine' is Barcelona (Spain). **a** A food cluster at the Barcelona Port Olimpic. **b** Banners on the façade of a building in the La Ribera district. Night-time noise is a big problem, not only in Barcelona. Even in postmodern society, the concept of the 'city as an entertainment machine' has its opponents. Source: Anna Kowalczyk 2015..................................	78
Fig. 3.1	Eating establishments at the Bródno housing project in Warsaw in the 1970s. Source: Information about facilities adapted from Bystrzanowski and Dutkowski (1975).........	100
Fig. 3.2	The concentration of restaurants and bars in the area close to the Avenida de Brasil in Madrid (Spain) (map A) and associated noise levels (map B). Source: *Concentración de actividades de ocio nocturno* and *Mapa de los Niveles de*	

List of Figures   xxxiii

	Ruido Actividad de Ocio en la zona de AZCA—Avenida de Brasil from the *Declaración de Zona de Protección Acústica Especial y Plan Zonal Específico de AZCA—Avenida de Brasil*, Boletín Oficial del Ayuntamiento de Madrid, Año CXIX, 26 de enero de 2015, Núm. 7.339, pp. 8–66, https://sede.madrid.es/csvfiles/UnidadesDescentralizadas/UDCBOAM/Contenidos/Boletin/2015/ENERO/Ficheros%20PDF/BOAM_7339_23012015134649429.pdf (Accessed 05 June 2017)..	101
Fig. 3.3	Eating establishments in downtown Madrid (Spain). This map was annexed to the *Declaración de Zona de Protección Acústica Especial y Plan Zonal Específico del Distrito Centro* approved by the city council in September 2012. Source: *Declaración de Zona de Protección Acústica Especial y Plan Zonal Específico del Distrito Centro*, http://www.madrid.es/UnidadWeb/Contenidos/Publicaciones/TemaMedioAmbiente/ZPAECentro/MemoriajustificativaZPAECentrodefinit.pdf (Accessed 05 June 2017)	102
Fig. 3.4	Areas with high noise levels in downtown Madrid (Spain). Highest levels are shown in red. The shaded area on the left is the Royal Palace, the Cathedral and their surroundings. Source: *Normativa de Plan Zonal Específico de la Zona de Protección Acústica Especial del Distrito Centro*, https://sede.madrid.es/UnidadesDescentralizadas/UDCBOAM/UGMedioAmbienteyMovilidad/ElementoBoletin/2012???/Anexo%20II%20Normativa%20ZPAE.pdf (Accessed 05 June 2017)	103
Fig. 3.5	Restaurant terraces. **a** The Old Town in Alanya (Turkey). Here, the terrace is an integral part of the restaurant. **b** A square in Venice (Italy). This is an example of the gastronomic invasion of public urban space. Source: Anna Kowalczyk 2010	103
Fig. 3.6	Three examples of food streets in tourist cities. **a** The main street in Santa Cruz de Tenerifa, Tenerife (Spain). **b** A food alley in Las Palmas, Gran Canaria (Spian). **c** A food alley in Funchal, Madeira (Portugal). In all cases streets are pedestrianised. Source: Anna Kowalczyk 2012, 2014 and 2015	104
Fig. 3.7	**a** The entrance to the Jetty 1905 Restaurant in Swakopmund (Namibia). The remains of the old pier (from 1905) were converted into a fashionable restaurant. **b** The entrance to the most luxurious restaurant in Żyrardów (Poland). This restaurant was opened in a building in the former textile	

	factory dating from about 1860. Its name, Szpularnia, suggests *szpulka* or spool, in English. On the right side is a sculpture of a pregnant woman designed to commemorate the women who worked in the nineteenth-century factory. Source: Anna Andrzej Kowalczyk 2016	105
Fig. 3.8	The *As Docas* zone in the Alcântara district of Lisbon (Portugal). This is an example of a food district (or food court) planned and implemented in a post-industrial area. Source: Information about facilities adapted from *Bares*, http://lourencolx.wix.com/docas#!__bar-pt (Accessed 15 July 2017), *15 aniversario*, http://lourencolx.wix.com/docas#!__mapa-n (Accessed 15 July 2017), *Restaurantes*, http://lourencolx.wix.com/docas#!__rest-pt (Accessed 15 July 2017)...	106
Fig. 3.9	The Alcântara district of Lisbon (Portugal). **a** The *As Docas* project showing restaurants and bars (centre), the marina (left), the railway line the old harbour (foreground) and the Ponte 25 de Abril. **b** Apartments and the chimney of the former factory, a reminder of the district's history. This is one example of the gentrification of post-industrial areas and neighbourhoods inhabited by working classes. Source: Anna Kowalczyk 2014	106
Fig. 3.10	Visitors to Oktoberfest in Munich (Germany) between 1980 and 2017. The fall in the number of visitors in 2001 was the consequence of the World Trade Center disaster in New York. In that year the festival was held between 22 September and 7 October. Source: *Anzahl der Besucher auf dem Oktoberfest in München von 1980 bis 2017 (in Millionen)*, https://de.statista.com/statistik/daten/studie/165511/umfrage/anzahl-der-besucher-auf-dem-oktoberfest-seit-1980 (Accessed 25 Aug 2017)...........................	113
Fig. 3.11	An example of the promotion of gastronomy in a city. This webpage is from the official dining guide for Whistler, British Columbia (Canada). The website promotes the culinary attractions of Whistler in the context of nature and lifestyle rather than urban space. Source: *Whistler dining and restaurant guide*, https://www.whistler.com/dining (Accessed 05 Sept 2017)......................................	113
Fig. 3.12	Booklets about gastronomy in relation to the urban space for Hong Kong Special Administrative Region of the People's Republic of China, the Neukölln district of Berlin (Germany), and Buenos Aires (Argentina). Source: *Hong Kong District Food Guide. January–March 2002*, Dimension Marketing	

	Ltd., Victoria; *Neuköln kulinarisch. Das Beste von Pommes bis Schampus. Genießen und Feiern, Kultur und Kinos, Hotels und Pensionen*, Bezirksamt Neuköln von Berlin, Berlin; *Cafés y Bares Notables de Buenos Aires. Localización*, Secretaría de Cultura, Subsceretaría de Patrimonio Cultural, Direccíon General de Patrimonio – Secretaría de Desarrollo Económico, Subsceretaría de Turismo, Direccíon General de Promoción y Desarrollo Turístico, Buenos Aires...............................	114
Fig. 3.13	An example of the promotion of gastronomy in the city using the web. Webpage of the official dining guide for Edinburgh, Scotland (United Kingdom). The first photo (top left) and the photo on the bottom right promote the culinary attractions of Edinburgh in the context of the urban space. Source: *Food and drink*, http://edinburgh.org/things-to-do/food-and-drink (Accessed 05 Sept 2017).............................	115
Fig. 5.1	Internal and external factors influencing the gastronomic attractiveness of a city or town........................	126
Fig. 6.1	Slow Food in practice. Farmers market in the vicinity of the cathedral in Barcelona (Spain). The inscription above the stalls rads *Productes tradicionals catalans* (**a**). Stall with homemade cheese at the Boxhagener Platz in Berlin-Friedrichshain district (Germany) (**b**). Source: Anna Kowalczyk 2015; 2014...............................	140
Fig. 6.2	Museums related to gastronomic heritage. The Madeira wine museum Blandy's Wine Lodge in Funchal, Madeira (Portugal) (**a**). The chocolate museum, Heindl SchokoMuseum in Vienna (Austria) (**b**). Teaware museum —'Flagstaff House—Museum of Tea Ware' in Victoria (Hong Kong Special Administrative Region of the People's Republic of China) (**c**). Source: Anna Kowalczyk 2009......	141
Fig. 6.3	Distribution of restaurants and bars in Madrid (Spain) with selected regional Spanish cuisine in 2013. **a** Andalusian. **b** Asturian. **c** Basque. **d** Galician. **e** Castilian..............	143
Fig. 6.4	Restaurants offering Spanish regional cuisine in central Madrid (Spain). **a** Established in 1911, the Andalusian restaurant Restaurante Tablao Villa-Rosa at Plaza Santa Ana 15. **b** Established in 1933, the Castilian (*cocina castellana*) restaurant the Casa Paco at Puerta Cerrada 11. **c** The Casa Labra restaurant is famous for *soldaditos de Pavía* (battered cod strips wrapped in red pepper) commemorating the political events in Madrid in 1874 and dissolving of	

	Parliament by hussars under the command of General Pavía. The other example of restaurant offered *cocina castellana* in the centre of Madrid. Source: Anna Kowalczyk 2015........	146
Fig. 6.5	Eating establishments in the Old Town of Warsaw (Poland) in 2018...	147
Fig. 6.6	Three restaurants serving traditional Polish cuisine in the Old Town in Warsaw (Stare Miasto) in 2018. Kamienne Schodki, which existed during the Communist period (**a**). Stolica restaurant (former Senator), which changed the name in 2017. Under the name of the restaurant is added Polish cuisine (Kuchnia polska in Polish) (**b**). New established restaurant Gospoda Kwiaty Polskie (**c**)............................	148
Fig. 6.7	The Zapiecek-Polskie Pierogarnie restaurant chain. Leaflet about restaurants offering traditional Polish cuisine in a part of Warsaw (Poland) visited by domestic and foreign tourists in 2018. The information on the leaflet is somewhat misleading, as it can be understood that restaurants have been operating since 1913. The map is also fraught with a mistake, because in reality the top part of the map is on a scale of about 1:10,000, and the bottom part is roughly from 1:25,000 to 1:30,000 (**a**). The Zapiecek-Polskie Pierogarnie outlet at Freta Street in New Town area of Warsaw (**b**). Source: Leaflet from Andrzej Kowalczyk collection, photo Andrzej Kowalczyk 2018..................................	149
Fig. 6.8	Two restaurants owned by Ms. Magda Gessler (a food celebrity in Poland) in the Old City of Warsaw (Poland). Tourists in front of the Polka restaurant (**a**). Total renovation of the restaurant U Fukiera at the Old Town Square (**b**)......	150
Fig. 6.9	Restaurants in the Old Town in Warsaw (Poland) attract tourists by referring to their historical heritage. The Bazyliszek in the Old Town Square is marked out by its legendary dragon (**a**). The Pod Samsonem restaurant at Freta Street offering dishes of Polish Jews (**b**).................	151
Fig. 6.10	Eating establishments in the Old Town in Warsaw (Poland) that attract customers by referring to their historical heritage in 2018. The signboard with name 'Pączek w maśle. Cukiernia warszawska' is similar to signs at stores in Warsaw before World War II. Further to the right (behind the grocery store chain Carrefour Express) is an ice cream shop with traditions dating back to 1968 (**a**). A similar signboard to the previously described is the inscription above the Restauracja Zapiecek, which has been operating since 1960 (**b**).........	152

Fig. 6.11	Stands selling obwarzanki in Kraków (Poland) in 2017. Source: Map by Magdalena Kubal-Czerwińska.	154
Fig. 7.1	Spatial distribution of eating establishments in Saint Petersburg 1894–2017. Source: Kostantin A. Kholodilin and Leonid E. Limonov 2018.	165
Fig. 7.2	Shift in the city centre of Saint Petersburg, 1869–2017. Source: Kostantin A. Kholodilin and Leonid E. Limonov 2018	166
Fig. 7.3	Growth in the number of eating establishments in Warsaw's central borough of Śródmieście between 1994 and 2017. Source: Rokicka-Donica (1995), Kaczorek (2002), Derek (2013) and Marta Derek's own research.	167
Fig. 7.4	Concentrations of eating establishments in Warsaw's central borough of Śródmieście in 2017. Source: Marta Derek's research.	170
Fig. 7.5	Growth in eating establishments in the Old Town of Kraków between 1990 and 2017. Source: Magdalena Kubal-Czerwińska's own research; Brzozowska and Maciuszek (1996), Wójcik (2007), Piziak (2011), Piziak and Zając (2012).	173
Fig. 7.6	Distribution of gastronomic facilities in the Old Town of Kraków in 1983, 2000, 2008, 2012 and 2017. Source: Piziak (2011), Magdalena Kubal-Czerwińska's own research	173
Fig. 7.7	The distribution of eating establishments in the Żoliborz district of Warsaw (Poland) in 1975. Source: Adapted by Andrzej Kowalczyk from Bystrzanowski and Dutkowski (1975).	175
Fig. 7.8	The distribution of eating establishments in the Żoliborz district of Warsaw (Poland) in 2017. Source: Makarewicz (2018) and fieldwork by Dawid Celiński-Jakubowicz and Konrad Frączek.	177
Fig. 7.9	Restaurants and bars in Podskali (Prague) in 2009	178
Fig. 7.10	Restaurants and bars in Podskali (Prague) in 2017. Source: Fieldwork by Jakub Šmída.	178
Fig. 8.1	Eating establishments on the Bródno housing estate in Warsaw (Poland) typical of the pre-1989 period. The Kacperek restaurant dates from the 1970s (**a**) and is different from the cafés (**b**) from the same decade, and milk bar with the same name (Kacperek). The only remaining milk bar from pre-1989 period on the Bródno housing estate (**c**)	184
Fig. 8.2	New food outlets on the Bródno housing estate in Warsaw (Poland) from the late 1980s and early 1990s. The outlet at	

	Wyszogrodzka Street offers fried chicken (**a**). The banner at the outlet on Skrajna Street displays information about fried chicken, zapiekanka and chips (**b**)	185
Fig. 8.3	The influence of foreign cuisine in gastronomy on the Bródno housing estate in Warsaw (Poland). Pizza Dominium on Kondratowicza Street (**a**). The Vietnamese bar Kim Nam at the crossroads of Rembielińska Street and Kondratowicza Street (**b**). The food outlet was probably built in the late 1990s, but the apartment house in the background was only completed in 2012. This new apartment block was built on a plot of land where there was a one-storey pavilion from the early 1970s with a grocery store and a pharmacy	186
Fig. 8.4	The distribution of eating establishments in the Bródno housing estate in Warsaw (Poland) in 2018	187
Fig. 8.5	Eating facilities along the main streets of the Bródno housing estate in Warsaw (Poland). Oriental bars border Kondratowicza Street. In the background are apartment houses for the lower middle class dating from the 1970s (**a**). Casual restaurants and bars along Świętego Wincentego Street. In the background is the new Tivoli middle-class housing estate (**b**)	188
Fig. 8.6	The most famous restaurants on the Bródno housing estate in Warsaw (Poland). The Italian-owned restaurant, Bella Napoli (**a**), is situated close to the town hall. The Serbian–Montenegrin restaurant, Bałkańska Dusza, is located in a detached family house on the edge of the Zacisze neighbourhood (**b**)	189
Fig. 8.7	Gastronomy at the CH Targówek (later CH Atrium Targówek) shopping centre in the Bródno housing estate in Warsaw (Poland). Distribution of eating establishments in 1997 (**a**). McDonald's restaurant to the north in 2018 (**b**). The McDrive service is on the left, the main restaurant is in the centre, with the terrace (front)........................	190
Fig. 8.8	Changes in the distribution of eating establishment in the CH Atrium Targówek shopping centre in Warsaw (Poland) 2002–2017	190
Fig. 8.9	Distribution of the Da Grasso chain pizza parlours in Warsaw (Poland) in July 2017. The areas bound by red line are the downtown area (Śródmieście borough)...................	192
Fig. 8.10	Pizza parlour at Gilarska Street in the Zacisze neighbourhood in Warsaw-Targówek borough	193
Fig. 8.11	Distribution of eating facilities in the Tarchomin and Nowodwory neighbourhoods in Warsaw (Poland) in 2018....	195

Fig. 9.1	Fashionable restaurants in the suburbs of Victoria (Hong Kong Special Administrative Region of the People's Republic of China). **a** Restaurants on the waterfront at Stanley. **b** Jumbo Kingdom, or the Jumbo Floating Restaurant, in Aberdeen was established in 1976 by Dr. Stanley Ho, one of the richest people in East Asia. The restaurant can accommodate up to 2,000 diners. Source: Anna Kowalczyk 2011....................................	201
Fig. 9.2	Distribution of eating establishments in the Piaseczno district (Poland) in 2016....................................	206
Fig. 10.1	Eating establishments in Siedlce (Poland) in 2017. Source: Adapted from Próchnicka (2017, p. 51), underlay map by Andrzej Kowalczyk....................................	215
Fig. 10.2	Sempre restaurant in the co-called Abraham's Small House in Gdynia. Entrance to the restaurant (**a**). Interior of the restaurant (**b**). Source: Mateusz Kaczmarek 2018...........	216
Fig. 10.3	Eating establishments in revitalised buildings at Ustka (Poland) in 2017. Restaurants in new tourist facilities at Ustka (**a**). The restaurant Pod Dębem in the wattle-and-daub house in the old part of Ustka (**b**). Source: Małgorzata Durydiwka 2018....................................	217
Fig. 10.4	Eating establishments in the buildings after revalorisation and restructurisation. Interior of eating establishments in former coal mine Guido in Zabrze (Poland). Source: Małgorzata Durydiwka 2016....................................	218
Fig. 11.1	Differences in the distribution of restaurants and bars in Madrid by ethnic cuisine in 2013. **a** Chinese cuisine. **b** Japanese cuisine. **c** Italian cuisine. Source: data gathered and mapped by Andrzej Kowalczyk, basing on information published in *Páginas Amarillas/Páginas Blancas 2014. Madrid Capital/Zona Centro, Consulta la guía de restaurantes más completa*, 1//11/2013, Connecti S.A.U. Impeso por Ediciones Informatizadas, S.A., Madrid, pp. 427–433	228
Fig. 11.2	Eating establishments serving ethnic cuisines in Amsterdam between 1975 and 2018. Source: *Horeca in Amsterdam: minder cafés, meer restaurants*, 2005; personal communication with Onderzoek, Informatie en Statistiek of the City Hall in Amsterdam; https://www.iens.nl/restaurant+amsterdam (Accessed 10 May 2018)	230
Fig. 11.3	Distribution of eating establishments serving ethnic cuisines compared to non-ethnic establishments and those	

	serving Polish cuisine in Śródmieście borough (Warsaw) in 2017. Source: Marta Derek, underlay map by Andrzej Kowalczyk 234
Fig. 11.4	Distribution of eating establishments serving different ethnic cuisines in Śródmieście borough (Warsaw) in 2017. Source: Marta Derek, underlay map by Andrzej Kowalczyk 234
Fig. 12.1	Ethnic restaurants and bars in Warsaw visited by students of the Faculty of Geography and Regional Studies of University of Warsaw (Poland)........................ 244
Fig. 13.1	Geographical distribution of eating establishments run by Strefa Kulinarna s.c. and their suppliers in Warsaw (Poland) in 2017 250
Fig. 13.2	Two types of transport used to deliver supplies to restaurants. Divundu (Namibia)—a truck used to supply long-distance supplies between cities, in this case to KFC's restaurant in Katima Mulilo (**a**). Warsaw (Poland)—a cargo van used to transport supplies over shorter distances (**b**) 253
Fig. 13.3	Geographical distribution within Warsaw (Poland) of Menu Express sp. z o. o. suppliers in 2017. Source: Author based on data from Menu Express sp. z o. o...................... 256
Fig. 13.4	Geographical distribution within Warsaw (Poland) of Menu Express sp. z o. o. customers in 2017. Source: Andrzej Kowalczyk based on data from Menu Express sp. z o. o...... 258
Fig. 13.5	Distribution of contractors in the supply chain for Menu Express sp. z o. o. in Warsaw (Poland) 259
Fig. 14.1	New dining hotspots in the urban space. In Wrocław (Poland), the former railway station (Dworzec Świebodzki, which closed in 1991) has been a popular meeting place for foodies for several years (**a**). In some urban agglomerations people eat and meet friends in parks and other green areas within the city, such as the Parc de la Ciutadella in downtown Barcelona (Spain) (**b**). Source: Anna Kowalczyk 2017, 2015 266
Fig. 14.2	Two types of spatial relations between restaurants and their clients. Type A (the old model): customers search for a restaurant (**a**). Type B (the new model): restaurants search for customers (**b**). Type B represents the process described as home food delivery................................. 268
Fig. 15.1	Eating establishments at the biggest shopping centres and galleries in Warsaw (Poland) in 2017. Source: Authors' own research.. 282

Fig. 15.2	Dining establishments at shopping centres in Warsaw (Poland). The Bierhalle chain restaurant in the Arkadia mall (**a**) and Złote Tarasy mall (**b**)	286
Fig. 15.3	Clusters of eating establishments at shopping centres in Costa Calma, Fuerteventura Island (Spain) in 2013	286
Fig. 15.4	Classification of the spatial distribution of eating outlets within shopping centres	289
Fig. 15.5	**a–c** The distribution of eating establishments (shown in red) and corridors (yellow) in the Galeria Mokotów shopping centre in Warsaw (Poland) in 2017	289
Fig. 16.1	Distribution of eating establishments concentrated in the Śródmieście borough (Warsaw) parks and along the Vistula river in 2017. Source: Authors' own research, underlay map by Andrzej Kowalczyk	298
Fig. 16.2	The Breakfast Market in Żoliborz borough in Warsaw (Poland) in 2017. Picknique at the Breakfast Market. Note the effects of trampling (**a**). More than a breakfast—Brazilian food at the Breakfast Market (**b**). Source: Monika Bartman 2017	301
Fig. 16.3	Vistula River in the centre of Warsaw (Poland) in 2017. Recently renovated left bank with a number of eating establishments faces the 'wild' right bank of the river (**a**). Floating bars and restaurants allow to stay as close to the river as possible (**b**). Source: Monika Kordowska 2017	304
Fig. 17.1	Street food stalls offering traditional and non-traditional meals. Permanent food stall with Chinese traditional delicacies in the Mong Kok district of Kowloon (Hong Kong Special Administrative Region of the People's Republic of China) (**a**). Semi-mobile food stall with hotdogs, drinks and chips and sausage in Divundu (Namibia) (**b**)	314
Fig. 17.2	Ways of offering street food. Mobile hawker in Lahore (Pakistan) (**a**). Semi-mobile hawker with clams at Ölüdeniz-Belceğiz (Turkey) (**b**). Semi-mobile push-cart with fresh juices in Marrakesh (Morocco) (**c**). Source: Anna Kowalczyk 2017	314
Fig. 17.3	Locations of stalls. Outskirts of Rehoboth (Namibia) (**a**) and the old part of the Havana (Cuba) (**b**). Source: Katarzyna Kowalczyk 2015	315
Fig. 17.4	Permanent food stalls. Food stall selling *bak kwa* in Macao (Macao Special Administrative Region of the People's Republic of China) (**a**). Food stall with dumplings at Tai O (Hong Kong Special Administrative Region of the People's Republic of China) (**b**). Source: Anna Kowalczyk 2011	315

Fig. 17.5	Ways of offering street food. Semi-mobile push-carts in Cienfuegos (Cuba) (**a**) and at Stanley (Hong Kong Special Administrative Region of the People's Republic of China) (**b**). Source: Katarzyna Kowalczyk 2015.	316
Fig. 17.6	Ways of offering street food. Semi-permanent vending table with chestnuts at Piazza di Spagna in Rome (Italy) (**a**). Semi-mobile push-cart with iced almonds at Bodrum harbour (Turkey) (**b**). Source: Anna Kowalczyk 2013	316
Fig. 17.7	Food trucks during the Slow Weekend festival organized in Warsaw at Soho Factory (Mińska Street) in June 2015. La Chica Sandwicheria (**a**), Carnitas Food Truck (**b**) and Pepe Crepe (**c**). Source: Anna Kowalczyk 2015	320
Fig. 17.8	The Tuk Tuk Thai Street Food truck at the Asian Street Food Fest in Warsaw on 27 June 2015 taken at 12.20 am (**a**) and 1.17 pm (**b**). Source: Anna Kowalczyk 2015	320
Fig. 17.9	The main clusters of food trucks in Warsaw in 2015–2018. Source: Adapted from Staniak (2017, p. 36) and a survey conducted by Andrzej Kowalczyk.	321
Fig. 17.10	A food truck park in Kraków (Poland). Food trucks at Isaac Square on the plot between the old buildings of Kazimierz, which was an independent city inhabited by the Jewish community and is now a tourist attraction. Source: Magdalena Kubal-Czerwińska 2017	323
Fig. 18.1	Means of transport used to deliver food from restaurants to clients in Warsaw (Poland). A van belonging to an independent Greek restaurant (**a**). Scooters belonging to Dominos Pizza chain (**b**).	331
Fig. 18.2	Location of eating establishments offering Polish and Italian cuisine in downtown Warsaw (Poland) supplying customers via Pyszne.pl (May 2018). Source: Adapted from https://www.pyszne.pl/jedzenie-na-telefon-warszawa (Accessed 01 May 2018)	335
Fig. 18.3	Location of eating establishments offering Japanese and Chinese cuisine in downtown Warsaw (Poland) supplying customers via Pyszne.pl (May 2018). Source: Adapted from https://www.pyszne.pl/jedzenie-na-telefon-warszawa (Accessed 01 May 2018)	336
Fig. 18.4	Outlets of the leading pizza chains in Poland. A pizzeria belonging to Telepizza on Drobnera Street in Wrocław (**a**). A pizzeria belonging to Dominos Pizza on Słomińskiego Street in Warsaw (**b**). Source: Anna Kowalczyk 2017	337
Fig. 18.5	Delivering food to homes in Warsaw. Biker working for Pyszne.pl with his orange bicycle (**a**). Cyclist working for Uber Eats (**b**)	338

Part I
Theoretical Aspects of Studies on Gastronomy in Urban Space

Chapter 1
Relations Between Gastronomy and the City

Andrzej Kowalczyk and Marta Derek

Abstract The purpose of this chapter is to establish the framework for analysing the relationship between gastronomy and urban space, notably from the historical perspective. It begins by defining the role of gastronomy in the premodern period, i.e. before the industrial revolution, and identifies four of the largest concentrations of taverns, inns, pubs, etc. found in cities prior to the eighteenth century. It then explores gastronomy and urban space in the modern period, especially the influence of the capitalist economy, new colonial conquests, the development of science and the introduction of new technologies, which took place mainly in the nineteenth century. The remainder of the chapter focuses on an analysis of gastronomy in the postmodern period (called also post-industrial; Pacione 2005), which starts in middle of the twentieth century. Drawing on a number of examples, it proposes models of the main concentrations of gastronomy found in the inner city, in the outer city, in the suburbs and in tourist cities. It highlights that although the functions performed by gastronomy are largely independent of the historical period, their role in the urban space has changed as the restaurant sector has become increasingly important in the context of the overall economy of a city.

Keywords Gastronomy · Urban agglomeration · Concentration of gastronomy · The inner city · The suburbs · The outer city · The tourist city

---

A. Kowalczyk (✉) · M. Derek
Department of Tourism Geography and Recreation, Faculty of Geography and Regional Studies, University of Warsaw, ul. Krakowskie Przedmieście 30, 00-927 Warsaw, Poland
e-mail: akowalczyk@uw.edu.pl

M. Derek
e-mail: m.derek@uw.edu.pl

© Springer Nature Switzerland AG 2020
A. Kowalczyk and M. Derek (eds.), *Gastronomy and Urban Space*,
The Urban Book Series, https://doi.org/10.1007/978-3-030-34492-4_1

1.1 Introduction

Gastronomy, understood as a sector of the economy and as a form of land use, took on a broader scale in ancient Rome. Until then food was consumed at home within the family circle in traditional rural communities. Eating out can therefore be considered an element of a so-called city lifestyle and, from a historical perspective, is closely linked to urbanisation and industrialisation. Restaurants, cafés and bars are visited by both residents and visitors (including tourists), which means that it is often difficult to precisely define who is using them. Although their gastronomic functions are largely independent of the historical period, their role in the urban space has changed as the sector has contributed more to the overall economy of a city.

1.2 Pre-modern Period

The first eating establishments were taverns, inns, etc. that catered to travellers. It was only later that they began to be visited by residents. They probably appeared first in the cities and towns of ancient China, Mesopotamia and Egypt, or in Indus Valley civilizations. They later appeared in ancient Greece, where they were known as *kapileia* (wine bars), or *lesche* and *phatnai* (offering food). O'Gorman (2007a) notes that the works of Xenophon and Thucydides refer to *katagogion* (inns); these are probably the oldest references to facilities that offered both accommodation and dining facilities. Gastronomy in urban areas expanded during the Roman period (O'Gorman 2007b). Archaeological excavations have shown that there were 158 bars and restaurants in ancient Pompeii (Ellis 2004, 2008). According to O'Gorman (2007b), there was a cluster of hospitality facilities in the city centre. Ellis (2008) suggests that establishments were located in the most profitable areas, as more than two-thirds of intersections were bordered by a bar. According to Fant (2009) these *thermopolia* had a counter that was open to the street and resembled street-side bars. It is no wonder that Seneca wrote 'I don't care to live among bars' (Fagan 2006, p. 373), because they were perceived as places of excessive drinking, gambling, fighting and prostitution. However, not only in Pompeii traces of the role of gastronomy in urban space in the cities of ancient Rome were found. Archaeological research indicates that it was similar in Herculaneum and Rome (Parham 2015, p. 98) (Fig. 1.1).

In medieval Europe, taverns and inns were the primary catering establishments. Most, but not all, served alcohol, and food was typically served to single men (Mac Con Iomaire 2013). For instance, in medieval Britain there were three types of facilities: ale houses, taverns and inns. The first served beer; taverns also served wine; and inns also provided accommodation. Ale houses were particularly popular:

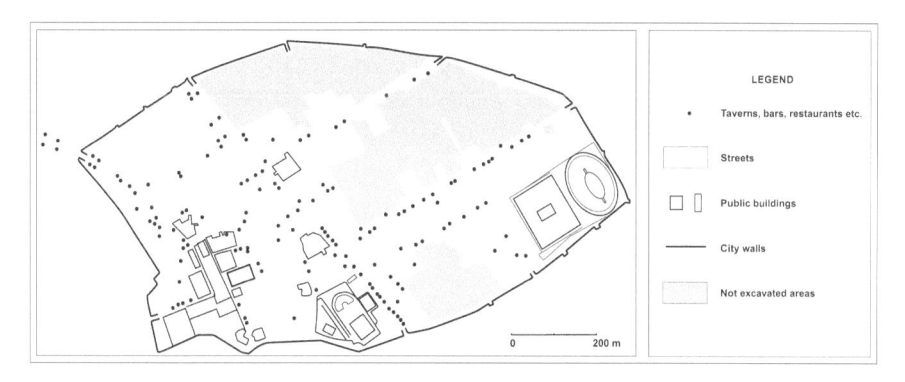

Fig. 1.1 The distribution of taverns, bars, etc. in Pompeii (Roman Empire). Source: Andrzej Kowalczyk after Ellis (2008, Fig. 42)

at the end of the sixteenth century, the three major cities in Derbyshire had 70, 69 and 38, respectively (Hart 1879). Their popularity continued into the next century and in 1660s Dublin (with a population of around 4,000 families) there were 1,180 ale houses and 91 so-called public brewhouses. Taverns were at the top of the hierarchy of gastronomy facilities. The first were probably opened in London and Oxford in the fourteenth century. They served wine or other alcohol, often as an accompaniment to food (Duensing 2014). Another type of eating establishment followed in the seventeenth and eighteenth centuries: so-called 'ordinaries' that were focused more on food than drink (Duensing 2014). Towards the end of the seventeenth century, coffee and tea houses appeared (Duensing 2014). Their popularity increased during the eighteenth century, as coffee and tea became fashionable among the upper classes. The same process could be observed across continental Europe—especially in France, Prussia, Austria and Poland. Around the same time, new types of dining facilities emerged. For instance, in 1738, in Dublin there were five groups of public dining or drinking establishments: taverns, eating houses, chop houses (steakhouses), coffee houses and a chocolate house (Mac Con Iomaire 2013).

> A coffee house is a public place that specialises in serving coffee and other refreshments and sometimes provides informal entertainment. According to Adamson (2004, p. xiv) the first coffee house was opened in Constantinople (now Istanbul) in 1475, although other sources suggest that coffee houses only became popular in the Ottoman Empire in the sixteenth century. Coffee appeared for the first time in the Christian part of Europe in the seventeenth century. The first coffee house was probably established in Venice in 1629, and in 1650s the first coffee houses were set up in England: in Oxford (1650), and in London (1652). In the following decades the expansion of coffee houses in London was so big that, by 1675, their number had grown to more

than 3,000. The first coffee house in Paris opened in 1672, and in Vienna in 1683, and North America's first coffee house appeared in Boston in 1676. Source: Adamson (2004), Coffeehouse (2017).

Indoor dining establishments were less widespread in other parts of the world. For example, in the Middle East, the equivalent of taverns was rare. In medieval times, the inhabitants of Cairo usually ate food bought from stalls and street vendors (Lewicka 2005). Between the thirteenth and sixteenth centuries the number of eating places ranged (according to the source) from 10,000–12,000 or from 20,000–24,000. Other data comes from Baghdad, where the Tigris River was bordered by wine houses or taverns, and shops selling roasted meat (Lewicka 2005). It is interesting to note, however, that neither of these two cities had the equivalent of what we would now call a restaurant (Lewicka 2005). At the same time, in India, cooked food was sold in catering shops or restaurants (*dukānān-i-ṭabbākhān*) (Siddiqi 1985), which were commonplace in market squares. Those located in old bazaars in cities like Delhi were among the most frequently visited food places (Kothiyal 2012). Finally, in Southeast Asia and China, street food was popular. Parham (2015, p. 98) speaks about 'lively food-centered streets' using the example of Kaifeng, China, already in Song period (960–1127).

According to French sociologist and geographer, Henri Lefebvre '…medieval society – that is, the feudal mode of production, with its variants and local peculiarities – created its own space. Medieval space built upon the space constituted in the preceding period, and preserved that space as a substrate and prop for its symbols; it survives in an analogous fashion itself today'. (Lefebvre 1991, p. 53). This statement is true also in the case of gastronomy.

Some regularities can be observed regarding the location of gastronomy facilities. These relate to cities and towns of similar size, with a similar geographic location, and inhabited by people belonging to particular cultural circles. Figure 1.2 shows that up until the eighteenth century, the largest concentrations of taverns, inns, pubs, etc. were in the following areas:

- around city or town squares (or marketplaces) and along the main streets;
- near ports (if the city or town was located by the sea or a river);
- in the surroundings of a religious sanctuary (church, mosque or other temple);
- on both sides of city gates (just inside the city or town, and just beyond its walls).

Markets have existed since ancient times and can be divided into two types: outdoor and indoor. Outdoor markets include marketplaces and market squares, street markets and so-called floating markets, which are popular in East and South East Asia. In the first three, goods are sold in open spaces in public squares or along streets. Street vendors do not have permanent stalls. In floating markets, goods are sold from boats. The defining characteristic of outdoor markets is there is no durable construction. This is unlike indoor markets, which are housed in specially

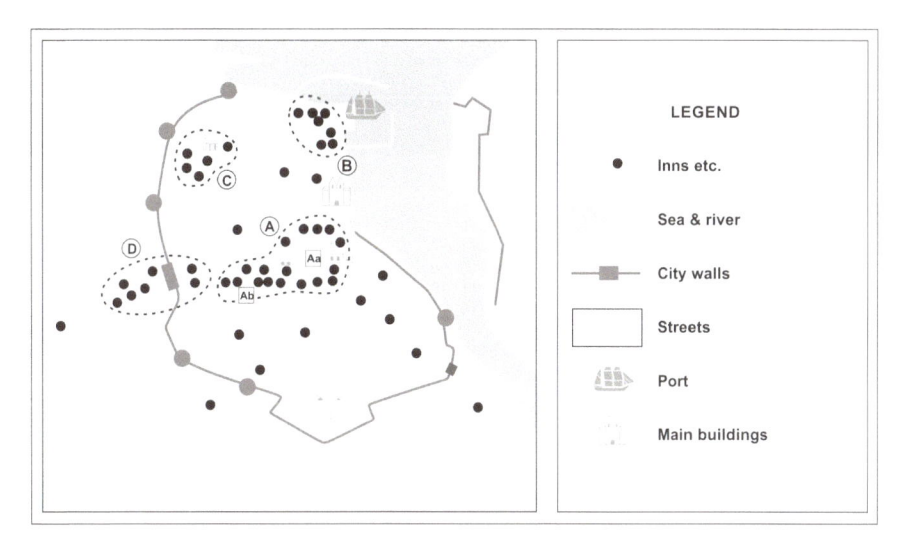

Fig. 1.2 Schematic map of a historical city or town showing the concentration of gastronomy facilities: A—surrounding the marketplace (Aa) and main street (Ab); B—merchant port; C—religious sanctuary; D—city gate

constructed halls. The first examples in Europe are Les Halles in Paris, which opened in 1100 and Covent Garden Market in London, which opened in 1265 (Adamson 2004).

Few other European market halls predate the nineteenth century. One example is the Sukiennice (literally 'Cloth Hall') in Kraków (Fig. 1.3), which dates back to the Middle Ages. Originally traders bought and sold textiles, spices, wines and salt, but there were also several food stalls. There are still restaurants and bars in the Sukiennice, but many of the rooms are now souvenir shops for tourists. This is not

Fig. 1.3 Indoor markets (market halls) in European cities. Sukiennice market hall on the main market square in Kraków (Poland, 1556–1559, renovated in 1875–1879) (**a**). The entrance to the Mercat de Sant Josep de la Boqueria on Las Ramblas in Barcelona (Spain, 1840–1853) (**b**). Source: Anna Kowalczyk 2006, 2015

Fig. 1.4 The Naghshe Jahan Traditional Restaurant on Naghshe Jahan Square in the old part of Isfahan (Iran). The entrance to the former *caravanserai* (**a**) and courtyard with restaurant on the 2nd floor (**b**). For the wealthy residents of Isfahan, this restaurant is a "place" in the meaning of Yi-Fu Tuan's (1975, 1979) and Edward Relph's (1976) concepts (more on these concepts in Sect. 2.2.4)

the case for La Boqueria in Barcelona (Spain). Although the present building was built in the mid-nineteenth century, the first mention of it dates back to 1217, when tables were installed near the city gates, mostly selling meat. There are still many food-related stalls today and, because it lies on the main street of Barcelona, many tourists who visit eat there (Fig. 1.3).

In cities and towns, from ancient times to the Middle Ages, taverns, inns and other facilities were located close to the market (Fig. 1.2, Aa). Taverns were often found in the courtyards of houses around the marketplace and served both visitors and local residents. In Islamic countries, their main function was to provide hospitality for travellers, rather than serve food (Bryce et al. 2013). These caravanserais still exist in Islamic cities and towns, but most have become luxury restaurants and bars catering for (foreign) tourists. One example is the Naghshe Jahan Traditional Restaurant, in the centre of Isfahan (Iran). Located next to the historic Sheikh Lotfollah Mosque (built in 1619), it regularly attracts foreign tourists (Fig. 1.4).

Another example is the teahouse that dates from the tenth century at Neyshabur (Iran). It is situated in the courtyard of the revitalised *caravanserai*, which nowadays includes art galleries, exhibitions of traditional crafts, etc. (Fig. 1.5).

> King Abbasi's *caravanserai* attracts Iranian tourists who visit Neyshabur to see the tomb of the famous Persian poet and theologian Abū Ḥamīd bin Abū Bakr Ibrāhīm (known by his pen-name as Farīd ud-Dīn or ʿAṭṭār, c. 1145–1221) and the artist Mohammad Ghaffari (known as Kamal-ol-Molk, 1848–1940), which are situated on the edge of town. The caravanserai and the teahouse inside it also attract overseas tourists, who visit Neyshabur for its archaeological excavations, which are the site of the massacre of the

Fig. 1.5 King Abbasi's *caravanserai* at Neyshabur (Iran). Courtyard and entrance (**a**). Interior of the teahouse (**b**). Traditional stone containers for cooking and serving the Iranian stew called *dizi*, *abgoosht* or *ābgusht* (**c**)

population by the Mongols in 1221 (among those killed was Abū Ḥamīd bin Abū Bakr Ibrāhīm) and turquoise mines. In 1970, the American band Santana released an instrumental track titled *Incident at Neshabur*, which is dedicated to memory of the victims of the 1221 massacre, on their album *Abraxas*.

In Europe, teahouses, coffeehouses and similar dining establishments appeared in indoor markets and along the main street (Fig. 1.2, Ab) and in and around the main market. Traders were also the main customers of taverns located near the city gates (Fig. 1.2, D). Dining places near churches, mosques, religious pagodas, etc. were mostly frequented by pilgrims (Fig. 1.2, C). Taverns and inns located in the area closest to the port were visited mostly by sailors, merchants and dockworkers (Fig. 1.2, B). They usually offered cheap meals and alcoholic beverages. These taverns or inns were often combined with brothels.

1.3 Modern Period

At the end of the Middle Ages the cities and towns in Europe, the Middle East and North Africa, and Asia were similar in terms of their socio-economic functions.[1] However, in the modern period the situation began to change. Geographic discoveries and colonial conquests in the sixteenth–seventeenth centuries, coupled with rapid economic development, led some European cities (especially ports) to grow faster than they had before. Consequently, the spatial arrangement and types of eating establishments became increasingly diverse. For example, in some places, administrative or commercial functions dominated, while neighbourhoods were inhabited by different social classes with different incomes. Up to this point, catering had mainly been limited to taverns and similar institutions offering a limited range of dishes. Moreover, their clientele was limited to the lower social classes as the upper classes generally had their own cooks.

It was only in the seventeenth–eighteenth century that things began to change. Gastronomic diversity appeared with the emergence of restaurants and coffee houses. Taverns began to be frequented not only by those who had no other option, but also to maintain social contacts. In Britain, the economy grew rapidly following the country's colonial expansion in North America and the Caribbean. The process accelerated with the advent of the Industrial Revolution, beginning in 1760, which led people to migrate from rural to urban areas and rapidly increased the urban population. Urban development also began to affect gastronomy, mostly in terms of its quality, and it was at this time that catering establishments began to be frequented by the upper classes: notably, the aristocracy, new social groups such as the *bourgeoisie* and a growing number of officials. By the end of the eighteenth century, changes were also taking place in Paris, Vienna, Berlin, St. Petersburg and other major continental cities, notably following the end of the Napoleonic wars in 1815.

> According to Brillat-Savarin (1994, p. 267, after Mac Con Iomaire 2013, p. 232) there are certain aspects of the restaurant that distinguishes it from other eating establishments: 'A restaurateur is a person whose trade consists in offering to the public an ever ready feast, the dishes of which are served in separate portions, at fixed prices, at the request of each consumer. The establishment is called a restaurant, and the person in charge of it the restaurateur. The list of dishes, bearing the name and price of each, is called the carte or bill of fare, while the record of the dishes served to the customer, together with the relevant prices, is called the carte à payer or bill'.
> Source: Mac Con Iomaire (2013, p. 232) (Fig. 1.6).

---

[1] It is difficult to say anything about the situation regarding gastronomy in the cities and towns of the pre-Columbian Americas in the context of the main topic of this book.

Fig. 1.6 a Restaurante Sobrino de Botin in Madrid (Spain). This restaurant, which opened in 1725, is among the oldest restaurants in Europe. **b** Antico Caffè Greco in Rome (Italy). This coffee house was established on Via dei Condotti 86 in 1760 and is one of the oldest coffee houses in Europe. It is a one of the must-see tourist sites in Rome. Its many guests have included historic figures such as the poets Goethe, Byron and Keats, composers Liszt, Wagner and Mendelssohn, writers Stendhal and Andersen, the playwright Ibsen and last, but not least, the adventurer Casanova. In October 2016 a cup of espresso cost about €7. These restaurants are other examples of "places" (Tuan 1975). Source: Anna Kowalczyk 2015

A specific example comes from Dublin (Ireland), where public dining changed significantly between 1700 and 1900. Restaurants, coffee houses and tea houses increased in popularity at the expense of taverns (Mac Con Iomaire 2009, 2013). This change was particularly evident in the latter part of the nineteenth century. Between 1850 and 1860 the number of taverns grew from 134 to 174, although the number of inns remained almost the same. Between 1870 and 1900 the total number of catering facilities declined from 86 to 61 (Table 1.1). Similar changes were observed in the number of hotels and other accommodation establishments (Mac Con Iomaire 2013).

Table 1.1 illustrates these changes. According to Mac Con Iomaire (2013, p. 238), coffee houses evolved into clubs in the eighteenth century and, in a similar manner, towards the end of the nineteenth century many taverns evolved into either restaurants or gin palaces. Similar processes occurred in other European countries as the result of the shift from a pre-capitalist to capitalist society. Perhaps the most important aspect, however, in the context of the urban space, is that these

Table 1.1 Changes in the structure of dining facilities in Dublin (Ireland) 1850–1900

Year	Taverns and inns	Dining rooms	Refreshment rooms	Total
1870	74	12	–	86
1880	33	20	19	72
1890	25	23	22	70
1900	24	20	17	61

Source: Adapted from Mac Con Iomaire (2013, p. 237, Table 1)

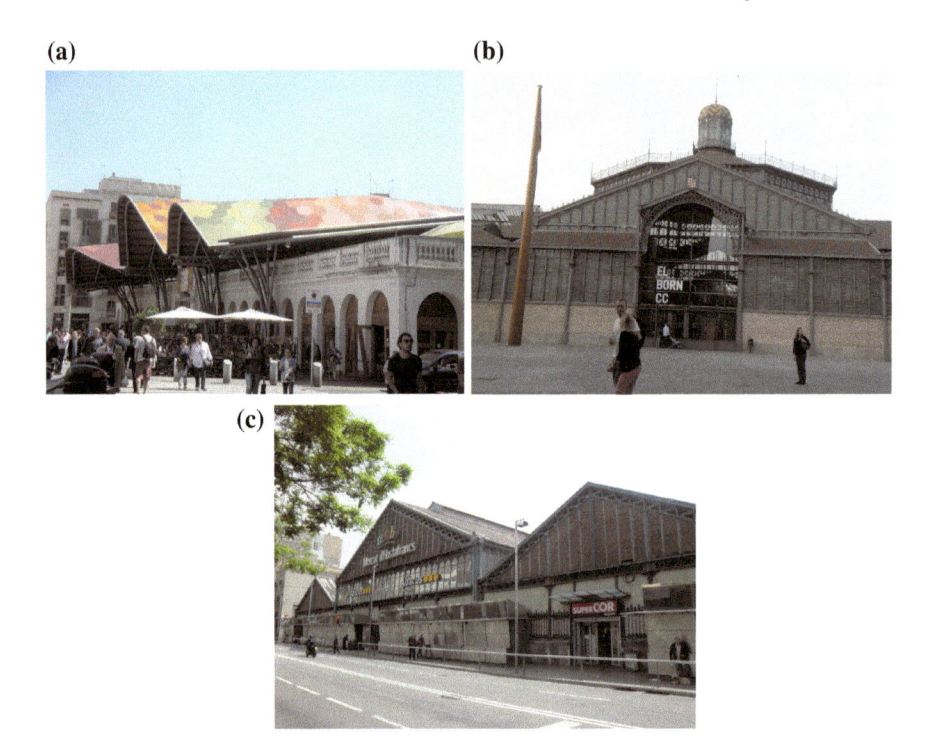

Fig. 1.7 Market halls in the central part of Barcelona (Spain). **a** Mercat de Santa Caterina (built in 1848) in the area of Santa Caterina. **b** Mercat del Born (built in 1878) in the area of La Ribera. **c** Mercat d'Hostafrancs (built in 1888) in the area of Hostafrancs. The revitalisation of the Mercat de Santa Caterina was completed in 2005 and it now hosts many restaurants and bars. The Mercat del Born was restored in 2013. Source: Anna Kowalczyk 2015

establishments began to appear in prestigious neighbourhoods, initially in areas near important public buildings and districts inhabited by the rich. This process lasted throughout the nineteenth century.

By the middle of the nineteenth century, the spatial expansion of cities meant that existing forms of selling and buying food were no longer adequate. In order to limit the chaotic and uncontrolled development of new marketplaces, authorities began to build market halls to provide food to local residents. They were the most rational and efficient means of controlling the quality, hygiene and prices of food in cities. Very soon they become the symbol of a modern urbanity (Lee 2009). This state of affairs persisted into the twentieth century and World War I. Examples appeared across Europe in cities such as Berlin, Vienna, Prague, Barcelona, Madrid, Lisbon, Stockholm, London and Rome (e.g. Atkins et al. 2007; Crespi-Vallbona and Pérez 2015; Parham 2012, 2015) (Fig. 1.7).

Figure 1.8 shows the distribution of the major market halls in Barcelona. Most were built in the late nineteenhh and early twentieth century. Typically, they are clustered in districts that were partly established in the Middle Ages but were more extensively developed in the nineteenth century, showing that their location was

1 Relations Between Gastronomy and the City

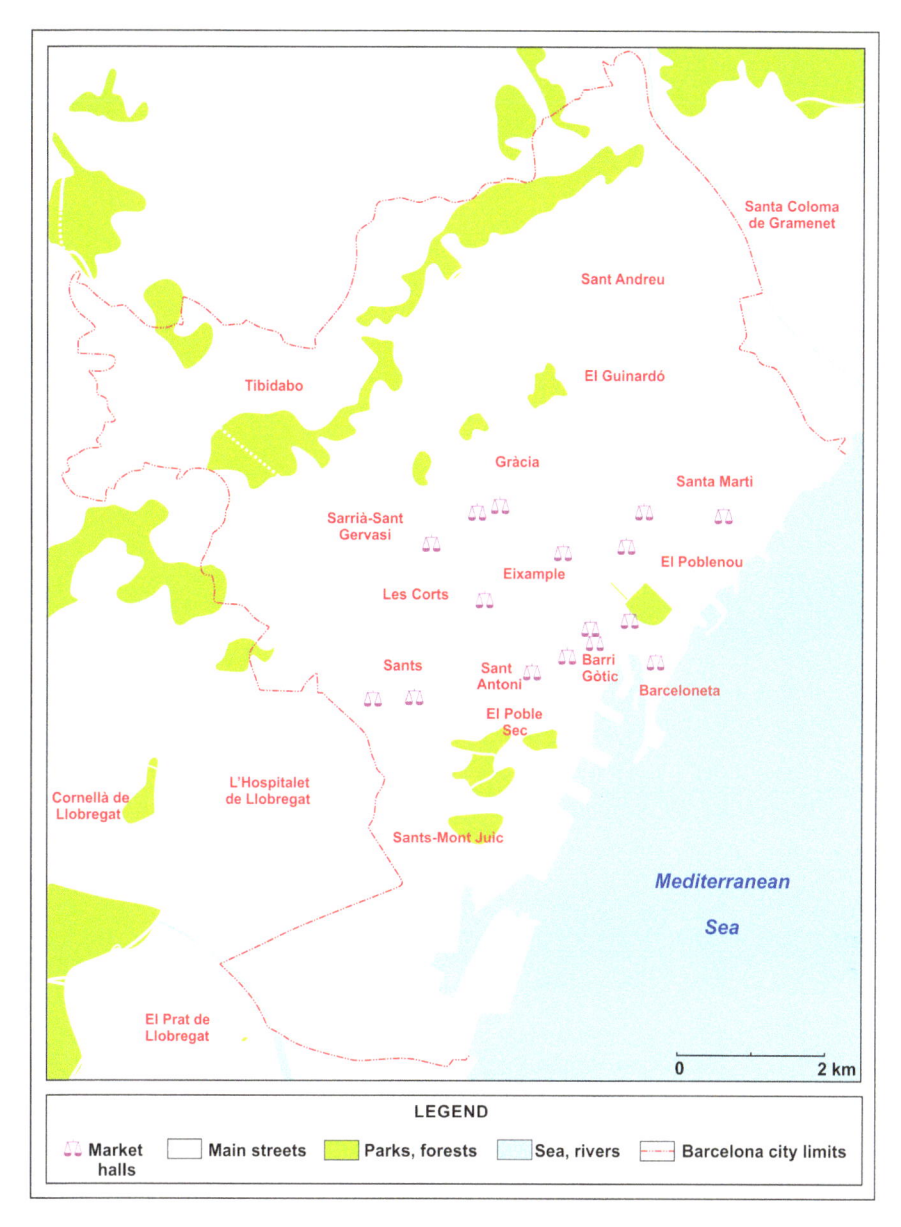

Fig. 1.8 The main roofed market halls (*mercat* in Catalan language) in Barcelona (Spain)

closely related to the distribution of the population. There are none in peripheral districts, partly because of the lower population density, but mainly because they had begun to be replaced by supermarkets and hypermarkets.

Table 1.2 shows the expansion of market halls in Germany between 1860 and 1900. During the same period, a similar process was happening in Madrid (Spain).

Table 1.2 Market halls in Berlin (Germany) in 1900

No.	Name	District	Year	Surface (m^2) and stands	Remarks
1	Zentralmarkthalle I Zentralmarkhalle II	Berlin-Mitte	1886 1893	16,079/ 1336 13,281/ 776	On both sides of Karl-Liebknecht-Str. (formerly Kaiser-Wilhelm-Str.); part of the building was destroyed in 1944 and the rest in the 1960s
2	Lindenhalle	Berlin-Mitte	1886	9114/ 746	Between Friedrichstr. and Lindenstr; destroyed during the Second World War, in 1962–1965 replaced by the Berlin Flower Market
3	Zimmerhalle	Berlin-Mitte	1886	4843/ 409	Zimmerstr.; since 1910 it has served as offices
4	Markthalle IV	Berlin-Mitte	1887	3778/ 364	Dorotheenstr.; closed in 1913
5	Markthalle V	Berlin-Tiergarten	1887	2538/ 231	Magdeburger Platz; destroyed during the Second World War, dismantled in 1956
6	Ackerhalle	Berlin-Mitte	1888	3546/ 344	At the corner of Invalidenstr. and Ackerstr.
7	Markthalle VII	Berlin-Kreuzberg	1888	4701/ 410	Legiendamm
8	Markthalle VIII	Berlin-Friedrichshain	1888	5070/ 532	At the corner of Andreasstr. and Krautsstr
9	Eisenbahnhalle	Berlin-Kreuzberg	1891	3296/ 300	Between Pücklerstr. and Eisenbahnstr
10	Arminiushalle	Berlin-Moabit	1891	4810/ 425	Between Arminiusstr., Bremerstr., Bugenhagenstr. and Jonasstr.; the only market hall that continues to operate since its opening
11	Marheinekehalle	Berlin-Kreuzberg	1892	2808/ 278	At the corner of Zossener Str. and Bergmannstr.; damaged during the Second World War, demolished in 1950, rebuilt

(continued)

Table 1.2 (continued)

No.	Name	District	Year	Surface (m²) and stands	Remarks
12	Markthalle XII	Berlin-Gesundbrunnen	1892	4198/ 273	At the corner of Grüntalerstr. and Badstr.; closed in 1898
13	Markthalle XIII	Berlin-Prenzlauer Berg	1892	5095/ 393	Wörtherstr., closed in 1910
14	Markthalle XIV	Berlin-Gesundbrunnen	1892	4066/ 352	Dalldorfer Str. (later Schönwalder Str.)

Source: Spiekermann (1999, p. 181), *Market halls in Berlin*, https://en.wikipedia.org/wiki/Market_halls_in_Berlin (Accessed 01 June 2017)

The mercado de la Cebada was founded in 1870. In 1875 the mercado los Mostenses was opened, in 1876 the mercado de Chamberí, and in 1882 the mercado de la Paz. Between 1913 and 1916 the mercado de San Miguel was built, and in 1934 the mercado de Olavide (Fidel 2007). From the very beginning, market halls have provided gastronomic functions (mainly street food). Nowadays, many of these facilities are being modernised (Figs. 1.7 and 1.9). Studies conducted in Madrid have shown that after renewal 'these spaces are the trending places to go to and people gather to them either during their lunch breaks, after work or over the weekend, creating a new social and food trend within the city, with its own physical space and identity. These new marketplaces not only attract locals, but also tourists' (Heathcote Sapey 2017, p. 80).

The modern era, which started in about 1500, lasted at least until the First World War (Barzun 2000). Huge changes which took place at the end of the nineteenth century, such as processes of democratisation and development of technology, were influencing global development until the middle of the twentieth century. The so-called Second Industrial Revolution, dated between 1870 and 1914, caused the growth of electricity, petroleum and steel, and used it to create mass production. Technological systems (telegraph, railroad, gas and water supply, sewage systems) were adopted worldwide, and were probably the first factors which enabled globalisation. These changes have also influenced food: thanks to the development of industrial farming and transportation, food started to be produced in larger quantities than ever before (Steel 2013). All these changes were stopped by the First World War. Than, after the end of the First World War, 'farseeing observers predicted the likelihood of another and it became plain that western civilization had brought itself into a condition from which full recovery was unlikely. The devastation, both material and moral, had gone so deep that it turned the creative energies from their course, first into frivolity, and then into the channel of self-destruction' (Barzun 2000, p. 712). In 1950 still fewer than 8% of British households had a fridge (Steel 2013). But at that time (i.e. the middle of the twentieth century) a shift

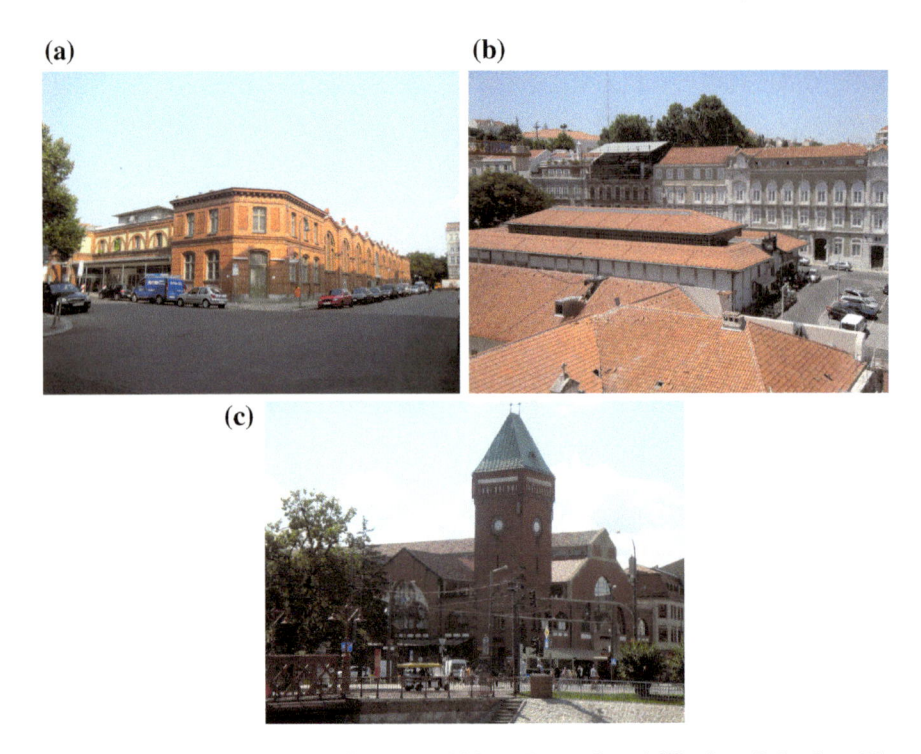

Fig. 1.9 Market halls in Berlin (Germany), Lisbon (Portugal) and Wrocław (Poland). **a** The Arminiushalle in Berlin-Moabit (built in 1891). **b** Mercado de Santa Clara (built in 1877) in the area of São Vicente (on the eastern edge of Alfama). Since 2011, the building has been home to the Centre of Culinary Arts (Centro das Artes Culinárias). **c** The Hala Targowa at Piaskowa Street in Wrocław (formerly Markthalle Nummer 1) was built in 1906–1908. It was established in order to organise street trading in the city centre. In addition to being a typical market place, it is now a hotspot for foodies. Source: Anna Kowalczyk 2014, 2017

towards postmodernity started taking place, which is called the contemporary historical period (e.g. Gregory et al. 2009).

The changes that took place in the United States have not been discussed, as trends followed those seen in Europe. The situation changed in the 1920s, however, with the appearance of fast-food bars and restaurants. Although this new type of establishment was partly the result of new technologies in the food processing industry and organisational changes in the service sector, it was mainly due to the growth in popularity of personal automobiles. The new establishments were usually situated on access roads into the city, or close to major junctions, or at expressway exits. From this moment on, innovation propagated in the opposite direction—from the United States to the rest of the world.

1.4 Postmodern Period

At the end of the nineteenth century even the biggest cities in Europe were usually small and spatially compact, and could be treated as a whole. The situation is different for contemporary cities, which show considerable internal diversity, not only demographic and socio-economic but also spatial. This diversity can be attributed to several factors. Since the middle of the twentieth century, diversification in the internal structure of cities has increased. They are becoming increasingly ethnically diverse and their economic, cultural and scientific functions extend far beyond their borders. All of this has happened in the context of globalisation, which can be considered as both cause and effect.

The location of restaurants, bars, cafés and other dining facilities will be discussed separately for the inner city, outer city and suburbs. The division of urban agglomerations into inner city and outer city has been used in geography and social sciences for many decades. Though, both concepts have been understood differently depending on time (Schiltz and Moffitt 1971; Downs 1997; Doucet 2010).

In this book, the term 'inner city' is understood as the equivalent of the term 'core'. In 1964 Wagner defined 'core' as '…an area of intensive land use characterized by offices, retail sales outlets, consumer services, hotels, theaters, and banks. This area is generally very compact and tends to be limited by pedestrian walking distances. This part of the downtown area is the hub of a city's mass transportation facilities, and tends to be more adequately served by mass transportation than any other section of the city' (Wagner 1964, p. 39). Three features of 'inner city' may be distinguished: proximity to the centre, historic character and mixed uses (Doucet 2010; Kährik et al. 2016). This approach is also used in urban planning (e.g. eThekwini inner city local area plan 2016).

In the past 150 years or more, however, the inner city has experienced 'a spoiled identity' (Gregory et al. 2009, p. 384), which resulted in its stigmatisation. In the 1980s, when social exclusion and deprivation in the inner city were at the centre of policy debates (especially referring to many declining industrial centres), urban regeneration became a significant policy direction in many European states. This regeneration, in turn, has led to 'a massive middle-class make-over of many inner-city districts' (Gregory et al. 2009, p. 384) within the process of gentrification (Górczyńska 2015; Hamnett 2003; Lees 2000; Smith 1979, 1996).

Another term that needs to be defined is the concept of 'outer city'. It is equivalent to the term 'frame', understood as '…area surrounding the core and involving less intensive land use' (Wagner 1964, p. 39). The main difference between 'inner city' and 'outer city' lies in the share of workplaces: most of them are located within the inner city, while the residential function is more developed in the outer city. Eating services are developed in both types of areas. In Seattle, for example, already over a half of a decade ago the share of eating and drinking places in the 'core' in relation to the whole of the city was 29.2%, and the share of 'frame' was 27.1% (Wagner 1964, p. 41, Table 2).

Both 'inner city' and 'outer city' are elements of traditional bid-rent theories (see Sect. 2.2.2 for more details). They are also distinguished by, e.g. Roulac (2003, p. 383, Exhibit 6, p. 384, Exhibit 7), who also adds a concept of 'connected suburb', understood as areas situated outside urbanised areas. Although suburbs are areas where people, housing, industry, commerce and retailing spread out beyond traditional urban areas, they are still connected to cities by commuting, and constitute increasingly common features of urban regions (Gregory et al. 2009). For these reasons they are also included into analysis in this book.

Finally, cities that mainly depend on tourist as an economic function will also be discussed separately, because their characteristics distinguish them from multifunctional cities.

The reason for introducing the concept of 'outer city' in this book is particularly important in the context of cities of Central and Eastern Europe (in the last decades also in East Asia). Until recently, in cities of the so-called socialist countries there were no suburbs similar to those in agglomerations in North America (i.e. with detached and semi-detached houses with a garage, a garden in front of it or behind the house, etc.). After the Second World War in Warsaw, Prague, Budapest and East Berlin (and earlier in Moscow, Kiev or former Leningrad) new housing estates were built with multi-apartment buildings, often with 10 or more floors, around the historic centre of the city.

These large housing estates were often built in areas that were previously a typical countryside with agricultural functions. Sometimes an old rural building was left as an enclave (e.g. in Prague, Budapest), but generally the former land use was demolished and multi-storey residential buildings were built instead of countryside cottages (e.g. in Warsaw). Due to its origins and specific land use, it is problematic to call these estates typical suburbs. Such settlements, very often gated estates, began to arise *en masse* only after political and economic transformations beginning in 1989. Therefore, in this book this zone is called 'suburbs', and the zone built up by large housing estates before transformation is referred to as 'outer city'. A part of the city called 'outer city' (or 'frame') corresponds to the term 'zone of transition' used by Burgess (1925) or with a much newer concept of 'outer-inner city' (Millington 2012).

Madrid (Spain) is a prime example of the presence of companies involved in gastronomy in particular areas of the city. The data in Table 1.3 and presented in Fig. 1.10 show that in the city centre and downtown there are 2 companies registered as restaurant businesses per 100 inhabitants, and almost 6 firms per 100 ha. For the whole of Madrid, the corresponding figures are 0.59 and 0.299. The second zone is formed by the Chamberí and Salamanca districts, and possibly (based on the number of registered gastronomy activities per 100 ha) Tetuán. They are directly adjacent to the centre and can be considered as part of the inner city. There are also

1 Relations Between Gastronomy and the City

Table 1.3 Registered activity in the gastronomy sector in Madrid (Spain) on 1 January 2017

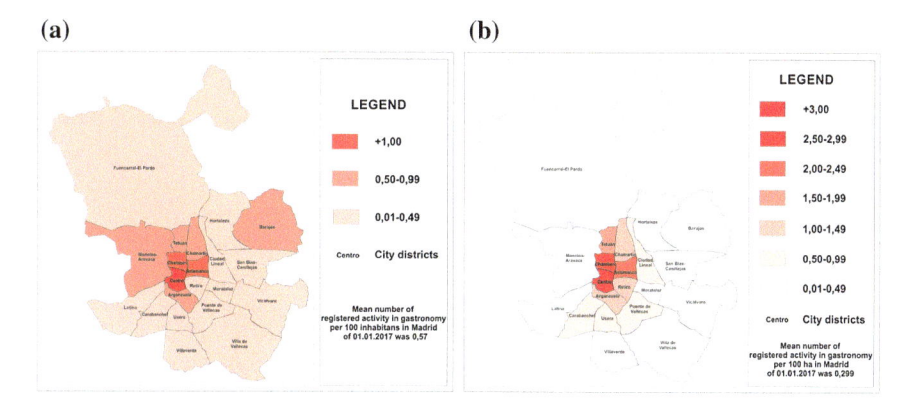

Districts	Gastronomy	Population	Area (hectare)	Number per 100 people	Density (per hectare)
Arganuzela	787	151,520	646	0.52	1.218
Barajas	328	46,264	4190	0.71	0.078
Carabanchel	1107	242,000	1405	0.46	0.788
Centro	2896	132,644	523	2.18	5.537
Chamartín	1001	142,610	918	0.70	1.090
Chamberí	1279	137,532	468	0.93	2.732
Ciudad Lineal	1010	212,431	1143	0.48	0.884
Fuencarral-El Pardo	923	235,482	23,781	0.39	0.039
Hortaleza	656	177,738	2742	0.37	0.239
Latina	891	234,015	2542	0.38	0.351
Moncloa–Aravaca	764	116,689	4653	0.65	0.164
Moratalaz	300	94,607	610	0.32	0.492
Puente de Vallecas	936	227,195	1497	0.41	0.625
Retiro	582	118,559	546	0.49	1.066
Salamanca	1240	143,244	539	0.87	2.301
San Blas	660	153,411	2229	0.43	0.296
Tetuán	1011	152,545	537	0.66	1.883
Usera	571	134,015	778	0.43	0.734
Vicálvaro	218	69,800	3527	0.31	0.062
Villa de Vallecas	374	102,140	5145	0.37	0.073
Villaverde	554	141,442	2018	0.39	0.275
MADRID	18,088	3,165,883	60,437	0.57	0.299

Fig. 1.10 Distribution of registered activity in the gastronomy sector in Madrid (Spain) in 2017. **a** Per 100 inhabitants and **b** per hectare showing differences between downtown, inner city and outer city areas

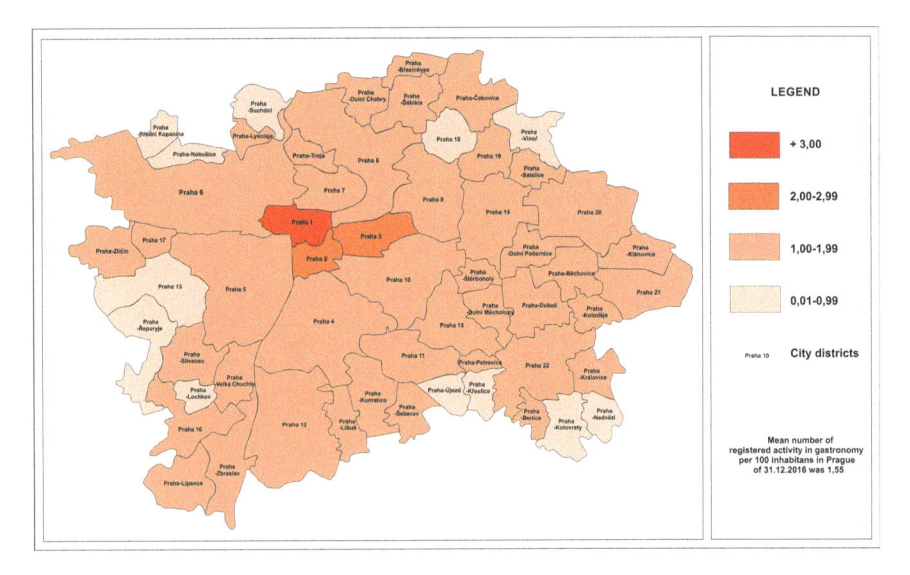

Fig. 1.11 Registered activity in the gastronomy sector in Prague (Czech Republic) in 2016 (by district) per 100 inhabitants

relatively large numbers of companies per 100 ha in Arganzuela, Chamartín and Retiro, which can also be considered part of the inner city. Although two other districts (Barajas and Moncloa–Aravaca) show a slightly higher than average number of companies, they should be considered as part of the outer city.

The spatial distribution of gastronomy functions in Prague (Czechia or the Czech Republic) at the end of 2016 is shown in Figs. 1.11 and 1.12. There are several identifiable regularities. First, the highest concentration of companies is found in the central part of the city. Per 100 inhabitants, the largest index is in the Prague 1 district, which includes the Old Town, Malá Strana (Lesser Town) and Hradčany, which lie on the other side of the Vltava River. These areas of Prague are not only the core of downtown but are also most frequented by foreign tourists. However, compared to Madrid, differences in the economic functions associated with gastronomy are less visible. Although Prague 1, Prague 2 and Prague 3 districts have a higher concentration of activity, differences between the inner and outer city are less clear-cut.

A summary of the calculated differences in the distribution of restaurant businesses in Prague is shown in Table 1.4. The results show that the number of companies is very highly correlated with the population distribution. The more populated parts of the city are home to more firms connected with the restaurant business. It is worth noting, however, that the Pearson's correlation coefficient between registered companies and their number per 100 inhabitants is lower (r = 0.553). Combining this with the high correlation coefficient between the

1 Relations Between Gastronomy and the City

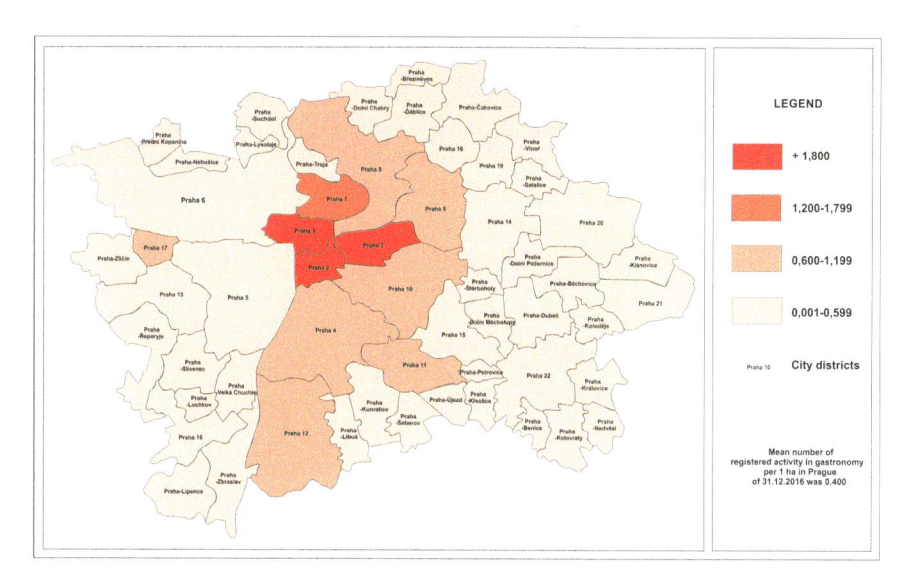

Fig. 1.12 Registered activity in the gastronomy sector in Prague (Czech Republic) in 2016 (by district) per hectare

Table 1.4 Linear correlation coefficients between registered activity in the gastronomy sector (Y) in Prague (Czech Republic) (by district, n = 57) and selected independent variables (X1, X2, X3 and X4) in 2016

Variables		Pearson correlation coefficient (r)	Coefficient of determination (R^2)	Percentage of variation in Y explained by X
X1	Population	0.898	0.806	80.6
X2	Area	0.556	0.309	30.9
X3	Restaurants/ 100 persons	0.553	0.306	30.6
X4	Restaurants/ 1 ha	0.760	0.578	57.8

Note The significance level for Pearson's correlation coefficients for all variables is 0.001

number of registered companies and their number per hectare (r = 0.760) it is possible to infer the spatial concentration in parts of the city that are less densely populated. This applies to the downtown districts of Prague 1 and 2, where there are 3.348, and 3.384 registered firms per hectare, respectively (the average for Prague is only 0.400).

Trends that are similar to those of Madrid and Prague can be found in other large cities. Although the intensity of the gastronomy function may vary from one city to another, the division into three zones (downtown, inner and outer city) is common to all of them.

1.4.1 The Inner City

The inner city often contains several concentrations of catering services, although whether this is the case largely depends on the size of the city. In larger cities that play a central role in the settlement system, the central part is often even more diverse and several other zones can be identified.

The main concentrations for gastronomy in the inner city are (Fig. 1.13):

A—historical city;
B—commercial centre;
C—financial centre;
D—theatre district and entertainment;
E—administrative buildings;
F—selected tourist attractions;
G—ethnic minorities district;
H—post-industrial revitalised area;
I—railway station;
J—public park;
K—riverside (or waterfront);
L—sport and recreation;
M—university.

Treating the city as a continuum, it is unsurprising to find that in contemporary cities and towns one of the main centres for gastronomy is the historical core (zone A in Fig. 1.13). This phenomenon has been confirmed by several studies. For example, in 1993, 25% of survey respondents who visited the Old Town in

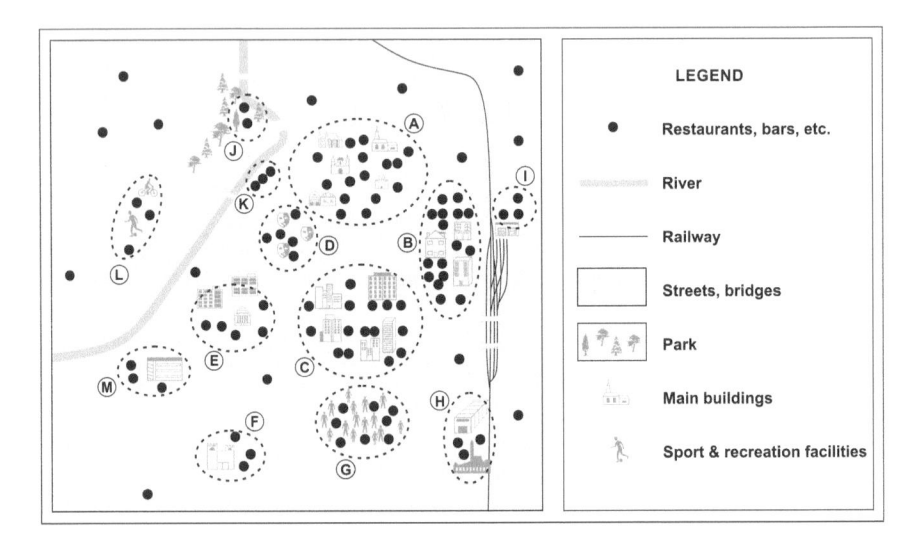

Fig. 1.13 Main concentrations of gastronomy in the inner city

Regensburg (Germany) spent time at cafés and restaurants, while in 2010 this figure had risen to 76% (Monheim 2010, p. 21). There are several reasons for this. Firstly, this area—if it has been preserved—is often still home to restaurants and bars that have a historical tradition, most often dating back to the nineteenth–twentieth centuries, but sometimes the Middle Ages. Second, the historic buildings in this part of the city are important tourist attractions. The presence of tourists results in the creation of more restaurants, bars, cafés, ice cream parlours, etc., often in places where other services have not yet been established. This expansion, which is due to the growing number of tourists is seen on a very large scale in Barcelona, Venice and Prague. In the oldest part of Prague along the 2.5-km Royal Way (*Královská cesta* in Czech) there were 120 eating facilities (29.8% of all services) in the summer of 2015, with the greatest concentration at Nerudova Street at the foot of the former royal castle (Dumbrovská and Fialová 2016, p. 12). Third, administrative and commercial buildings and offices attract residents from other parts of the city. Fourth, some people still live in older neighbourhoods, and some restaurants and bars are linked to their demand.

The defining characteristic of gastronomy in zone A is its great diversity. Many restaurants and bars charge higher prices, partly because of the presence of tourists and partly to offset high rents. Another feature that defines restaurants and bars in the oldest part of town is that they are usually open until late at night. This can cause conflicts with local residents who complain about the noise, as has been seen in the historical areas of Barcelona, Prague and Venice. Although there may be organised culinary events in zone A, they are often aimed at tourists. It should also be noted that historic areas in cities often have narrow streets that were not designed to handle car traffic. This can make competition for road space (including parking and access) particularly acute and constrains the development of retail trades, including the restaurant business (Yen et al. 2015, p. 3) (Figs. 1.14, 1.15 and 1.16).

The next highest concentrations of catering services are found in the parts of the inner city which, if they form a coherent whole, are often called the central business district or CBD (zones B and C in Fig. 1.13). In European cities many of these restaurants and bars are located in buildings built around the turn of the nineteenth and twentieth centuries. Here, restaurants, bars and cafés are a part of the entrepreneurial activity, with a large share of small businesses, especially along streets where independent restaurants occupy most of the space. These catering establishments serve residents, business visitors and tourists. Many customers work nearby and use the restaurants for lunch, or after work. In Brisbane (Australia) the mean distance between the workplace and a restaurant (or café) is 490 m (Burke and Brown 2007, p. 25). As in zone A, gastronomy in these zones is very diverse in terms of the range of food and prices. But in many cases, the opening hours of restaurants and bars are shorter because they are adapted to office hours. This is especially true for facilities located in office buildings or their immediate vicinity. Some very expensive restaurants are used for business lunches, while others are much cheaper and serve employees. The opening hours for this type of facility generally run from early morning (usually between 8:00 to 10:00) until early evening (Figs. 1.17 and 1.18).

(a) (b)

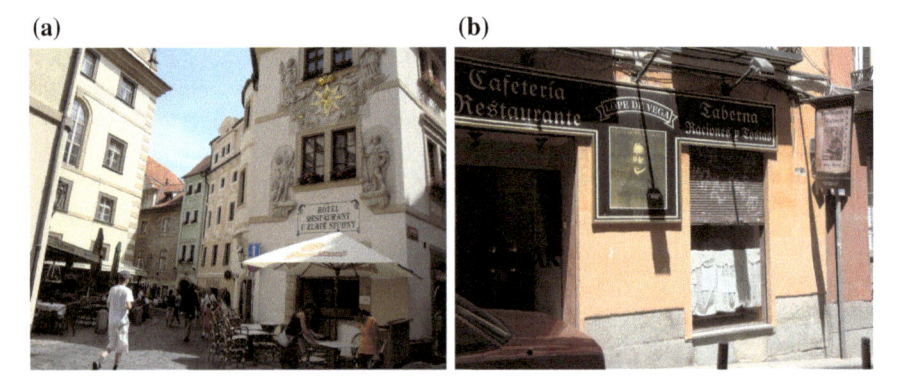

Fig. 1.14 a Eating establishments in the historical centre of Prague (Czech Republic). On the right-hand side is the restaurant U Zlaté studně, which is located in the Dům U Zlaté studně (built in seventeenth century in Renaissance style) at the junction of Seminářská and Karlová Streets. In the house to the right there is a fast-food bar, the café Clementin (green building) and the Hotel at Black Star restaurant. On the left side is one of the terraces of the café Clementin. b The Restaurante Lope de Vega in the old part of Madrid. This restaurant is situated on the Calle de Lope de Vega 37, not far from the museum Casa Museo Lope de Vega (Calle de Cervantes 11), where the famous Spanish playwright, poet and novelist, Félix Lope de Vega y Carpio (1562–1635) lived. Both cases refer to the concept of "place". Source: Anna Kowalczyk 2009, 2015

(a) (b)

Fig. 1.15 a Nerudova Street in the Malá Straná quarter of the old centre of Prague (Czech Republic). This street has the biggest concentration of restaurants and bars on The Royal Way (Královská cesta). b Roof terrace restaurant in one of the medieval houses on Nerudova Street (viewed from Hradčany castle). Source: Anna Kowalczyk 2006

Concentrations of restaurants, bars and cafés are often found in the area called the theatre district, referred to here as zone D. Such neighbourhoods are a central part of London, Antwerp, Amsterdam, Boston, Cleveland, Tokyo (Hibiya district) etc., but the most famous is New York City's theatre district in midtown Manhattan. The main characteristic of restaurants and bars in this part of the inner city is that they open in the afternoon (or early evening) and do not close until just before

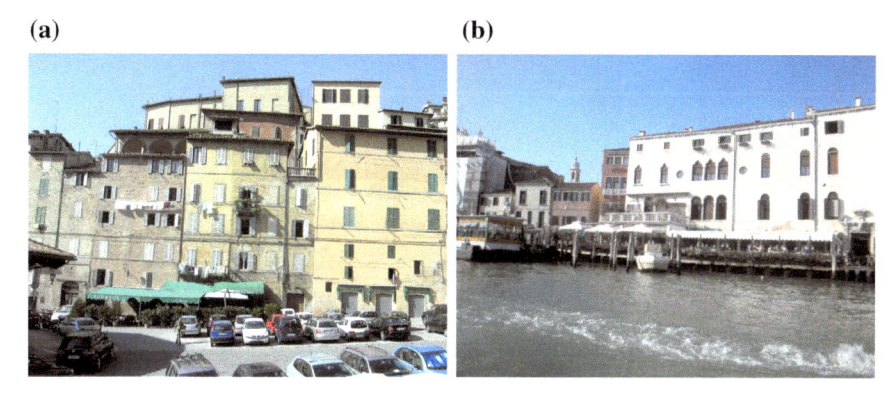

Fig. 1.16 Eating establishments in the historical parts of Italian cities. **a** *Taverne* with terraces in the old part of Sienna. **b** *Ristoranti* on the banks of the Canal Grande in Venice. In both cases, the dominant groups of visitors are overseas tourists. Source: Anna Kowalczyk 2004

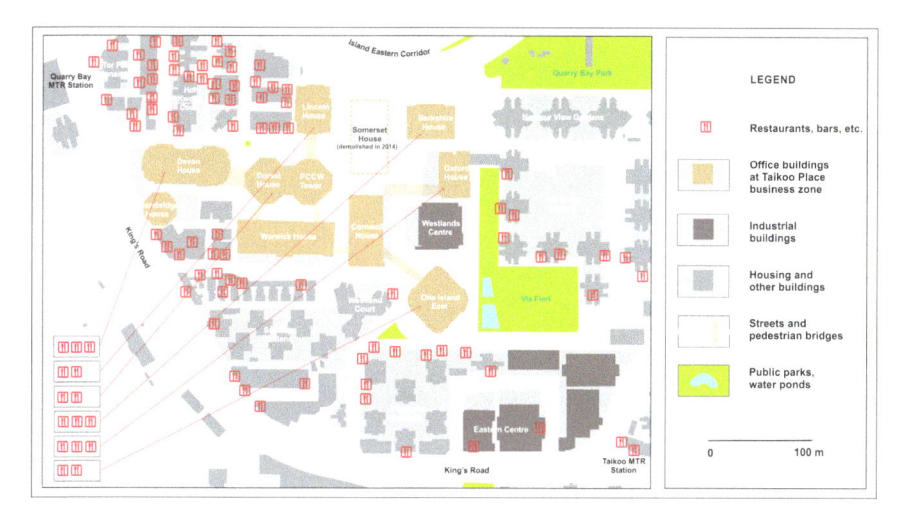

Fig. 1.17 Eating establishments in the Taikoo Place business zone in eastern Victoria (Hong Kong Special Administrative Region of the People's Republic of China) in 2017

dawn. Like restaurants, bars and cafés located in zones B and C, many facilities are housed in buildings dating from the nineteenth or early twentieth centuries. In general, they are expensive and visited by upper social classes and tourists (Fig. 1.14). A good example of this type of gastronomy are the restaurants, cafés and wine bars that can be found inside the Obecní dům (Municipal House) in the centre of Prague (Fig. 1.19).

Zone E gastronomy is very similar to that in zones B and C, although the opening hours are more restricted. This is because public administrations generally have shorter working hours than corporations, banks and insurance companies.

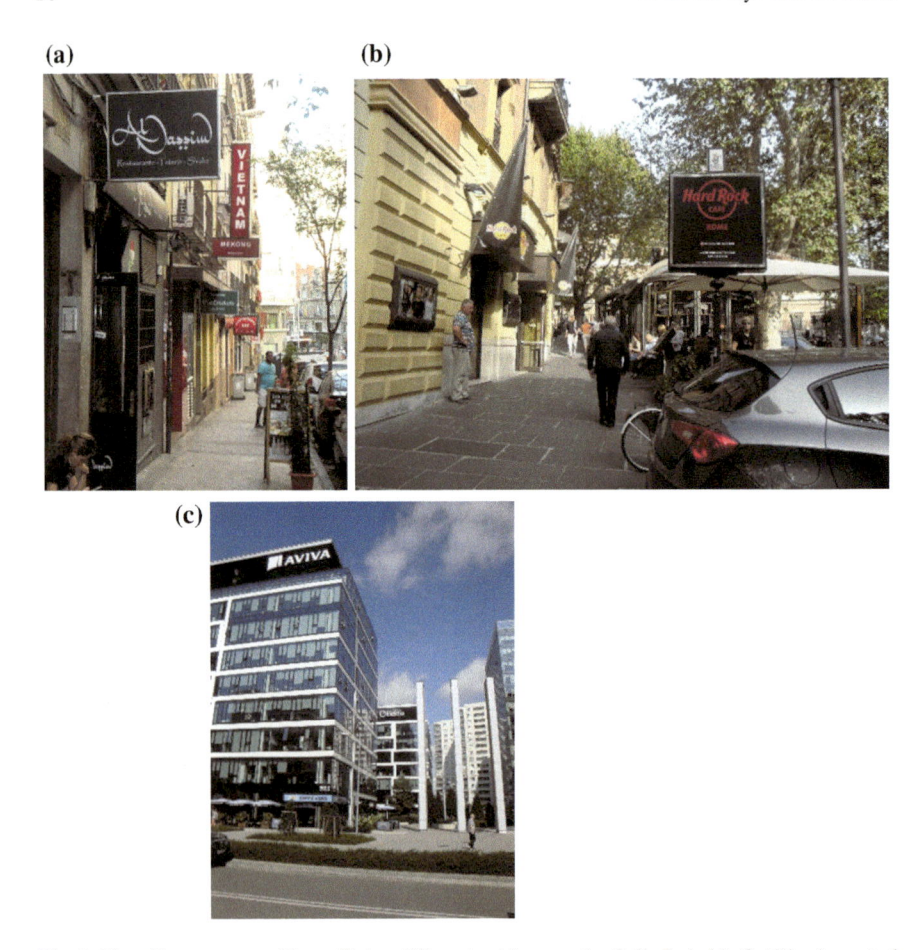

Fig. 1.18 **a** Restaurants and bar offering different cuisines on the Calle Isabel la Católica in central Madrid. On the right is Gran Via, the main street in the commercial centre. **b** The Hard Rock Cafe Rome restaurant on Via Vittorio Veneto, one of the busiest streets in central Rome. The Hard Rock café is a chain of themed restaurants that was founded in London in 1971. By June 2017 it had 181 bars worldwide. **c** Coffee bar operated by Green Cafè Nero, part of the Polish–British joint-venture chain in the Gdański Business Centre, a new office park in the inner city of Warsaw. Buildings in this business park mainly house the headquarters of transnational corporations from Poland and the Central Eastern European region. Source: Anna Kowalczyk 2015

A good example of zone E is the EUR (Esposizione Universale di Roma) district in Rome, where there are many ministries and other state offices.

In some large cities, tourist attractions are situated beyond the historical core, although still in the inner city. In Barcelona, for example, tourists are attracted to the church La Sagrada Família and Parc Güell both of which are some distance from Las Ramblas and the Barri Gòtic in the historical part of the city. The same is true of Berlin, where the palace and park of Charlottenburg are several kilometres

Fig. 1.19 a The main entrance to *Obecní dům* (Municipal House) in Náměstí Republiky square, in the centre of Prague. The building, which opened in 1912, is in the Art Nouveau (*Jugendstil* in German, *Secessiosstil* in Austrian; *Modernismo* in Spanish) architectural style. The ceremonial halls on the first floor house the works of Alphonse Mucha (1860–1939), the painter and decorative artist. The building also includes five restaurants and cafés, and is used as a concert hall (with 1200 seats) and ballroom. **b** Restaurant tables in front of the *Stadttheater* in Baden bei Wien (Austria). **c** Eating establishments close to the theatre in Wrocław (Poland). On the right is Café Kulisy in the building of the Teatr Polski theatre. In the background are the bars Mekong and Hot Seven, which are situated under the railway viaduct. Source: Anna Kowalczyk 2009, 2013, 2017

away from other tourist attractions. Accordingly, in some cities the cluster of dining establishments can be categorised as zone F. Usually this zone contains mid-range restaurants and bars, often casual dining and fast-food establishments, which are open at the same time as tourist attractions (Fig. 1.20).

The ethnic minorities zone (G in Fig. 1.13) was present in ancient and medieval cities and towns, and is becoming increasingly important. The specific issues related to ethnic cuisine are discussed elsewhere in this book. It should be pointed

Fig. 1.20 a Restaurant Parc Güell near the main entrance to Parc Güell, one of the main sightseeing spots in Barcelona. **b** Café Lennon on Plaça de John Lennon in the Gràcia quarter in Barcelona. This café is located some 600–700 m from La Sagrada Família and Parc Güell. Source: Anna Kowalczyk 2015

out here, however, that ethnic gastronomy is not exclusive to the central part of cities, as is the case for the Chinatown districts of New York, San Francisco and Tokyo. It should also be noted that ethnic gastronomy is not only limited to areas where immigrants live (another topic that is developed later in this book). Many are located in other parts. In Madrid, for example, one of the streets in the commercial centre has adjacent Arabian and Vietnamese restaurants (Fig. 1.18a). In general, restaurants and bars serving ethnic cuisines have long opening hours (sometimes from early morning to late night) and prices are low (Fig. 1.21).

A new phenomenon in contemporary cities is the development of catering functions in what were once industrial areas (former ports and railways). As industry collapsed in large cities (in Europe, in the 1960–1970s), large areas of land became available and were redeveloped and revitalised. When these post-industrial areas were in the city centre, development projects often incorporated a gastronomic function. Zone H (Fig. 1.13) generally includes high-quality (and highly priced) restaurants and bars, for several reasons. Firstly, these redevelopment projects are usually carried out by real estate developers who want to recoup their costs. Second, in contemporary cities these urban spaces have become fashionable among richer residents who consider them part of their lifestyle. Third, redeveloped areas are attractive to tourists. Fourth, companies that are situated in post-industrial buildings (often IT or design companies) offer high wages, and their employees often frequent these restaurants, bars and cafés. They also visit their friends who live in loft apartments located in former industrial facilities, more often to drink and socialise, than to eat (Fig. 1.22).

Unlike zone H, areas near, and even inside railway stations started to become gastronomy clusters in the nineteenth century (zone I). The role of restaurants and

Fig. 1.21 **a** A Cartouxina restaurant at the Rua das Farinhas 7 in the Mouraria quarter of the old part of Lisbon. This small, family owned restaurant offers exotic African meals from the Republic of São Tomé and Príncipe. Although this is a typical example of an ethnic restaurant, it is not located in a part of the city where ethnic minorities live. **b** Mustafa's Gemuese Kebab in the Kreuzberg district of Berlin. This street food bar has gained cult status (note the long queue) among residents and visitors to Berlin. It is situated in an area that has a high percentage of immigrants (in 2010, 10.9% of residents were of Turkish origin). Source: Anna Kowalczyk 2014

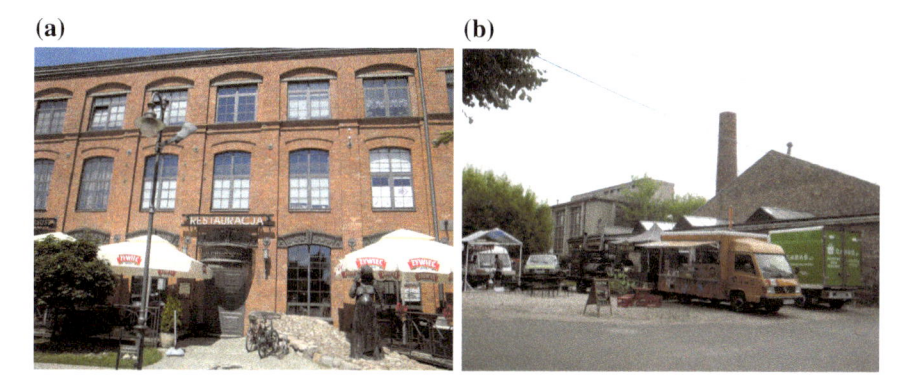

Fig. 1.22 **a** Restaurant in the building of the former textile factory at Żyrardów (Poland). An example of an eating establishment in a redeveloped industrial area. **b** Food trucks in the revitalised SOHO Factory in Warsaw (Poland). In this post-industrial zone, a few casual dining restaurants have opened, which sometimes host culinary events (e.g. food truck meetings). Source: Anna Kowalczyk 2015

bars in this area (which generally serve low-priced dishes) is similar to taverns in the port districts of medieval towns. They are mainly frequented by visitors and tourists and are characterised by their long opening hours (sometimes almost 24 h, with a very short closure for cleaning) (Fig. 1.23).

Gastronomy in public parks (zone J) dates back to the nineteenth, or possibly even the eighteenth century. It started with the largest and most elegant parks of Paris, Vienna, London, Berlin and Madrid, which provided cafés and pastry shops

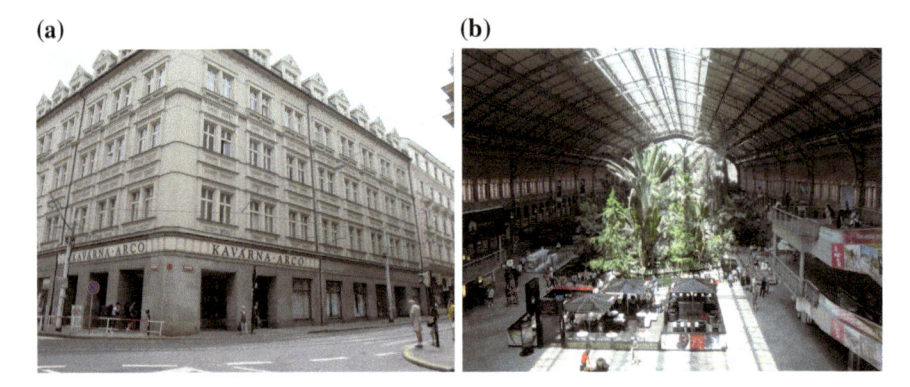

Fig. 1.23 a Café Kavárna Arco opposite the Masarykovo nádraží (Masaryk's Railway Station) in Prague (Czech Republic). This coffee house was one of the places that was often visited by the author Franz Kafka. **b** Café bar inside Atocha Railway Station (Estación de Atocha) in Madrid. Source: Anna Kowalczyk 2009, 2015

to the park's visitors. The situation still prevails today, and the characteristic feature of establishments located in parks is that they are mostly visited during the summer and at weekends. It is also worth highlighting that various gastronomy events—food truck meetings, culinary festivals and competitions, etc.—are often organised in zone J. These events often attract thousands of people, mostly made up of the city's residents (unlike those organised in zone A) (Fig. 1.24).

Figure 1.25 shows some examples of zone K, situated beside bigger rivers or lakes or the seaside. These shore areas offer different leisure facilities, including cultural institutions such as museums, art galleries and theatres. The combination of these facilities can create a waterfront. Examples of cities with a well developed zone K include London (London Docklands), Liverpool, Manchester, Barcelona and Genoa in Europe, Baltimore, Boston, New York, Toronto and Vancouver in North America, Cape Town in Africa, Singapore and especially Victoria and Kowloon (Hong Kong) in Asia, and Melbourne in Australia. In all of these cities the types of restaurants and bars range from the very expensive to fast-food catering. Because zone K fulfils a similar role to zone D, its restaurants and bars are usually open until late at night. They are regularly used during the week if zone K is close to the city centre, otherwise they are used mostly at weekends. According to some researchers, restaurants, bars and cafés located in this part of the city are generally considered to be prestigious and good places for socialising. Eating an elegantly served dish or drinking originally prepared coffee is regarded as a symbol of one's socio-economic status (Goss 1996, pp. 238–239).

Zone L is sometimes found in the inner city, but more often lies in the outer city. Its restaurants and bars are adjacent to recreational areas, sports and exhibition facilities, etc. Most catering establishments are either moderately expensive or cheap. Their opening times are closely linked to those of the nearby facilities, most often, Saturdays and Sundays.

1 Relations Between Gastronomy and the City 31

Fig. 1.24 **a** Biergart'l im Stadtpark in Vienna (Austria). This 170-seat restaurant and café is located in Stadtpark, the first public park in Vienna, which opened in 1862. At weekends, visitors can attend open-air concerts. **b** Restaurante Eleven in the Jardim Amália Rodrigues public park in the centre of Lisbon (Portugal). **c** The dining facility for visitors to Wrocław's (Poland) zoological garden. On the right is a signpost showing directions to attractions. In the background is the bar. Source: Anna Kowalczyk 2013, 2014, 2017

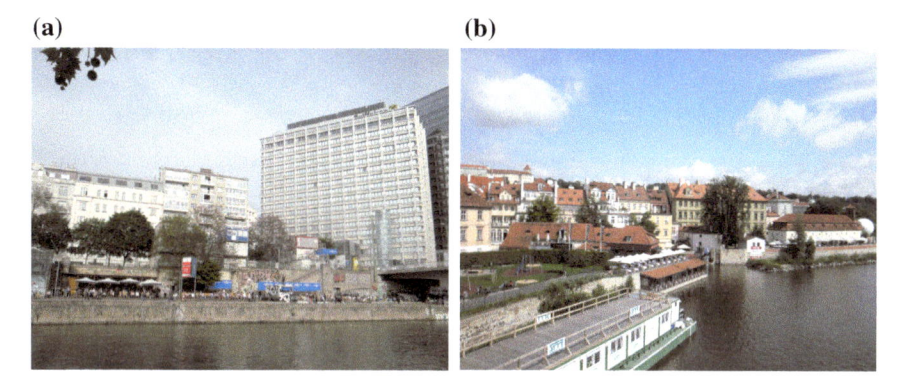

Fig. 1.25 **a** Bars along the Donaukanal in the centre of Vienna (Austria). **b** Restaurants on the bank of the Vltava River in the old part of Prague (Czech Republic). The main difference between these two examples is that bars and cafés on the bank of the Donaukanal are visited mostly by residents of Vienna, whereas in Prague high-priced restaurants are more tourist-oriented. Source: Anna Kowalczyk 2013, 2009

Finally, zone M includes universities, research institutes, etc. There are relatively few restaurants here, and the main types of catering establishments are bars, cafés and canteens, many of which are housed in faculty buildings or libraries. Their opening hours are tied to the hours of universities and libraries, and usually have a limited menu with low prices. In general, they are mostly visited by students and local staff.

The inner city zones described above may not all be present in every city, although most of them will be found in large cities. In some cases, there are exceptions, with a cluster of restaurants and bars in parts of the city other than those identified above. In the inner cities of Copenhagen and Vienna, for example, there are theme parks (the Tivoli and the Prater, respectively) within which (and nearby) there is a high concentration of catering establishments. In other cities, some zones (for example within and near a university) may appear in the outer city or even in the suburbs.

1.4.2 The Outer City

It can be difficult to establish a clear boundary between the inner and outer city. One distinction is when there are more people living in a given area than there are workplaces. Applying this definition to the outer city, there are usually six main zones (Fig. 1.26):

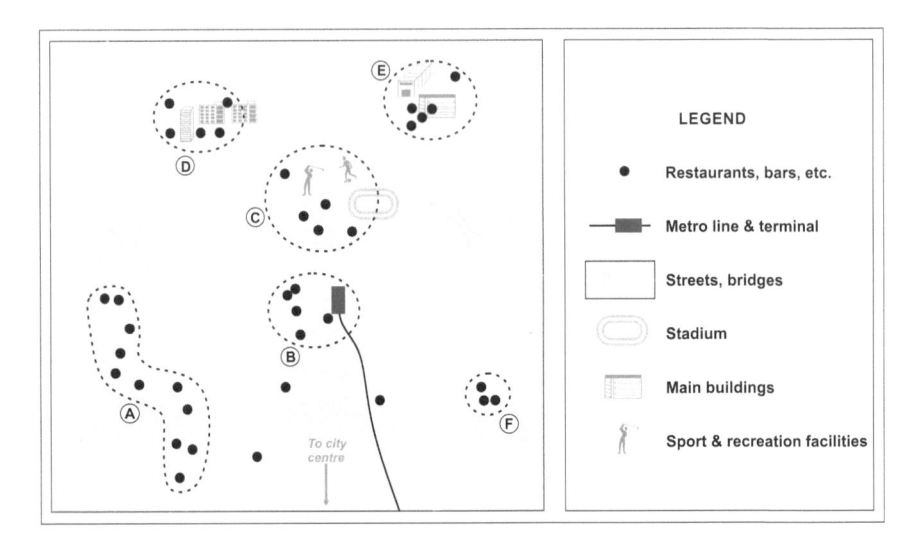

Fig. 1.26 The main gastronomy zones in the outer city

Fig. 1.27 a Fast-food bars at the urban rapid transit station (U-Bahn Linie U 6) Alterlaa in south-west Vienna (Austria). **b** The fast-food bar Konnopke's Imbiß in the district of Berlin-Prenzlauer Berg (Berlin, Schönhauser Allee 44B). The Konnopke's Imbiß was established in 1930 close to the Eberswalder Straße metro station on Linie-U 2. It is one of the most iconic places to eat for *Currywurst* lovers in Berlin. Source: Anna Kowalczyk 2013, 2014

A—along the main (or transit) street connecting the inner city to the suburbs;
B—around the terminus of the metro or other urban rail transit system (metro, light rail or *Stadtbahn*, tram, etc.);
C—in or near sport and recreation facilities;
D—in business parks;
E—in the commercial centre;
F—in residential housing estates.

The first zone (A) consists of restaurants and bars located along the main street that connects the city centre to suburban areas. They serve both local residents and visitors (who travel along the main street). In general, these restaurants and bars offer poorer quality food. Many belong to multinational companies like the fast-food chains, McDonald's, Pizza Hut, KFC, etc.

Zone B includes facilities located near to a terminus of the urban rail transit system. Most often, these establishments are fast food and street food bars (rather than restaurants) (Fig. 1.27) that serve people returning home after work or school. Very often their clientele comprises people who transiting from the urban rail system to buses and who do not have much time to eat. Usually, the dishes offered are cheap. The characteristic feature of these bars is their long opening hours (often from 6:00 to 00:00 or later), which are usually linked to the timetable of the adjacent transport facility.

The third concentration of catering establishments is in the vicinity of sport and recreation facilities (zone C). Sometimes clusters of this type of restaurants and bars are found near stadiums and sports halls. They usually offer a narrow range of cheap dishes. Zone D includes catering facilities that are mainly located in office

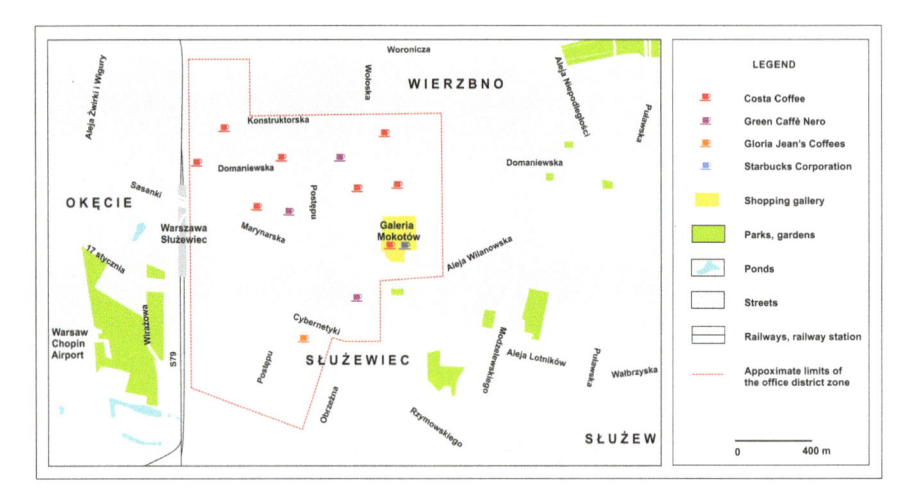

Fig. 1.28 Distribution of coffee chain shops in the Służewiec area (Mokotów borough) of Warsaw (Poland) July 2017

buildings. This concentration corresponds to zone C in Fig. 1.13b. The main difference between them is that in outer city business parks there are fewer high-quality restaurants. These restaurants and bars are generally open most of the day, from about 08:00 or 09:00 until late in the evening.

The concentration of dining services that are related to business and commercial functions in the outer city is exemplified by the Służewiec area, in the southern part of Warsaw. Research conducted in 2017 showed that there were 11 coffee shops belonging to international chains (excluding independent canteens and bars) located in office buildings housing the headquarters of foreign trade and financial corporations (Fig. 1.28). These establishments are owned by the British chain Costa Coffee, the British–Polish company Green Caffè Nero and the Australian chain Gloria Jean's Coffees. In the Galeria Mokotów mall, adjacent to the office blocks, is another Costa Coffee shop and one belonging to the Starbucks Corporation.

Zone E comprises restaurants and bars that primarily serve shopping centre customers (Fig. 1.29). These are rarely prestigious facilities; they mainly comprise mid-range restaurants and fast-food bars, and less formal restaurants. The latter are often visited by families at weekends.

The final concentration of dining establishments in the outer city is zone F. This is made up of restaurants, bars and cafés catering for residents of the housing estates that are located some distance from the outer city and are similar in character to the suburban areas of cities. These catering facilities are therefore described in more detail in the section that looks at gastronomy in the city suburbs.

(a)   **(b)**

Fig. 1.29 a The view from the car park of the Kentucky Fried Chicken fast-food restaurant between The Grove Mall and the MegaCentre in Windhoek (Namibia). This restaurant lies in the mall area in the Klein Kuppe district in the southern part of Windhoek. This district is inhabited mainly by the middle classes. **b** The entrance to another fast-food restaurant in The Grove Mall. This restaurant belongs to Wimpy SA (created in 1934 in the United States as Wimpy Grills), a multinational chain of fast-food restaurants headquartered in Johannesburg (South Africa) since 2007. Source: Anna Kowalczyk 2016

1.4.3 The Suburbs

Areas located on the outskirts of cities act as a transition zone between the city and the countryside. For this reason, the gastronomic infrastructure is a hybrid of the city-specific pattern and rural areas.

Nine types of gastronomy can be distinguished in suburban areas (Fig. 1.30):

A—in former independent towns, which have been merged into the urban zone;
B—inside and in the vicinity of a civil airport;
C—in business parks;
D—in commercial centres;
E—in new housing estates;
F—along the transit street connecting the city with airport, expressway, etc.;
G—in industrial and commercial parks, wholesale centres, etc.;
H—at expressway rest areas;
I—in recreational zones.

Zone A is the first type. These catering establishments are in towns that already existed but were later incorporated into the urbanised area. Most of the restaurants, bars and cafés are located here because these areas are larger than the surrounding villages and play a broader role in the local settlement system. This means that eating establishments in this zone serve the needs of residents, plus those of nearby villages and sometimes visitors (e.g. tourists, business travellers).

Zone B covers gastronomic facilities located at a civilian airport. They are generally located inside the airport terminal, but sometimes may lie outside (e.g. in

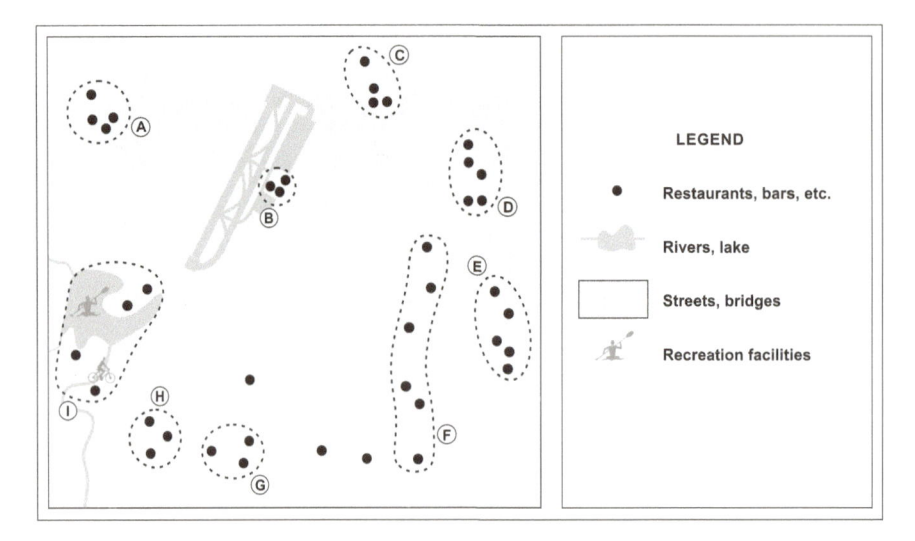

Fig. 1.30 Gastronomy in suburban areas

transit hotels that are often situated near larger airports). These establishments tend to be restaurants and bars with moderate prices and are open 24 h a day (Fig. 1.31).

Zone C concerns office or business parks. These parks are a relatively new phenomenon and are often adjacent to airports and main expressways (especially interchanges). As with the outer city (only more so), these business parks rarely

Fig. 1.31 a Entrance to the Gordon Ramsay Plane Food restaurant in Terminal 5 at Heathrow Airport in London (United Kingdom). **b** The food court, My City Helsinki, at Helsinki-Vantaa International Airport. The difference between these two dining facilities is clearly visible—the first is isolated from its surroundings, whereas the second is in the public space. According to Marc Augé (1995), both cases can be considered as "non-places" (read more in the Sect. 2.2.4). Source: Anna Kowalczyk 2011, 2016

include expensive restaurants and bars, cafés and canteens dominate. Zone D has the same characteristics as Zone E in Fig. 1.26. Therefore, there is no need to describe it. Zone E is specific to the suburbs. These suburban areas, adjacent to pre-existing towns (some may be cities, marked with an A in Fig. 1.26) are built on greenfield sites in new housing estates. It often takes time for these facilities to become established, as housing estates develop. Their main task is to satisfy demand from suburban residents, and they usually consist of bars and restaurants with low-priced meals. Opening times can vary because some (pubs and wine bars, in particular) function as places for residents to socialise after work.

As Zone F is similar to the cluster of restaurants and bars marked in Fig. 1.26 as zone A, it is not discussed further here. Zone G is also unique to the suburbs. This is because logistics and wholesale centres are generally only found in suburban areas. These parks often employ several thousand people and operate 24 h a day. There is therefore a need for employees to be able to eat in cheap local bars and canteens, because their workplace is usually located a considerable distance from towns or larger villages that have a well developed gastronomy infrastructure. Zone H also tends to be unique to suburban areas. The gastronomy clusters here are generally small and rarely include more than 3–5 restaurants and bars. As a rule, dining facilities at rest areas along expressways are open almost 24 h a day and offer a limited range of cheap (and usually fast) food.

Finally, zone I covers recreational and tourist areas that attract the urban population at weekends. As these areas are places of recreation for various socio-demographic and socio-economic groups, there are both high-priced restaurants and bars (with high-quality service), and fast-food bars. Some of the restaurants in zone I are famous for a particular cuisine and are therefore visited for this reason. These usually have the highest prices, which are often the same (or even higher) than similar restaurants in the city centre.

One example of a restaurant located in the recreation zone of a large city is Gaststätte Neu Venedig in Berlin (Fig. 1.32a). Its customers are primarily owners of neighbouring summer homes, bikers and weekend tourists with kayaks, boats, yachts and motorboats.

The restaurant Neu Venedig (official name Gaststätte Neu Venedig) is located on the Neu-Venedig estate at the south-eastern periphery of Berlin. This area was acquired by the city council in 1890, and became an administrative district of Berlin in 1920. The lands of the former Rahnsdorf estate included the swampy Spreewiesen, a flood plain of the Spree River. In 1925 the local authority approved development plans for part of the Spree meadows and the area was drained by canals from 1926 onwards. The area was initially called Neu-Kamerun (New Cameroon), but later renamed to Neu-Venedig (New Venice). Neu-Venedig has six islands, nine bridges and three pedestrian bridges. In 1928, the Neue Rahnsdorfer Terrain-Actiengesellschaft was commissioned to sell the 374 plots (of 600 m^2 each) to private owners.

Fig. 1.32 a Gaststätte Neu Venedig restaurant in the Köpenick district of Berlin (Germany). This is an example of a restaurant located in a former suburb of the city. It is closed on Tuesdays and can be entered either from the street, or directly from a boat. **b** The Grinzing-Heuriger Rudolfshof restaurant in Vienna (Austria). This is an example of a *Heuriger*: a tavern that is typical of the outer zone of Vienna and its surroundings. Source: Anna Kowalczyk 2014, 2013

Neu-Venedig subsequently became an estate with summer cottages for Berlin's lower middle class. There are no permanent residences because the area lies in a river flood plain.
Source: *Neu Venedig. Geschichte*, http://www.neu-venedig.de (Accessed 25 May 2017).

These types of eating facilities were often established on the outskirts of large European cities and in their suburban areas in the nineteenth and twentieth centuries. Examples include the restaurants and bars at Casa de Campo (Madrid), Parque Florestal de Monsanto (Lisbon), and Bois de Boulogne and Bois de Vincennes (Paris). Many of these facilities are seasonal, usually opening during the summer, but for the rest of the year they are either closed or only open at weekends. Another example of a restaurant (or wine restaurant) closely related to leisure is the *Heuriger*. They are mainly located outside the urban core, in the peripheries of the city. Areas with the largest numbers of outlets are former villages, which were amalgamated to Vienna at the end of the nineteenth century (e.g. Grinzing or Heiligenstadt; see Table 1.5). In these restaurants local winemakers serves new wine under a special licence in alternate months during the growing season. They are prevalent in the northern part of Vienna and are famous for their convivial atmosphere, which is result of enjoying young wine, traditional food and folk music in one place. *Heurigers* are often frequented by tourists, particularly visitors to Grinzing, part of the district of Döbling (Fig. 1.32b).

1 Relations Between Gastronomy and the City

Table 1.5 Selected *Heurigers* in Vienna in 2017

Area	Name	Address
Innere Stadt	Bitzinger's Augustinerkeller	Augustinerstraße 1
	Esterhazykeller	Haarhof 1
	Zwölf-Apostelkeller	Sonnenfelsgasse 3
Heiligenstadt	Mayer am Pfarrplatz	Pfarrplatz 2
	Feuerwehr Wagner	Grinzinger Straße 53
	Heuriger Muth	Probusgasse 10
	Heuriger Sirbu	Kahlenbergerstraße 210
	Werner Welser	Probusgasse 12
Grinzing	Alter Bach-Hengl	Sandgasse 7-9
	Altes Presshaus Grinzing	Cobenzlgasse 15
	Weingut Cobenzl	Am Cobenzl 96
	Weingut-Destillerie-Heuriger—Dr. Müller-Schmidt	Cobenzlgasse 38
	Passauerhof	Cobenzlgasse 9
	Der Reinprecht	Cobenzlgasse 22
Neustift	Fuhrgassl-Huber	Neustift 68
	Hengl-Haselbrunner	Iglaseegasse 10
	Kierlinger	Kahlenberger Straße 20
	Der Wiener Heurige Wolff	Rathstraße 44-50
	Zeiler am Hauerweg	Rathstraße 31
Jedlersdorf	Christ	Amtsstraße 10-14
Stammersdorf/ Strebersdorf	Göbel	Stammersdorfer Kellergasse 131
	Schilling	Langenzersdorfer Straße 54
	Wieninger	Stammersdorfer Straße 78
	Winzerhof Leopold	Stammersdorfer Straße 18
	Fritsch-Wanderer	Stammersdorfer Straße 76
	Schmidt	Stammersdorfer Straße 105
Mauer	Edlmoser	Maurer-Lange-Gasse 123
	Wiltschko	Wittgensteinstraße 143
	Zahel	Maurer Hauptplatz 9
	Heuriger Neuwirth	Maurer Lange Gasse 18
	Steinklammer	Jesuitensteig 28
Ottakring	10er-Marie	Ottakringer Straße 222-224
	Buschenschank Huber	Roterdstraße 5
Oberlaa	Zum Werkelmann	Laaerwald 218

Source: http://www.stadt-wien.at/lifestyle/essen-trinken/heuriger-in-wien-ein-kleiner-fuehrer-durch-die-heurigenorte.html (Accessed 25 May 2017)

Heuriger is the abbreviation for *heuriger Wein* (this year's wine) in the Austrian and Bavarian dialects of German. Originally, it was an open-air tavern on the winemaker's premises, where people would bring simple food and drink the new wine. In the renowned wine-growing areas of Vienna (Grinzing, Sievering, Neustift and Liesing) many of the restaurants that are oriented towards domestic, but mainly overseas tourists, have a rustic interior design.

Vienna's *Heurigers* play a significant role in the promotion of tourism. Information is provided in booklets published by the city's authorities, on official websites and in other promotional materials. Several have a long history. Feuerwehr Wagner at Grinzing, for example, was established in 1683, and at the Mauer Heuriger, the Steinklammer family that owns it has been serving wine and traditional Viennese cuisine since 1697. Similarly, the Passauerhof at Grinzing was established in the old monastery cellar in the twelfth century, and visitors to Zeiler am Hauerweg at Neustift can view the original wine press from 1730. It should also be mentioned that the renowned music composer, Ludwig van Beethoven drank at *Mayer am Pfarrplatz*, highlighting the important role of the *Heuriger* tradition in the city's cultural heritage.

1.4.4 Tourist Cities

As tourism developed in the eighteenth–nineteenth centuries, dining establishments —along with hotels—became a significant land user. At that time, tourism was essentially a privilege of the upper classes, and most restaurants and cafés had elegant interiors and served expensive food. In most cases, these luxury restaurants were situated in the vicinity of hotels, on promenades, piers, in parks, near spas and sources of mineral water, at concert halls, next to theatres and in other places visited by tourists. This may be seen as an archetype of what Finkelstein later called 'foodatainment', noting that 'ambience and the aestheticization of food have made eating-out into a fashionable performance, and, as such, part of the expanding, insinuating entertainment industries' (Finkelstein 1999, p. 131).

The same situation prevails today. There are different types of tourist cities and towns—seaside resorts, so-called hill stations (*stations climatique* in French), ski resorts (*stations de sports d'hivers*), pilgrimage centres, spa towns and many others —but changes in the urban space due to touristification are best seen in the coastal cities and towns that are primary targets for mass tourism. The scale of the phenomenon and the diverse needs of tourists mean that, in coastal cities, tourist infrastructure is often the most developed and diverse, as is the gastronomy.

1 Relations Between Gastronomy and the City

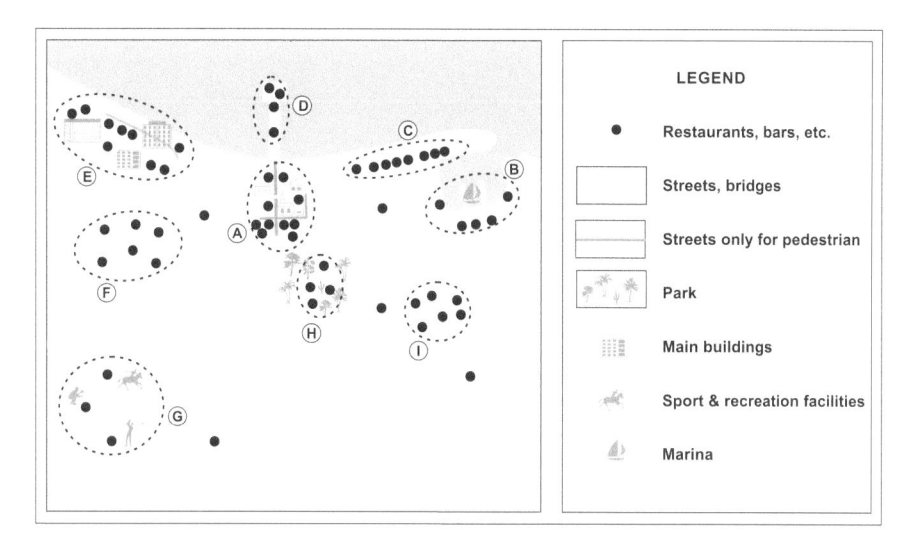

Fig. 1.33 The main zones of concentration of gastronomy at seaside tourist cities

The main concentrations of eating facilities in seaside tourist cities (shown in Fig. 1.33) are:

zone A—historical towns or villages that were later transformed into tourist cities and towns;
zone B—former ports (usually fishing) that were converted into marinas;
zone C—esplanades (promenades) along the beach;
zone D—piers;
zone E—areas with a concentration of hotels;
zone F—parts of the city with apartments for renting, second homes, etc.;
zone G—areas with recreation and sport facilities;
zone H—public parks;
zone I—residential areas, where local inhabitants live.

Since many of today's tourist cities have existed for a considerable time (although some developed from villages), they have often preserved their historic centre, which is one of their main attractions (zone A). Examples of such cities can be found in Europe, Latin America and Asia. The presence of tourists is the main factor driving business. This affects both prices (usually high) and meals, which often reflect the tastes of tourists. Figure 1.34 shows restaurants and bars located in the historical parts of Funchal (Madeira, Portugal), Arrecife and Puerta del Carmen (Lanzarote, Spain) and Lindos (Rhodes, Greece). These locations attract many tourists, who are interested in the historical area, and who often decide to eat lunch or dinner there. An excess of restaurants or bars in the older parts of tourist cities can lead to too much competition as each try to attract business.

Fig. 1.34 Examples of eating establishments in historical buildings situated in tourist cities. **a** Funchal, Madeira (Portugal). **b** Arrecife, Lanzarote (Spain). **c** Playa del Carmen, Lanzarote (Spain). **d** Lindos, Rhodes (Greece). Where a restaurant lies inside an historic building, this becomes an additional attraction for visitors. Source: Anna Kowalczyk 2009, 2010, 2011, 2016

Residents are rarely the main customers of restaurants in the historical centres of tourist cities. Therefore, the heart of tourist cities is the location for fast-food restaurants and restaurants serving traditional dishes (attractive to tourists). This is because chain restaurants are well-known to overseas visitors, something that is particularly important in countries that have an exotic cuisine. Some chain restaurants are also located in places that are not visited by the local population, except for those working in the tourist industry (hotel staff, bar and night clubs workers, tourist office staff and souvenir shop workers).

Many tourist cities include seaside districts, where there is a clear concentration of facilities near the harbour and in the vicinity of the seafront: along the promenade, beach, etc. (zones B, C and D). Figure 1.35 shows restaurants in the immediate vicinity of a harbour, which was formerly a fishing port, and now serves as a place for tourists to moor their boats. In both towns shown in this figure, the tourist function is located in the historic heart of the city.

In turn, Fig. 1.36 shows eating establishments in tourist resorts where the main attractions are the sea and sandy beaches. At Varadero, Agadir and Swakopmund—

1 Relations Between Gastronomy and the City

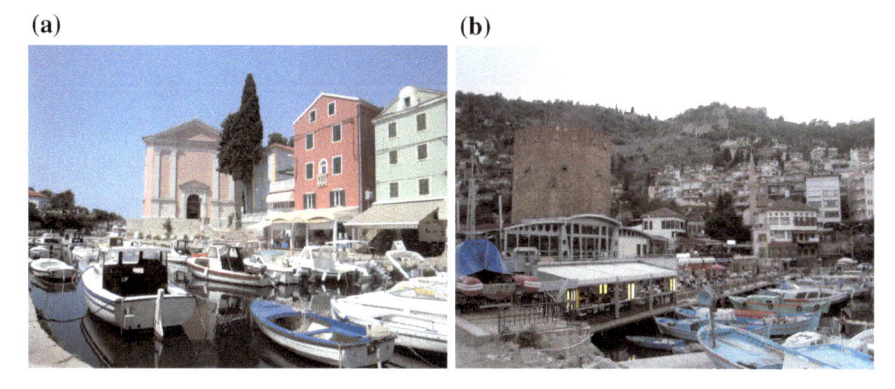

Fig. 1.35 In coastal tourist cities and towns eating outlets are an integral part of fishing ports and marinas. **a** Restaurant at the quay in Lošinj, Lošinj Island (Croatia). **b** Restaurants are mainly occupied tourists visiting Alanya (Turkey) harbour in the evenings. Source: Anna Kowalczyk 2016

Fig. 1.36 Tourist-oriented amenities on the seafront promenade in Agadir (Morocco) showing global fast-food restaurants. **a** Pizza Hut and McDonald's restaurants. **b** KFC. The location of these restaurants is an example of a gastronomic cluster. Source: Anna Kowalczyk 2017

neither of which has much of an historic centre—the historic part of the city is not regarded as a tourist attraction. The main attractions are the sea and the beach, and they are the most important factors in determining the location of restaurants, bars, etc. In some cases, such as Morro Jable (Fig. 1.37), an additional factor that supports eating facilities is the accessibility provided by local authorities. In most seaside towns in Spain, walkways that guaranteed access to the sea have existed for many years. They are protected by a special law, called the *Ley de Costas*, which was introduced in the 1980s (Fig. 1.38).

Fig. 1.37 Eating establishments in typical seaside tourist resorts. **a** Hotel restaurant terrace at Varadero (Cuba). **b** Cafés near the beach at Swakopmund (Namibia). In both **a** and **b**, the beautiful sea view is an additional attraction for visitors. **c** Restaurant with views of the sea and tourist boats in Alanya (Turkey). **d** Exterior of the restaurant at Bitez (Turkey) with views of the beach and sea (on the right) and the walkway that leads along the coast (on the left). Source: Katarzyna Kowalczyk 2015 (a), Anna Kowalczyk 2016 (b), 2010 (c and d)

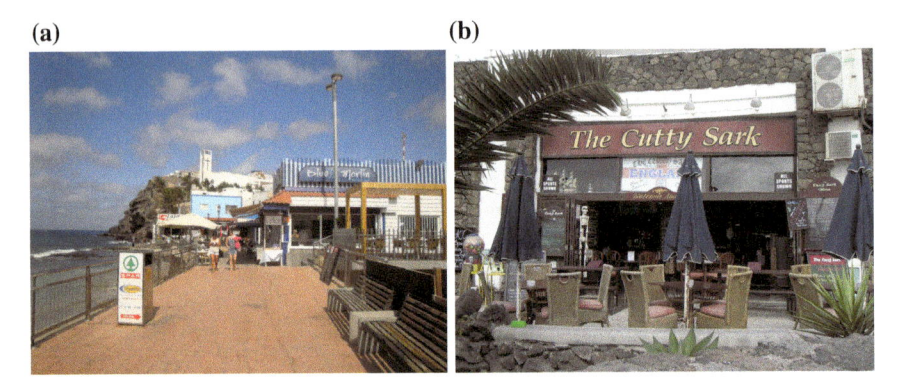

Fig. 1.38 **a** The esplanade at Morro Jable, Fuerteventura (Spain). **b** The English pub at Costa Teguise, Lanzarote (Spain). The sign for The Cutty Sark and the (English) flag beneath it are intended to attract tourists from the United Kingdom (particularly England). Source: Anna Kowalczyk 2010

1 Relations Between Gastronomy and the City

Fig. 1.39 Alanya (Turkey)—an example of restaurant location along the river. Such localization is another factor in attracting guests. Source: Anna Kowalczyk 2015

Customers are also attracted to riverside locations. In Alanya, for example, a dozen or so restaurants, serving a range of freshly caught fish dishes, have been established along the Dim River (*Dim Çayı* in Turkish) (Fig. 1.39).

Some people visit tourist cities for reasons other than gastronomy. Cities may be worth visiting purely for the local tourist attractions. For example, wine, brandy and cognac producers, etc. located in a charming part of town can also encourage visitors to take romantic walks, listen to music or go dancing. Although these types of establishment often lie in very different parts of tourist cities, they can form clusters. In Porto (northern Portugal), for example, wine stores and wineries can mainly be found along the Douro River, and a similar situation exists for sherry-drinking tourists in Jerez de la Frontera (Spain) and Cognac (France) (Fig. 1.40).

In seaside cities and towns, areas adjacent to hotels and rental apartments have their own concentrations of restaurants, bars, etc. (zones E and F). This is the case in places like Nice and Cannes (Côte d'Azur, France), Funchal (Madeira, Portugal) and Opatija (Croatia) (Fig. 1.41), although in all-inclusive resorts, many guests

Fig. 1.40 Gastronomy facilities do not have to be dining places. **a** Winery in the outskirts of Corfu town, Corfu Island (Greece). In addition to learning about wine technology and tasting, tourists can also buy wine. **b** Food court in Puerto de la Cruz, Tenerife (Spain), which is considered to be a good place for walking and taking photographs. Source: Anna Kowalczyk 2010, 2014

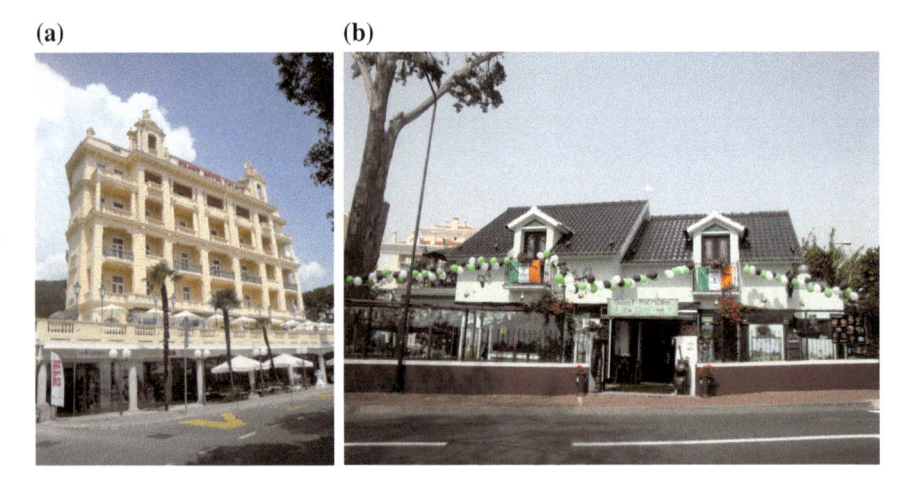

Fig. 1.41 a Restaurants and bars on the ground floor of the 4-star Grand Hotel Palace Bellevue in Opatija, Croatia. **b** The Hole in One pub in Funchal, Madeira. This Irish-style pub is located on Estrada Monumental opposite the 5-star CS Madeira Atlantic Resort & Sea SPA (which became the VidaMar Resort Hotel Madeira in July 2017). This was the world's first Crowne Plaza Resort (an American brand established in 1983 by the Holiday Inn Corporation) when it opened in 1999. Source: Anna Kowalczyk 2012, 2016

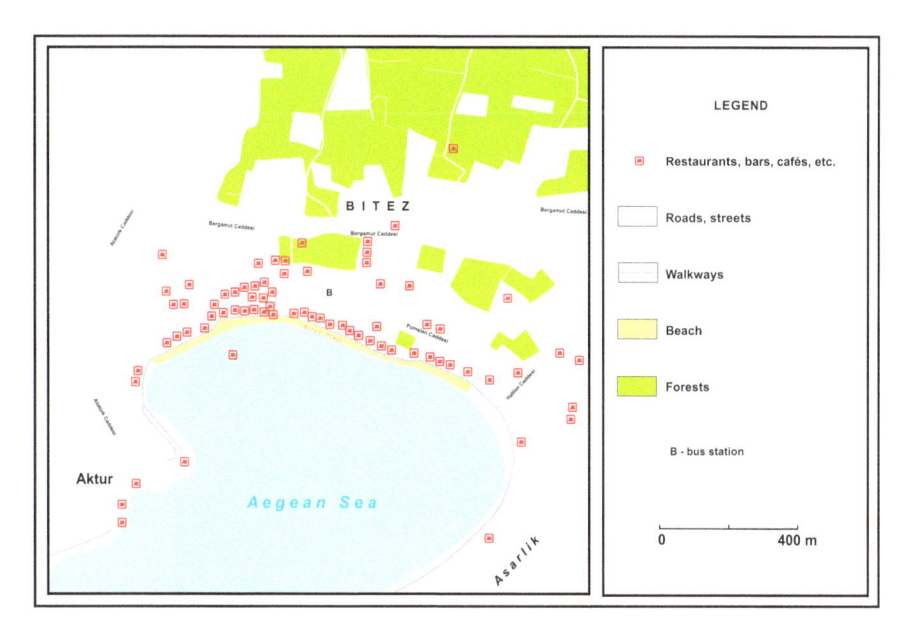

Fig. 1.42 Location of eating establishments in Bitez, Muğla Province (Turkey) in 2013. The linear arrangement along the beach, which is clearly visible, is typical of coastal tourist towns

only ever eat in the hotel. As a result, restaurants situated near these hotels can struggle to attract enough customers to make them financially viable. This problem occurs mainly in resorts in the Canary Islands, Madeira and the Balearic Islands (Fig. 1.42).

The prevalence of eating establishments in tourist cities makes them one of the most important parts of what is called the *touristscape*. It is not just their presence in the urban space that is important, but the visual (neon signs, banners), aural (music) and olfactory (smell of cooked, fried and baked foods) sensations that they contribute. Unlike their material presence, however, these intangible footprints of gastronomy are only a temporary element of a tourist city's landscape. Ferrari (2019) calls it *servicescape*, and defines it as 'a holistic and interdependent view of several physical elements that may affect customers' choices whenever they decide or not to return to a place of consumption' (Ferrari 2019, p. 161) (Fig. 1.43).

Fig. 1.43 Examples of the touristscape linked to gastronomy. **a, b** Maspalomas, Gran Canaria (Spain). **c** Lignano Sabbiadoro, Friuli-Venezia Giulia (Italy). In both cases photographs were taken during the winter season when restaurants were closed and tourists were absent. Source: Anna Kowalczyk 2015 (Maspalomas), 2004 (Lignano Sabbiadoro)

References

Adamson MW (2004) Food in medieval times, Food through history series. Greenwood Press, Westport, CT and London
Atkins PJ, Lummel P, Oddy DJ (2007) Food and the city in Europe since 1800. Ashgate, Burlington
Augé M (1995) Non-places. Introduction to an anthropology of supermodernity. Verso, London and New York, English translation by J. Howe
Barzun J (2000) From dawn to decadence. HarperCollins Publishers, New York
Brillat-Savarin J-A (1994) The physiology of taste. Penguin Books, London
Bryce D, O'Gorman KD, Baxter IWF (2013) Commerce, empire and faith in Safavid Iran: the caravanserai of Isfahan. Int J Contemp Hosp Manag 25(2):227–246

Burgess EW (1925) The growth of the city: an introduction to a research project. In: Park RE, Burgess EW, McKenzie RD, The city, The University of Chicago Press, Chicago and London, pp 47–62
Burke M, Brown AL (2007) Distances people walk for transport. Road Transp Res 16(3):16–29. https://research-repository.griffith.edu.au/bitstream/handle/10072/17867/49100_1.pdf?sequence=1. Accessed 1 June 2017
Crespi-Vallbona MC, Pérez MD (2015) Tourism and food markets: a typology of food markets from case studies of Barcelona and Madrid. Reg Mag 299(1):15–17
Coffeehouse (2017) https://en.wikipedia.org/wiki/Coffeehouse. Accessed 2 June 2017
Doucet B (2010) Inner cities, inner suburbs, outer suburbs: geographies, changing preferences. GeographyJobs.com, January 15, 2010. http://www.geographyjobs.com/articles/inner_cities_inner_suburbs_outer_suburbs_geographies_changing_preferences.html. Accessed 10 Feb 2019
Downs A (1997) The challenge of our declining big cities. Hous Policy Debate 8(2):359–408
Duensing SN (2014) Taverns, inns and alehouses? An archaeology of consumption practices in the City of London, 1666–1780. School of Arts, Languages and Cultures, Faculty of Humanities, The University of Manchester, Manchester, Ph.D. dissertation. https://ethos.bl.uk/Logon.do;jsessionid=27B84579AC96E03C712262C438AFC208?ordering=1. Accessed 1 June 2017
Dumbrovská V, Fialová D (2016) *Turistické okrsky a turistifikace v Praze. Případová studie Královské cesty*, "Studia turistica: online časopis pro vzdělávání v cestovním ruchu", roč. 7, č. 1 (červenec), pp 6–17
eThekwini Inner City Local Area Plan (2016) http://www.durban.gov.za/City_Services/development_planning_management/Documents/Inner%20City%20LAP%20Nov%202016.pdf. Accessed 10 Feb 2019
Ellis SJR (2004) The distribution of bars at Pompeii: archaeological, spatial and viewshed analyses. J Rom Archaeol 17(1):371–384
Ellis SJR (2008) The use and misuse of 'legacy data' in identifying a typology of retail outlets at Pompeii
Ferrari F (2019) Servicescape and gastronomic tourism. In: Kumar Dixit S (Ed) The Routledge handbook of gastronomic tourism. Routledge, Oxon and New York, pp 161–168
Internet Archaeology 24(4) (July), June 30, 2008, http://intarch.ac.uk/journal/issue24/4/7.2.html. Accessed 01 June 2017
Fagan GG (2006) Leisure. In: Potter DS (ed) A companion to the Roman Empire. Blackwell Publishing Ltd., Malden, MA and Oxford and Carlton, Vic., pp 369–384
Fant JC (2009) Bars with marble surfaces at Pompeii: evidence for sub-elite marble use. J Fasti Online. http://www.fastionline.org/docs/FOLDER-it-2009-159.pdf. Accessed 30 May 2017
Fidel E (2007) *Mercados de Madrid, cuestión de supervivencia*. Urban Idade. Memorias de las redes urbanas, 31 mayo 2007. https://urbancidades.wordpress.com/2007/05/31/mercados-de-madrid-cuestion-de-supervivencia. Accessed 2 June 2017
Finkelstein J (1999) Foodatainment. Perform Res 4(1):130–136. http://www.thankyouforcoming.la/wp-content/uploads/2013/03/FOODATAINMENT.pdf. Accessed 20 Jan 2019
Goss J (1996) Disquiet on the waterfront: reflections on nostalgia and utopia in the urban archetypes of festival marketplaces. Urban Geogr 17(3):221–247
Górczyńska M (2015) Gentryfikacja w polskim kontekście: krytyczny przegląd koncepcji wyjaśniających. Przegląd Geograficzny 87(4):589–611
Gregory D, Johnston R, Pratt G, Watts MJ, Whatmore S (eds) (2009) The dictionary of human geography, 5th edn. Blackwell Publishing Ltd., Chichester
Hamnett C (2003) Gentrification and the middle-class remaking of inner London, 1961–2001. Urban Stud 40(12):2401–2426
Hart WH (1879) List of the alehouses, innes, and taverns in Derbyshire in the year 1577. Derbys Archaeol J 1:68–80. http://archaeologydataservice.ac.uk/archives/view/daj/contents.cfm?vol_id=990. Accessed 1 June 2017
Heathcote Sapey F (2017) Today's marketplace. Reengaging society & the power of food. In: Vadini E (ed) Public space and an interdisciplinary approach to design, L'architettura delle città. J Sci Soc Ludovico Quaroni, 10, Edizioni Nuova Cultura, Rome, pp 73–86

Heuriger in Wien: Ein kleiner Führer durch die Heurigenorte. http://www.stadt-wien.at/lifestyle/essen-trinken/heuriger-in-wien-ein-kleiner-fuehrer-durch-die-heurigenorte.html. Accessed 25 June 2017

https://www.czso.cz/documents/10180/46014684/33012117q1s01.xlsx/95399516-d374-4b89-a5ca-accf9af11d6c?version=1.0. Accessed 20 June 2017

http://intarch.ac.uk/journal/issue24/4/images/figure42.html. Accessed 30 May 2017

http://www.madrid.es/UnidadesDescentralizadas/UDCEstadistica/Nuevaweb/Economía/EmpresasyLocales/censo/D2110317.xls. Accessed 20 June 2017

Kährik A, Temelová J, Kadarik K, Kubeš J (2016) What attracts people to inner city areas? The cases of two post-socialist cities in Estonia and the Czech Republic. Urban Stud 53(2):355–372

Kothiyal T (2012) Living with history. Deccan Herald. October 6, 2012. http://www.deccanherald.com/content/283426/living-history.html?iframe=true&width=90%&height=90%. Accessed 1 June 2017

Lee J (2009) The market hall revisited. Culture of consumptions in urban food retail during the long twentieth century. Linköping Studies in Art and Science, No. 497. Linköping University

Lees L (2000) A reappraisal of gentrification: towards a 'geography of gentrification. Prog Hum Geogr 24(3):389–408

Lefebvre H (1991) The production of space. Basil Blackwell, Inc., Oxford and Cambridge, MA, English translation by D. Nicholson-Smith

Lewicka PB (2005) Restaurants, inns and taverns that never were: some reflections on public consumption in medieval Cairo. J Econ Soc Hist Orient 48(1):40–91

Mac Con Iomaire M (2009) The emergence, development and influence of French haute cuisine on public dining in Dublin restaurants 1900–2000: an oral history. School of Culinary Arts and Food Technology, Dublin Institute of Technology, Dublin, Ph.D. dissertation, http://arrow.dit.ie/cgi/viewcontent.cgi?article=1011&context=tourdoc. Accessed 1 June 2017

Mac Con Iomaire M (2013) Public dining in Dublin: the history and evolution of gastronomy and commercial dining 1700–1900. Int J Contemp Hosp Manag 25(2):227–246

Market halls in Berlin. https://en.wikipedia.org/wiki/Market_halls_in_Berlin. Accessed 1 June 2017

Mercado de Santa Clara. https://pt.wikipedia.org/wiki/Mercado_de_Santa_Clara. Accessed 2 June 2017

Mercat del Born. https://en.wikipedia.org/wiki/Mercat_del_Born. Accessed 2 June 2017

Mercat d'Hostafrancs. https://ca.wikipedia.org/wiki/Mercat_d'Hostafrancs. Accessed 2 June 2017

Mercat de Santa Caterina. https://ca.wikipedia.org/wiki/Mercat_de_Santa_Caterina. Accessed 2 June 2017

Millington G (2012) The outer-inner city: urbanization, migration and 'race' in London and New York. Urban Res 5(1):6–25

Monheim R (2010) *Die Regensburger Altstadt – Shopping-Spezialisierung im Schatten zweier Einkaufszentren.* In Innerstädtische Shopping-Center, "Berichte des Arbeitskreises Geographische Handelsforschung", 28 (Dezember), pp 18–25

Neu Venedig. Geschichte. http://www.neu-venedig.de. Accessed 25 May 2017

O'Gorman, KD (2007a) Dimensions of hospitality: exploring ancient origins. In: Lashley C, Lynch P, Morrison A (eds) Hospitality: a social lens. Advances in Tourism Research, Elsevier, Oxford and Amsterdam, pp 17–32

O'Gorman KD (2007b) Discovering commercial hospitality in ancient Rome. Hosp Rev 9(2):44–52

Pacione M (2005) Urban geography. In: A global perspective, Routledge, London and New York

Parham S (2012) Market place: food quarters, design and urban renewal in London. Cambridge Scholars Publishing, Newcastle upon Tyne

Parham S (2015) Food and urbanism: the convivial city and sustainable future. Bloomsbury Publishing, London and New Delhi and New York and Sydney

Relph E (1976) Place and placelessness. Pion Limited, London

Roulac SE (2003) Strategic significance of the inner city to the property discipline. J R Estate Res 25(4):365–394

Schiltz T, Moffitt W (1971) Inner-city/outer-city relationships in metropolitan areas. A bibliographic essay. Urban Aff Q 7(1):75–108

Schofield J, Vince A (2003) Medieval towns: the archaeology of British towns in their European setting, 2nd edn. The Archaelogy of Medieval Europe, 1100–1600. Continuum, London and New York

Siddiqi IH (1985) Food dishes and the catering profession in pre-Mughal India. Islam Cult 59(2):117–174

Smith N (1979) Toward a theory of gentrification: a back to the city movement by capital, not people. J Am Plan Assoc 45(4):538–548

Smith N (1996) The new urban frontier: gentrification and the Revanchist City. Routledge, New York

Spiekermann U (1999) *Basis der Konsumgesellschaft. Entstehung und Entwicklung des modernen Kleinhandels in Deutschland 1850–1914*. Schriftenreihe zur Zeitschrift für Unternehmensgeschichte", Band 3, C. H. Beck, München

Steel C (2013) Hungry city: how food shapes our lives. Random House

The history of La Boqueria. http://www.citybitesbarcelona.com/category/the-history-of-la-boqueria. Accessed 30 May 2017

Tuan Y-F (1975) Place: an experiential perspective. Geogr Rev 65(2):151–165

Tuan Y-F (1979) Space and place: humanistic perspective. In Gale S, Olsson G(eds) Philosophy in geography. Theory and Decision Library, 20. D. Reidel Publishing Company, Dordrecht and Boston and London, pp 387–427

Turks in Berlin. https://en.wikipedia.org/wiki/Turks_in_Berlin. Accessed 5 June 2017

Wagner LC (1964) A realistic division of downtown retailing. J Market 28(3):39–42

Wimpy. https://en.wikipedia.org/wiki/Wimpy_(restaurant). Accessed 30 May 2017

Yen B, Burke M, Tseng C, Ghafoor MMT, Mulley C, Moutou C (2015) Do restaurant precincts need more parking? Differences in business perceptions and customer travel behaviour in Brisbane, Queensland, Australia. Paper presented at the 37th Australasian Transport Research Forum, Sydney, Australia, 29 September–2 October, 2015. http://atrf.info/papers/2015/files/ATRF2015_Resubmission_84.pdf. Accessed 1 June 2017

Chapter 2
Theories and Concepts Related to Gastronomy in Urban Space

Andrzej Kowalczyk

Abstract This chapter describes the functioning of the restaurant sector in the urban space based on spatially oriented and non-spatially oriented theories. The former are of particular importance, as they concern not only the spatial distribution of restaurants, bars, etc., but also consumer behaviour. Central-place theory, bid-rent theories and the socio-spatial structure of a city (such as Burgess' concentric zone theory, Hoyt's sector theory, and Harris and Ullman's multiple nuclei theory) are proposed as useful concepts to explain the concentration of eating establishments in urban areas. New approaches to the city and city planning, such as the concepts of the creative city, the compact city and the smart city, are also offered.

Keywords Central-place theory · Models of socio-spatial structure of the city · Berry · Space and place theory · Consumer behaviour · The creative city · The compact city · The smart city

2.1 Introduction

In the previous chapter, which discussed general patterns in the evolution of the distribution of gastronomy in urban space, attention was paid to the regularity of spatial patterns. These regularities suggest that it can be useful to consider the problem of the distribution of eating facilities in the urban space using general location theory (particularly retail location theory) as well as concepts from the domains of urban planning and tourism.

In this chapter, we establish the academic context for the study of gastronomy in urban space by providing an overview of the main conceptual approaches, theories

---

A. Kowalczyk (✉)
Department of Tourism Geography and Recreation, Faculty of Geography and Regional Studies, University of Warsaw, ul. Krakowskie Przedmieście 30, 00-927 Warsaw, Poland
e-mail: akowalczyk@uw.edu.pl

© Springer Nature Switzerland AG 2020
A. Kowalczyk and M. Derek (eds.), *Gastronomy and Urban Space*,
The Urban Book Series, https://doi.org/10.1007/978-3-030-34492-4_2

and concepts, which can be employed in exploring the subject. We illustrate them with examples of research done in the field of gastronomy. We focus mainly on the spatially oriented theories, but we also present behavioural and non-spatially oriented concepts, as well as some new approaches to the city planning which we find useful in the context of urban gastronomy.

An issue that will attract particular attention is the location of eating establishments. This is because the real estate agents' mantra, 'the most important factors for success are location, location and location', rings true in the gastronomy business (Farrell 1980, cited in: Law 1996, p. 123). Yen et al. (2015, p. 2) note that 'accessibility for its customers is fundamental to any restaurant business'. Further, the authors precise that the '…optimal restaurant location is understood to be that which is accessible to customers, near other restaurants, with high-quality transport connections with good frequency, and within convenient walking distance of the restaurant location' (Yen et al. 2015, p. 4). This statement is supported by earlier work carried out in Melbourne (Australia), which showed that clustering restaurants in the centre, and requiring customers to walk past businesses located in intermediate and peripheral areas helps to spread the benefits of being in the city (Reimers and Clulow 2004). Similar observations have been made in other cities.

2.2 Spatially Oriented Theories

2.2.1 Central-Place Theory

In the urban space, eating establishments tend to be clustered together rather than dispersed. This trend can be explained by Christaller's classical central-place theory, especially when we analyze the presence of clusters of restaurants and bars in the central business district (CBD). The theory defines *range* as the maximum distance that a consumer would be willing to travel to purchase a particular good offered at a given place. It means that the limit is determined by competition among businesses supplying the same good, as well as other factors.

Also central to the theory is the concept of the *market area*. In the context of gastronomy, O'Sullivan (2003) uses the simple example of market areas for pizza parlours and Tibetan restaurants. He assumes that both types of catering establishments produce 200 meals per day. If total demand for pizza is 10,000 meals per day, the region will have 50 pizza parlours (10,000:200). The total demand for Tibetan food, however, is only 200 meals per day, so a single Tibetan restaurant or bar can serve the whole region. Based on this example, O'Sullivan concludes that 'the market area is not determined by scale economies per se, but by scale economies relative to per capita demand'. In other words, where there are many potential consumers for a given type of meal, there should be more eating establishments in order to meet this demand. And where the number of clients is smaller the number of catering facilities will also be smaller.

Following on from Christaller's theory, American geographers Berry and Garrison (1958), formulated the terms *threshold* and *range*. They defined threshold as the minimum purchasing power necessary to support the supply of a good from a centre. In other words, threshold is the smallest size of community (potential customers) in which the function exists. According to empirical studies carried out in Washington State in the north-western United States, they calculated ranges for different types of economic activity and gastronomy in rural areas. This found that were 276 persons for restaurants and 282 for taverns (Berry and Garrison 1958, p. 150).

Berry (1963) went on to identify five centres of economic activity: (a) isolated convenience stores and street corner development, (b) neighbourhood business centres, (c) community business centres, (d) regional shopping centres and (e) central business centres or CBD (Berry 1963, p. 19). This classification was based on extensive research in the Chicago metropolitan area. Isolated convenience stores (the first, lowest level in the hierarchy) were defined as a commercial establishment that serves the occasional demands of local people. In gastronomy, this equates to the pub in the United Kingdom, the *trattoria* in Italy, the *hospoda* in the Czech Republic, the *kneipe* in Germany and the *taberna* in Spain. The neighbourhood business centre (second level) usually includes grocery stores, pharmacies, beauty shops and small restaurants (cafés or bars). The third level consists of community business centres, which serve city residents. In gastronomy they include the same services as the lower levels plus other dining facilities, such as a restaurant plus café, or a beer bar plus pizzeria. The next level refers to regional shopping centres. They serve people living in the bigger parts of the city, and usually include a range of different types of eating facilities—casual restaurants, fast-food bars, cafés, ice cream bars, etc. Some are located separately, but usually they are co-located in food courts. The highest level in the hierarchy is the CBD. In this part of the city there is almost every kind of service you can think of, which in terms of gastronomy, means restaurants, bars, pizza parlours, cafés, wine bars, etc., all concentrated in a large cluster that serves the entire city (and its visitors).

The notion of a catchment (market) area (similar to Berry's threshold) is fundamental to central-place theory (Fig. 2.1). In gastronomy, it is usually measured as the distance between the location of the eating facility and where potential consumers live. Catchment areas are delimited by transportation and land use in the surrounding area. They may be expressed in terms of the radius (or distance), surface area, size of the population served or the time needed to reach the central point of the area being served. In urban areas, there is a hierarchy of catchment areas for eating facilities. At the lowest level are small, local walk-up catchments. In Brisbane (Australia), for example, their radius was measured at around 1.5 km (Burke and Brown 2007, p. 25, Fig. 5b). Slightly larger are bicycle catchment areas, because cyclists can travel further than pedestrians in the same time. Next are public transport catchments, which are limited by routes connecting to the restaurant and the specifics of the transportation service (frequency, hours of operation, travel times). Probably on the same level or slightly higher (or wider in spatial terms) are catchments for cars, which are determined by street access and travel

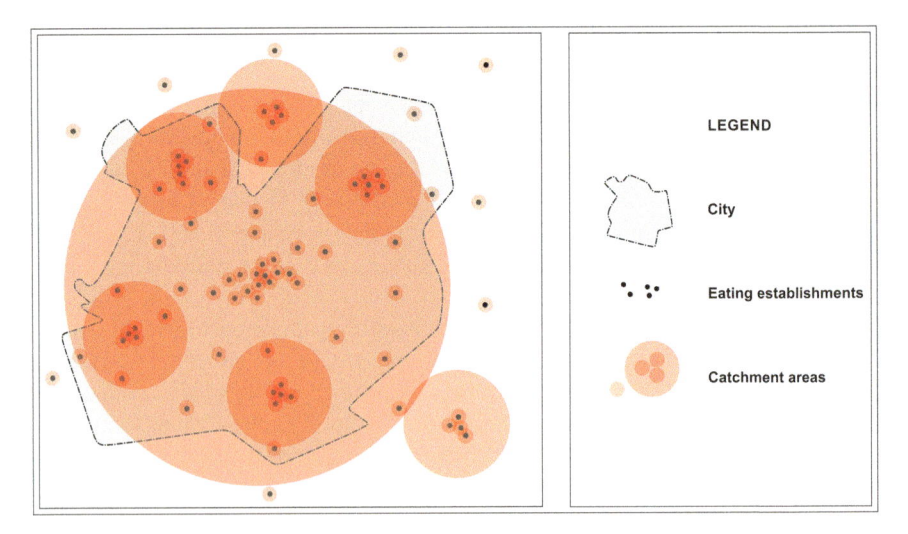

Fig. 2.1 Catchment areas of eating establishments in a hypothetical city according to central-place theory (Berry 1963)

times across the city and the availability of parking near the restaurant. Parking issues can be resolved by using taxis and taxi-like services (real-time ridesharing services, e.g. Uber) especially at night-time (Yen et al. 2015, p. 3). Although Christaller's concept is no longer used to describe the distribution of services, it was recently applied to the dining industry in the United States (Schiff 2015).

2.2.2 Bid-Rent Theory

The concept of bid-rent theory, close to the Ricardian theory of rent, explains why demand for land, and therefore land use patterns, vary across the urban area. This is an important concept in economic geography, in general, and in urban geography (and, of course, urban economics) in particular. In the model the city centre is assumed to be the most accessible and therefore most valuable location. Since for some land uses accessibility is of a greater importance than for others, they are prepared to pay higher rents for central locations (Pacione 2005). Gastronomy is one of such land uses.

Bid-rent theory was formulated by Alonso (1964). It states that different land users will compete with one another for land close to the city centre. Retailers who seek to maximize their profits are willing to pay more for land in the CBD, but less for land that is further away. The price and demand for land therefore change as the distance from the centre increases (Alonso 1964, p. 14). Figure 2.2 shows the general assumptions of the theory as they apply to gastronomy. Close to the centre, high land prices result in a proliferation of expensive restaurants. In surrounding

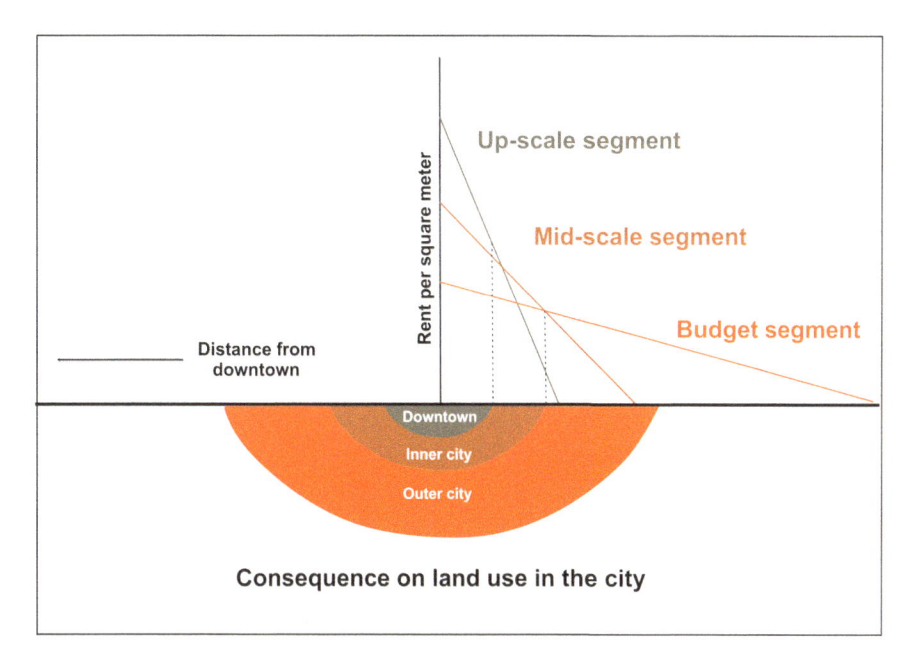

Fig. 2.2 Distribution of restaurants, bars, etc. in the urban space according to bid-rent theory

areas, the share of restaurants and bars in the mid-price segment increases. Finally, in the peripheral zone, the lower price of land means that there are more restaurants, bars and coffee shops, etc. in the budget segment. The concept has been used (together with central-place theory and concentric zone theory), for example, to investigate differences in land use in the Roath area of Cardiff, Wales (Narvaez et al. 2013, p. 089:16). This study identified clusters of eating outlets that mainly served locals.

According to bid-rent theory the optimal location for restaurants is highly centralized areas in the inner city with heavy foot traffic. Unsurprisingly, there are usually a variety of restaurants, bars, pubs, wine bars, cafés and other gastronomic facilities in these zones. They attract investors in the gastronomy business and, more generally, retail outlets and consumer services. The more accessible areas with the greatest concentration of customers are more profitable and top players compete for the most accessible areas. Others locate their restaurants or bars in rings around the city centre, creating a concentric zone model of urban gastronomy.

Some authors suggest, however, that real distance models (popular in the retail trade literature) are only partially effective in the context of gastronomy. They can be used to explain the location of pizza parlours, family oriented restaurants and some ethnic restaurants, but not gastronomy as a whole. Distance is only one of many different factors that affect the success of a particular restaurant location (Pillsbury 1987, p. 327). Other aspects will be analyzed further in this volume.

2.2.3 Models of Socio-spatial Structure

The uneven distribution of eating establishments in urban areas can be also explained by classical theories of the spatial-social structure of cities, i.e. concentric zone theory (Burgess 1925); sector theory (Hoyt 1939); and multiple nuclei theory (Harris and Ullman 1945). Burgess (1925) took the example of Chicago, and suggested that five concentric zones could be distinguished, including a more distant zone (30–60 min travelling time from the city centre). He called this a *commuter zone*, a *suburban area*, or a *satellite* city. His zonal modal was proposed as an ideal type, not as a representation of reality (Pacione 2005). Hoyt (1939) extended this classification by adding sectors that run from the city centre to its limits along the major transport routes. In contrast, Harris and Ullman's (1945) multiple nuclei model suggests that the city consists of identifiable parts that do not exhibit the regularity described in the other two theories. While the Burgess zonal model and, to a lesser extent, the Hoyt sectoral pattern suggest inevitable predetermining patterns of location, Harris and Ullman suggest that land use patterns vary depending on local context (Pacione 2005).

Figure 2.3 shows the distribution of gastronomy in urban areas using classical models of the socio-spatial structure of the city. Although the first two models are very general, they can be used for gastronomy. The number of restaurants and bars generally declines with increasing distance from the city centre (Burgess model) and there is often a greater concentration of catering establishments along the main transport routes into and out of the city centre (Hoyt model). However, in many

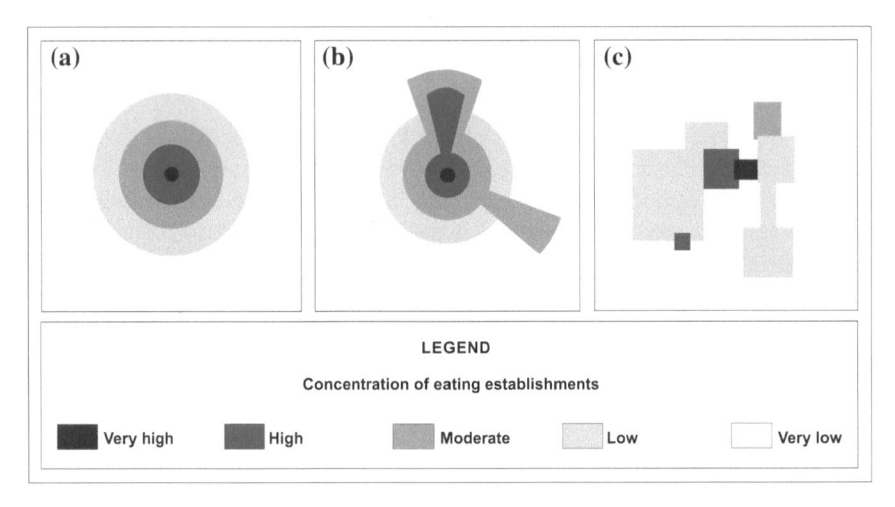

Fig. 2.3 Application of models of the city using Burgess' concentric zone theory (**a**), Hoyt's sector theory (**b**) and Harris and Ullman's multiple nuclei theory (**c**) to explain the concentration of eating establishments. Source: Adapted from 5 urban models. https://www.slideshare.net/ecumene/5-urban-models. Accessed 10 June 2017

cases, the third theory (Harris and Ullman model) is better suited to explain the concentration of gastronomy, e.g. in situations where eating establishments are close to stadiums, in the vicinity of tourist attractions, inside (and around) shopping centres, etc.

The applicability of these models depends on the geographical location, size and function of cities and towns. In towns, the situation often closely resembles the concentric model, while in large cities the multiple nuclei model is more appropriate. In both cases, however, gastronomy is located along the main streets and transit roads (in line with sector theory). Geographical location has the greatest impact on the distribution of gastronomy outlets in coastal cities and towns, cities located along large rivers or in mountain valleys. Sometimes the distribution of eating establishments resembles a radial star-shaped system, with most restaurants and bars along streets, especially in towns and smaller cities and at transportation nodes.

Pillsbury's (1987) study of Atlanta, Georgia, revealed larger clusters of restaurants in the city centre, in local sub-centres and, above all, along major transport routes. The results confirmed the assumptions of central-place theory as well as Hoyt's sector theory, although the study was based on socio-demographic and socio-economic diversification within urban areas and on differences in quality of housing. Pillsbury's results in Atlanta in the early 1980s support Berry's (1959) earlier study in Spokane, Washington State. Berry noted that services exist along four types of communication arteries and described the concentrations of services along streets and roads as *ribbons*. He divided them into four categories: (a) highway-oriented ribbons, (b) urban arterial-oriented ribbons, (c) new suburban ribbons, and (d) traditional shopping streets. With respect to gastronomy, his survey showed that restaurants tended to be concentrated along highway-oriented ribbons and bars along urban arterial-oriented ribbons (Berry 1959, p. 147).

One of the strongest criticisms of the classical models of urban land use referred to their economic bias and neglect of cultural influences on urban land use patterns (Pacione 2005). Some further works aimed to provide concepts more relevant to contemporary urban society.

Vance's urban realms model was a refinement of Harris and Ullman's model. The main thesis of this model is that the CBD and inner city areas are not as important to urban areas as they were in earlier models. This can be attributed to the growing popularity of car ownership and the development of expressways. *Realms* (in other words, the catchment area or zones of influence) are focused on suburban centres and the centre of the city. The urban realms model was created using the example of the San Francisco Bay Area. It proposes that some functions of the CBD can be moved to the periphery, leading to a decline in the importance of the city centre. The best example of such a situation is the transfer of commercial functions from the CBD to new shopping centres (malls) in the suburbs. The same is true of

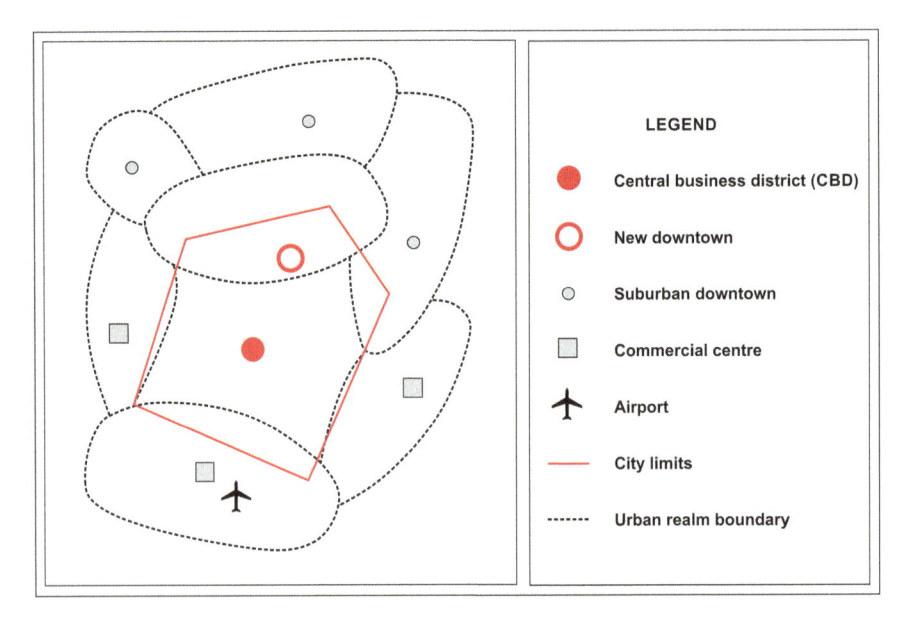

Fig. 2.4 Vance's urban realms model. This model shows that contemporary cities have several functional centres: the CBD (sometimes in crisis), the new downtown, suburban downtowns and commercial centres. There are eating establishments in all of these locations. Source: Adapted from Chen (2015)

offices that are moved from the downtown area to new centres in the outer city. As a result, the realms of the main city and suburban downtowns create a huge metropolitan area where the role of the CBD declines in relative terms. Four major types of centres make up realms (the CBD, the new downtown, suburban downtowns and commercial centres) together with the airport, although the airport does not have its own realm but is located within the area (realm) served by the nearby commercial centre (Fig. 2.4).

While the Vance's model highlights the importance of the new downtown, local suburban downtowns, shopping malls in the outer city and the airport, the Kearsley (1983) model is much more important in terms of the location of eating facilities in urbanized areas. Although some suggest that it is a derivation of the concentric zone model, it also includes elements of the sector model and the multiple nuclei model (Fig. 2.5).

Classical theories and models of the inner structure of the city suggest that the distribution of restaurants, bars and other dining facilities is the result of a complex interaction between traditional and non-traditional location factors. Traditional factors include demand (connected with the socio-demographic and economic character of the population), the price of land and spatial accessibility.

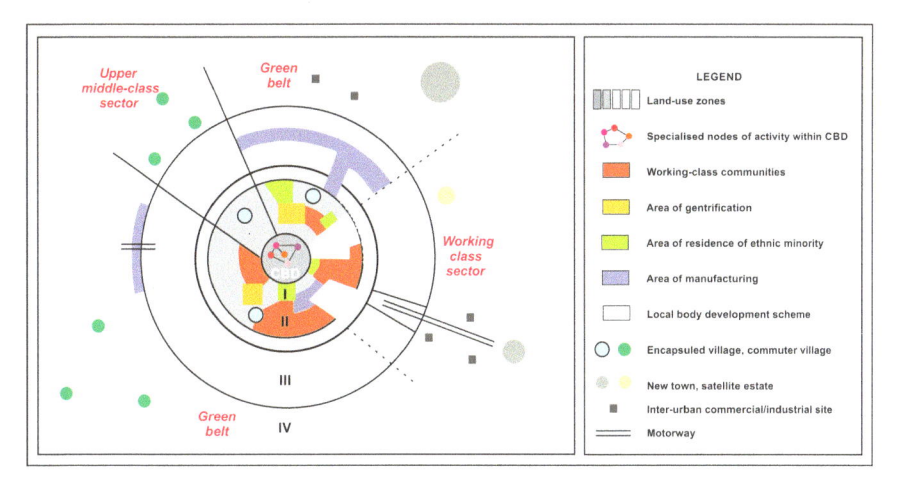

Fig. 2.5 Kearsley's model of the city. This model is much more detailed than that of Vance. It includes districts with ethnic minorities, gentrification zones (where expensive and fashionable restaurants are located) and post-industrial areas. It also shows the development of gastronomy in suburban areas, notably new towns and inter-urban commercial sites. Source: Adapted from 5 urban models. https://www.slideshare.net/ecumene/5-urban-models. Accessed 10 June 2017

Non-traditional factors include the type of food, the chef's reputation, etc. Pillsbury (1987, p. 326) asks 'What exactly constitutes a good location? A careful analysis of restaurant histories, locations and success rates indicates that the concept of good location is difficult to define. (…) Good location certainly is more than a constant stream of people passing a store site or a large population of potential clientele in the immediate vicinity.'

The second group of concepts that can explain the location of restaurants, bars, etc. in the urban space comes from urban planning. They are generally associated with smaller parts of the city and are more normative than descriptive.

The festival market place concept (FMP) deserves special attention. This concept has been a leading strategy for the revitalisation of downtown American cities since the 1970s (Gravari-Barbas 1998, p. 263). The main idea involves the application of planning solutions designed to make downtown areas look like a European city centre, as they have evolved historically. The concept was first applied in Boston (the Quincy Market project in 1976) and Baltimore (the Harborplace project, finished in 1980). It was later used as a major tool during the revitalisation of downtown areas in the largest metropolitan areas in the United States, including Atlanta (Underground Atlanta), Boston (Faneuil Hall), Chicago (Navy Pier), New Orleans (Jackson Brewery and Riverwalk), New York (South Street Seaport), San Francisco (Pier 39 and Ghirardelli Square) and many others. The fundamental principle of the FMP is a mix of shops, restaurants, cultural facilities and public areas to energize the urban space. As previously noted, there are now many eating

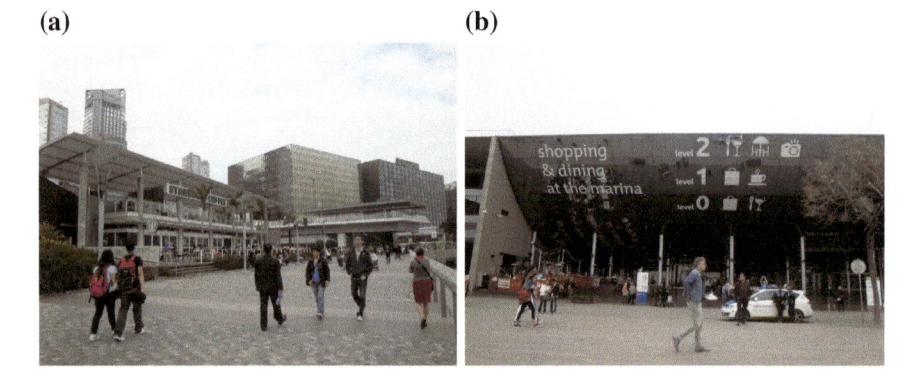

Fig. 2.6 Eating establishments are an integral part of revitalisation programmes. **a** Kowloon (Hong Kong Special Administrative Region of the People's Republic of China)—Avenue of Stars in the post-industrial zone created from former railway yards. **b** *Barcelona Centro Commercial Maremagnum* at Port Vell in Barcelona (Spain)—formerly a commercial sea port. Source: Anna Kowalczyk 2009, 2013

establishments in zones that were revitalized using FMP principles: in July 2017 there were 12 restaurants and bars in the Harborplace area (Baltimore),[1] seven at Underground Atlanta (Atlanta),[2] 13 at Station Square (Pittsburgh),[3] 26 at Riverwalk (New Orleans)[4] and 59 at Faneuil Hall (Boston).[5] In the geographical literature on FMP projects in the United States, this phenomena was noted by Gravari-Barbas (1998, pp. 270–271). The example of Baltimore was also reflected by Harvey (1989, 1990), who wrote that the social space undergoing changes are defined differently and acquires new meanings (Harvey 1990, p. 422).

Downtown areas revitalized using the FMP concept can be found not only in the United States, but also many other countries across the globe. In many cases, it has been applied to parts of cities that lie close to the sea, by a lake or beside a large river. Revitalized zones are often part of a larger programme of waterfront regeneration. Examples include London Docklands, the waterfronts at Liverpool (Albert Dock, King's Dock, etc.), Cape Town (Victoria & Alfred Waterfront), Singapore (Marina Bay Waterfront), Victoria and Kowloon (Wan Chai Waterfront, Tsim Sha Tsui), Barcelona (Port Vell), Genoa (Porto Antico) and Lisbon (Alcântara and Oriente) (Fig. 2.6).

---

[1]Store Directory (2017a).
[2]Dine at Underground Atlanta (2017).
[3]Store Directory (2017b).
[4]Dine (2017).
[5]Dining directory (2017).

The area that was designed as the festival market place is Victoria and Alfred Waterfront in Cape Town (South Africa). The entire area covers 123 ha, but in the next years will expand by new areas. The Victoria and Alfred Waterfront project covers the former port area and the name comes from the Alfred Basin (built in 1870) and Victoria Basin (completed in 1905). Since in the following decades the commercial port moved east, the area became an post-industrial 'brownfield'.

In 1988 Victoria and Alfred Waterfront (Pty) Ltd. was established as a company responsible to redevelop the historic docklands around the Victoria Basin and Alfred Basin. According to the plan it should be a mixed-use area. It means that the project included retail (including gastronomy), culture, recreation, tourism and residential functions with a working harbour at its centre (Our history 2019). This last element was especially important as it referred to the historical heritage of the area. This, in turn, meant that the process of revitalizing of the V&A Waterfront would be in realized according to the place-making concept (Behera 2017). It practice it meant that the most important '…planning motivation for the project was the re-establishment of physical links between Cape Town and its waterfront in order to create a quality environment; a desirable place to work, live and play; and a preferred location to trade and invest for Capetonians and visitors' (van Zyl 2005). The first facilities related to retail, gastronomy and entertainment began operating in 1991–1992 in the Victoria Wharf Shopping Centre. In the following decades, further facilities were put into use, and in 2017 the zone named Silo District was completed (Victoria and Alfred Waterfront consists of 9 structural units).

From the very beginning, Victoria and Alfred Waterfront has become an important tourist attraction of Cape Town. As Pirie wrote, between 1999 and 2003 much more tourists visited the Victoria and Alfred Waterfront than the famous Table Mountain (Pirie 2007).

Victoria and Alfred Waterfront is very important for the economy of Cape Town. Every day nearly 22,000 people come to work at the Victoria and Alfred Waterfront, and up to 180 000 visitors come here every day in summer season (V&A Waterfront Fact Sheet 2018). Of the total visitors 60% are Capetonians, 17% are South African and 23% international (Overview 2019). Such a large number of Cape Town visitors to the Victoria and Alfred Waterfront means that it is perceived as a place by Capetonians in the sense described by Edward Relph, Yi-Fu Tuan and other prominent humanistic geographers. A similar opinion is expressed by Ferreira and Visser (2007, p. 241), who wrote that '…the place-making success of the V&AW has led to economic development that should be read at different levels of analysis'. This means that the Victoria and Alfred Waterfront project has been successful both in terms of humanistic geography (as place) and neo-positivist geography (as a 'growth machine'). In other words, regardless of the research approach we represent, Victoria & Alfred

Waterfront project should be considered as a successful implementation of the festival market place concept.

In autumn 2018, there were 93 eating outlets on Victoria and Alfred Waterfront, with as many as 63 in the so-called Quays District area (including 42 in Victoria Wharf Shopping Centre) (Restaurants for whatever your taste 2019). In addition, 37 food stands were located in the V&A Food Market, which is housed in a former port power plant (Fig. 2.7). Although the restaurants and bars located at Victoria and Alfred Waterfront offer cuisine from around the world, the people of Cape Town and tourists come here mainly to taste fish and seafood dishes. This is confirmed by survey which showed that the main purpose of presence at Victoria and Alfred Waterfront was 'gastronomic activity' (Oosthuizen, p. 6). Similar results were obtained by survey organized in 2012, which showed that for 53% percent of Capetonian respondents the most important reasons for visiting Victoria and Alfred Waterfront were to eat at restaurant or bar (de Villiers 2016, p. 72).

The problem of locating catering services in the urban space has also been described using urban tourism concepts. In particular, the *tourist business district* (TBD) and the concept of a *recreational business district* (RBD), which are related to the idea of the city as an entertainment machine, described later. Getz defines a TBD as the concentration of visitor-oriented attractions and services in a city, which is similar to the RBD that is often associated with seaside resorts. Although similar, TBDs and RBDs differ in their form and function (Getz 1993, pp. 583–584). One important distinguishing factor is the position of gastronomy relative to other functions. According to Getz, in the RBD eating services (catering and beverages)

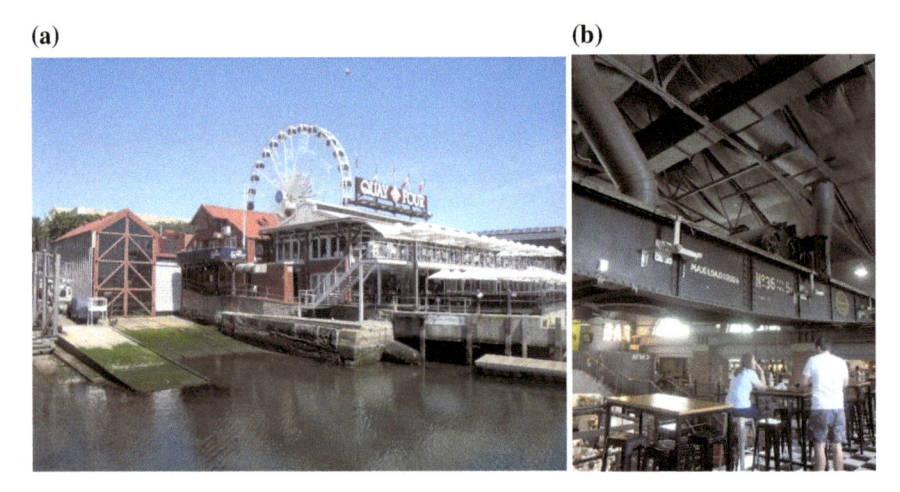

Fig. 2.7 Gastronomy at Victoria and Alfred Waterfront in Cape Town (South Africa). **a** Casual restaurants on Quay Four; **b** V&A Food market in the old power station (1882)

occupy a separate and leading position among other functions. In the case of the TBD, they are not separate and are included among other business and visitor services (Getz 1993, p. 586, Table 1). Such a distinction is questionable. The rapid development of gastronomy in recent years has made the catering industry one of the leading attractions—often for both residents and visitors—in many city centres. Examples include the Alfama district in Lisbon (Portugal), Vieux Lyon in Lyon (France) and Lan Kwai Fong or SoHo in Victoria (Hong Kong Special Administrative Region of the People's Republic of China). Examples of concentrations of RBD eating establishments can also be found in typical tourist destinations that are not officially towns. Often, their distribution resembles the letter T as they are concentrated along the coast and the main street (or streets) leading to the beach (Fig. 2.8).

Another theory, similar to the FMP and TBD and RBD concepts, is the notion of the *festivalisation* of culture. Here, various types of outdoor events are regarded as a permanent element of contemporary urban space and have an impact on the local economy, the everyday life of the city's population, the image and brand of the city, etc. (Cudny 2016, pp. 157–159). This concept can be used to explain why many city councils are ready to subsidise gastronomic events and use their culinary heritage in marketing, including tourist promotion. One of the best examples of this is the Munich authorities' support for *Oktoberfest*, which brings together 5–6 million people, including many domestic and foreign tourists (Hall et al. 2003, pp. 331–332).

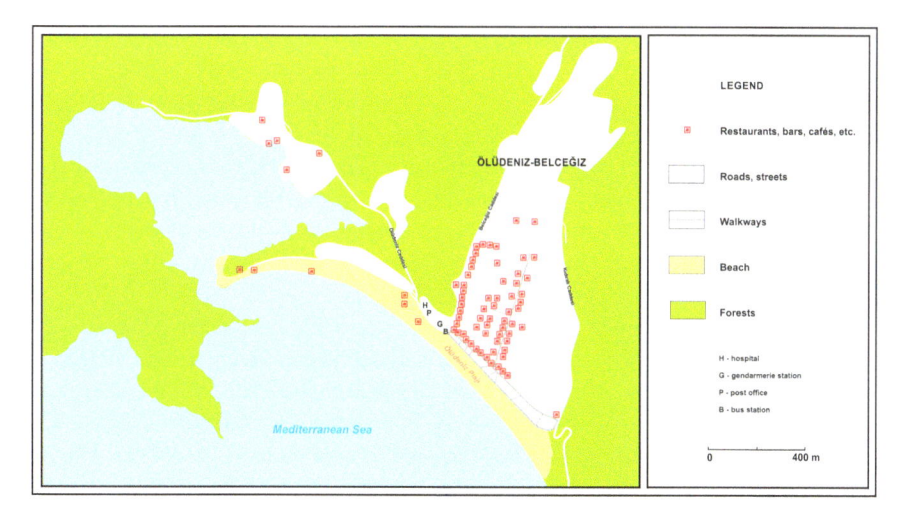

Fig. 2.8 Eating establishments (open to the public) at Ölüdeniz-Belceğiz in Muğla Province (Turkey) in 2011. Although Ölüdeniz-Belceğiz is not a town (it is a part of the Hisarönü—Ovacık—Ölüdeniz-Belceğiz tourist zone) it is a good example of the typical distribution of restaurants, bars, etc. in tourist cities and towns. Dining facilities are concentrated along the seaside and main streets (usually transformed into pedestrian zones) leading to the beach

2.2.4 Space and Place Theory

Starting from the late 1970s, the concepts of space and place are one of the most commonly used concepts in human geography. As Agnew notes, '...space and place are about the "where" of things and their relative invocation has usually signaled different understandings of what "where" means, it is best to examine them together rather than separately' (Agnew 2011, p. 316). Sack, in turn, emphasizes that 'Place implies space, and each home is a place in space. Space is a property of the natural world, but it can be experienced. From the perspective of experience, place differs from space in terms of familiarity and time.' (Sack 1997, p. 16).

One of the most prominent human geographers, representing the so-called humanistic geography, Yi-Fu Tuan, states that 'Space and place together define the nature of geography'. (Tuan 1979, p. 387). In the same text, the wrote that 'The study of space, from the humanistic perspective, is thus the study of a people's spatial feelings and ideas in the stream of experiences' (Tuan 1979, p. 388). Starting from this premise, in the next sentences the author concludes that the space understood in this way should be called a 'place'. According to him 'Place (…) is more than location and more than the spatial index of socio-economic status. It is a unique ensemble of traits that marits study in its own right' (Tuan 1979, p. 409).

So, in the further part of this section, both concepts are given more attention, given that they are fundamental to urban geography also in the context of gastronomy.

Space

In the history of the development of geographical thought, the twentieth century was the epoch of space (Gregory et al. 2009). In geography, understood as a social science, dealing with space dominates neo-positivist geography (Cresswell 2013, p. 7), whose main theories regarding urban space services are described in Sect. 2.1. At this point, Cresswell's opinion should be quoted. According to him 'Spatial scientists were not very interested in how people related to the world through experience; they tended to think of people as objects or rational beings' (Cresswell 2008, p. 135). However, the following section will discuss concepts close to structuralist geography and humanistic geography.

One of the most prominent geographers dealing with the former urban space is undoubtedly David Harvey. In one of his early (and best known) works Harvey wrote '...space is neither absolute, relative or relational in itself, but it can become one or all simultaneously depending on the circumstances'. (Harvey 1973, p. 13). Interestingly, the same opinion Harvey expressed more than 30 years later (Harvey 2004, pp. 2–5). Using the examples of New York City and Mexico City Harvey (2008, p. 38; 2012, p. 23) noted that local authorities reshaping the city are favours to developers and visitors. The latter case is directly related to gastronomy, because in the downtown of the cities visited by tourists, prices in restaurants are often so high that they practically make it impossible for residents to use them. This problem is becoming increasingly significant in Prague (Czech Republic), Kraków and

Warsaw (Poland), in the historical city centres some restaurants and bars are almost exclusively visited by tourists. This phenomenon can be treated as one of the manifestations of globalization, which is associated with another problem. Many of these restaurants and bars belong to global corporations (McDonald's, KFC, Pizza Hut), or are located in buildings bought by global real estate institutions. This means that the profits from catering activities go above all to international corporations, which is described in details by Harvey. In one of his recent books, he shows that globalization is conducive to the development of big cities, but at the same time is a threat to them (Harvey 2012, chapter 'The urban roots of capitalist crises').

At about the same time as David Harvey, the French sociologist (and geographer) Henri Lefebvre became interested in the concept of space in the context of the city. Like Harvey, he represented a Marxist approach. In 1968 he published the work 'Le droit à la ville', which in English is entitled 'The right to the city'. As Purcell writes 'Lefebvre does not see the right to the city as an incremental addition to existing liberal-democratic rights. He sees it as an essential element of a wider political struggle for revolution'. (Purcell 2013, p. 142). According to some contemporary critics of Lefebvre's work, you can simultaneously see that '...He was also a hopeful romantic, looking back to medieval spatial and civic forms to find evolutionary possibilities'. (King 2019, p. 23).

In his manifesto from 1968, he writes that 'The *right to the city* cannot be conceived of as a simple visiting right or as a return to traditional cities. It can only be formulated as a transformed and renewed *right to urban life*'. (Lefebvre 1996a, p. 158). At the same time, Lefebvre is in favour of urban planning (Lefebvre 1996a, pp. 153–154) and a holistic view of social life in a city where human experience is included.

According to Lefebvre, space in the context of the city has its '...long *history of space*, even though space is neither a "subject" nor "object" but rather a social reality—that is to say, a set of relations and forms'. (Lefebvre 1991, p. 116). In the same book Lefebvre writes that his concept of space includes what he calls 'perceived space', 'conceived space', and 'lived space'. The first one refers to the objective, concrete space people encounter in their environment. 'Conceived space' he described as mental constructions of space (image). And 'lived space' means the complex combination of 'perceived' and 'conceived' spaces. In other words, it represents an individual's actual experience of space in everyday life. In the cited book, Lefebvre (1991, pp. 352–353) also distinguishes between *space of consumption* and *consumption of space*, what means an unproductive form of consumption. Both distinctions can be important when dealing with gastronomy, because it is made up of material and intangible content and forms. [At this point, it should be added a personal reflection that from the point of view of geography (not sociology or political science), it is the book 'The production of space' ('La production de l'espace', published in French language in 1974) which is more useful in the context urban space, rather than the manifesto 'The right to the city'.]

According to Purcell, Lefebvre in his works '...imagines and advocates a new urban politics, what I call an urban politics of the inhabitant.' (Purcell 2002, p. 100).

An example of such an approach in relation to gastronomy can be civic initiatives regarding organizing outdoor feasting, promoting the idea of Slow Food, etc. This phenomenon is described in Sect. 16.4 on the example of events in Warsaw named 'The Breakfast Market'.

Another very important issue in the works of Henri Lefebvre is the concept of *oeuvre* (Lefebvre 1996b, pp. 65–66), which relates to the terms 'attractivity' or 'attractiveness' commonly used today in urban context (see Chap. 5).

In conclusion, the main Lefebvre views on urban space quoted above, means that his approach to the question of urban space takes into account human experiences, the role of the time, etc. This, in turn, means, that his opinions are very similar to those of researchers dealing with the 'place'. Celine Jeanne (2016) thinks similarly. Comparing the views of Harvey and Lefebvre, she titled a fragment of her considerations 'Spatial determinism versus practices in place-making', pointing out that while dealing with urban issues at the same time (late 1960s and early 1970s), the first author represented the approach of treating the city in terms of 'space' and the other represented an approach that is close to the reality of the 'place' concept, albeit in a Marxist perspective (Jeanne 2016).

According to Gregory et al. (2009), contemporary theorizations of space in human geography share four following features: the integration of time and space; the co-production of time and space; the unruliness of time–space; and the porousness of time–space.

Place

In the humanistic geography 'place' refers to how people perceive a certain piece of space and what it means for them. This way of thinking is close to the phenomenology which tries to find the essential features of individual experiences.

As Yi-Fu Tuan wrote 'Geographers have approached the study of place from two main perspectives: place as location, a unit within a hierarchy of units in space; and place as a unique artefact. Thus we have a growing literature on "central-place" theory on the one hand, and on the other a small body of work devoted to depicting the unique character of individual places, mostly towns and cities.' (Tuan 1975, p. 151). For Tuan (1975, p. 152) 'Place is a center of meaning constructed by experience. Place is known not only through the eyes and mnd but also through the more passive and direct modes of experiences, which resist objectification'. A little farther Tuan (1975, p. 156) writes that 'Cities are places and centers of meaning par excellence', and finally comes to the conclusion that 'Chatting with neighbours on the stoop, going to the drugstore for a milk shake, emerging into glare of sunlight from the dark cavern of a movie house or bar, fresh neon colour in a wet night, and the thick Sunday newspapers—these experiences are too commonplace to sit for portraits.' (Tuan 1975, p. 157). At this point, it is worth noting that there are references to gastronomy in the quoted passage. According to Tuan, time is a very important element which constitutes 'place'. This is why, for example, signboards that can be found at an entrance to the restaurant or bar include not only its name, but also year when it was established.

Another geographer who, at about the same time as Yi-Fu Tuan, began to deal with the concept of 'place', is Edward Relph. His sense of the concept of 'place' is slightly different. He refers to physical setting, activities and meanings as the '… three basic elements of the identity of places'. (Relph 1976, p. 47). According to him these three components are deemed to be interrelated and inseparably interwoven in experience. As Withers (2009, p. 640) writes, 'Where Tuan defined place in relation to space: space as an arena for action and movement, place as about stopping, resting, becoming, and be coming involved, Relph emphasized a more experiential notion of place, and drew upon Edmund Husserl's work in phenomenology in doing so. Place in this sense had an almost spiritual dimension, having to do with dwelling, with being in the world'. (Whithers 2009, p. 640). It means that for Relph the essence of place '…does not therefore come from locations, nor from the trivial functions that places serve, nor from the community that occupies it, nor from superficial or mundane experiences. (…) The essence of place lies in the largely unselfconscious intention' (Relph 1976, p. 43). Such a statement brings even closer the views of Relph to phenomenology.

It should be recalled here that the phenomenological approach in human geography was already promoted by Carl Ortwin Sauer, who in 1925 wrote that 'The task of geography is conceived as the establishment of a critical system which embraces the phenomenology of landscape …' (Sauer 1925, p. 25). Although it may be assumed that Sauer understood the term 'phenomenology' differently (this may be indicated by using the phrase 'critical system') than Edmund Husserl, Martin Heidegger or Maurice Merleau-Ponty, it must be noted that he introduced it to geography nearly 50 years before Relph. At this point, it should be mentioned that the phenomenological approach, although rare, is also used in tourism research (see Cohen 1979; Kowalczyk 2014).

Returning to Relph, he stated in the preface of his book that his research method is 'a phenomenology of place' (Relph 1976, pp. 4–7). He described this statement in terms of three components: (a) the geographical location (place's physical setting); (b) place's activities, situations, and events; and (c) the individual and group meanings (experiences and intentions) in regard to the place.

An important role in Relph's deliberations on the subject of the place is the concept of *insideness* and *outsideness* (Seamon and Sowers 2008). He understood *insideness* as the degree of attachment, involvement, and concern that a person or group has for a particular place. As Seamon and Sowers wrote 'On the other hand, a person can be separate or alienated from place, and this mode of place experience is what Relph calls *outsideness*' (Seamon and Sowers 2008, p. 45).

Although Tuan and Relph (and Anne Butimer) can be considered precursors of the humanistic approach in geography, this direction has aroused and aroused the interest of many other geographers. Many of their considerations concern the concept of a place.

Among the geographers dealing with the concept of place in relation to cities, Tim Cresswell should be mentioned. In one of his works, he resembles the concept of 'meaningful location' by Agnew, who distinguished three aspects of 'place': (a) location, (b) locale, and (c) sense of place. The first aspect is related to the geographical location, the second is defined as '...material setting for social relations—the actual shape of place within which people conduct their lives...', and the third one as '...subjective and emotional attachment people have to place.' (Cresswell 2004, pp. 132–133). Another problem raised by Cresswell is the issue of authenticity versus the inauthenticity of the place. Referring to the views of Relph (1976), he considers the manifestations of culture which are often referred to as 'inauthentic' places and are described as 'McDonaldization' or 'Americanization' (Cresswell 2004, p. 174). At the same time, he refers to the French anthropologist, Marc Augé, who distinguished the category of 'non-places' (*non-lieux* in French), including such sites as airports, motorways, gasoline stations, hotels belonging to hotel chains, supermarkets, etc. (Augé 1995, pp. 94–106).

In another of his works Cresswell writes that 'To some geographers the humanistic conception of place, which has been the predominant understanding of place since the 1970s, is simply too fixed, too bounded, and too rooted in the distant past'. (Cresswell 2009, p. 176). He further recalls Doreen Massey, who notes that '...a notion of place where specificity (local uniqueness, a sense of place) derives not from some mythical internal roots nor from a history of isolation—now to be disrupted by globalization—but precisely from the absolute particularity of the mixture of influences found together there'. (Massey 1999, p. 22). In her concept of place, Massey writes about a 'progressive sense of place', a 'global sense of place', and about an 'extrovert sense of place'. She shows the numerous connections of 'place' with external factors, as other cultures, international relations, global market, etc. All these factors result from greater mobility of people and a larger scale of flows of goods and services. One of the more important arguments for such understanding of the term 'place' is tourism. More attention to this problem is given in the chapters 'Culinary attractiveness of a city—old and new destinations' (Chap. 5), 'Ethnic cuisine in urban space' (Chap. 11), and 'Tourist experience and change in culinary tastes' (Chap. 12).

The concepts of space and place are very useful in gastronomy research, especially in research carried out in qualitative approach. Referring to the views of Yi-Fu Tuan or Edward Relph, we can more easily understand the phenomenon of sentiment of London's or Leeds' residents to the nearest pub; people's visits from a *quartiere* of Florence or Siena in the local *trattoria* and so on. In turn, the category 'lived space' given by Henri Lefebvre allows to understand the role of information at the entrance to the restaurant about the year in which it was established. At this point, one should not reject the 'inauthentic' category of the 'place' (created by Edward Relph), because for tourists spending holidays in India, even a bar offering Indian cuisine in the centre of Warsaw (Poland) becomes a 'place'.

2.3 Behavioural Theories

We have already seen that the physical location of a dining establishment within the competitive urban space may be a major factor in its success or failure. Location can also be described using spatial behaviour models.

The first type is producer behaviour models. Pred (1967) defined the principles that entrepreneurs follow when choosing a particular place to do business. His first assumption was that while the choice of location is based on well-defined criteria, decision-makers (e.g. owners of independent bars or fast-food restaurants) are not completely rational. They use two important criteria: the availability of locational information (because such information is required to make an optimal decision); and the ability to use that information. Pred constructed a behavioural matrix with one axis representing the available information and the other the ability to use it. He concluded that most location decisions are suboptimal, but acceptable, in that they are still profitable. A profitable location refers to a spatial margin of profitability; in other words, a set of locations (often in the same area) where income from an activity is higher than the costs (purchasing land, rent, staff salaries, etc.) of that location. Many locational decisions are considered to be suboptimal because, even if the decision maker has all of the necessary information to hand, it is still not guaranteed that the chosen location will be profitable. This theory applies particularly to decisions regarding the location of dining services in urban areas where there is competition from other restaurants, changing consumer preferences, changing business rents, etc. All of these factors can cause a new food establishment to struggle financially, and even close shortly after opening, with their owners moving to another place or even going bankrupt. It is worth reiterating that many location decisions are not based solely on economic reasons. 'Soft' factors are also often used (family relations, recreational attractiveness, etc.). Timmermans (1986) used the producer behaviour model in the broad context of catering services to explain the location of economic activity in the town of Meijel (The Netherlands), taking into account other variables, including the presence of restaurants and bars.

The behaviour of entrepreneurs when choosing where to locate restaurants was explored in Taiwan, notably Japanese restaurants in Taipei. The first was founded in 1988 then, several years later, the restaurant's managers decided to open another in one of four possible locations (A, B, C and D) in different parts of the city (Tzeng et al. 2002, pp. 178–179). They based their decision on a combination of quantitative and qualitative data. Quantitative data were: rental costs, transportation costs, convenience of the mass transportation system, available parking, pedestrian volume, number of competitors and the size of the commercial area. Qualitative data included the intensity of competitors, the convenience of garbage disposal and sewerage capacity. The study found that transportation was the most important factor. Although sites B and D were both regarded as good locations, site B was in an expanding commercial centre at the intersection of two subway lines (Tzeng et al. 2002, pp. 181–182).

The second type of model is consumer behaviour models. It is very important to understand the spatial behaviour of consumers for any service activity, because it is often a major factor in the success of a restaurant or bar. Decision-makers use it when deciding on the location of eating establishments. The first analysis of geographic behaviour of consumers was linked to central-place theory. However, it was later noted that customer behaviour is largely linked to distance. The desire to reduce distance is a central assumption of central-place theory. Huff (1963) found that where consumers go depends on the distance (travel time) and the size (surface area) of the facility. Huff's model referenced the classical gravity concept of the retail trade area as formulated in Reilly's breaking point model. Responding to other authors, Huff noted that the 'vast majority of market analysts (…) have used the gravity model to delineate intra-urban retail trade areas have naively assumed that (…) intra-urban trade movements would possess the same value within urban areas'. He proposed including specific details about the shop, arguing that customers were willing to cover different distances depending on the type of store (Huff 1963, p. 86; 1964, p. 37). His mathematical model included expected average annual expenditure for a given product by consumers, and the annual amount budgeted by consumers for a given product or product class (Huff 1966, pp. 294–295). However, his models were applied to the retail business, and are not suited to explaining the behaviour of consumers using dining facilities. For this, it is necessary to include several others factors, including transport modes and costs, and consumer characteristics (gender, age, ethnicity, profession, income, etc.). This need was noted by Nakanishi and Cooper (1974) who developed their own, more mathematically advanced, Multiplicative Competitive Interaction model.

A similar approach was developed by Cadwallader (1975), although his theory appears rather unsophisticated in comparison. Cadwallader came up with the concept of the *behavioural environment*, based on the assumption that the consumer behaves according to the perceived rather than the real environment: 'A fundamental axiom of the behavioural approach is that an individual's behaviour is based upon his perception of the environment, and not upon the environment as it actually exists' (Cadwallader 1975, pp. 339–340). Using an example of 53 households from West Los Angeles, Cadwallader explained consumer behaviour using just three variables: store attractiveness (measured by speed of checkout service, range of goods sold, quantity of goods sold and prices), distance and information (Cadwallader 1975, p. 341, 344–346). In a similar vein, Timmermans (1979, p. 46) used distance and attractiveness of the place (measured by the number of employees) to explain consumer behaviour in Eindhoven (The Netherlands).

From the perspective of gastronomy in urban space, the most important factor in consumer behaviour research is the role played by location. A literature review by Longart (2015) suggests that most authors who analyze the motives for a consumer's choice of a particular restaurant ignore the effect of location. The exceptions are Auty (1992), Upadhyay et al. (2007) and Cannarozo-Tinoco and Duarte-Ribeiro (2012), who rank location as the fourth (out of 15), fifth (out of six) and sixteenth (out of 28) most important factor (Longart 2015, pp. 52–59 and 75). For Kivela (1997) location is particularly important for fast-food restaurants and bars. The question of location was studied in Seoul by Park et al. (2007, pp. 1135–1137) who devised a mobile application using an expectation maximization algorithm for selecting a restaurant based on aspects such as availability of parking and distance from the consumer, together with attributes such as type of restaurant, price level, socio-demographic characteristics of users, etc. In Taiwan, researchers found that customers were using the newest information technology to search for restaurants, including parking options (Lo et al. 2011).

The problem of car parking and accessibility has also investigated in Turkey (Kincaid et al. 2010). It is worth noting that earlier studies in the United States have shown that some restaurant chains, most notably Taco Bell Corporation, have worked with designers and planners to improve access and determine the size and layout of parking bays (Jaynes and Hoffman 1994, p. 165). The importance of access to restaurants has also been examined in India. Diners surveyed in Gwalior (Madhya Pradesh) rated parking higher than either ambience, prices or the chef's reputation when choosing a restaurant (Upadhyay et al. 2007, p. 10). The provision of car parking by restaurants in Toronto (in 1999) and Seoul (in 2011) was also found to have a significant positive relationship with the average amount spent (Susskind and Chan 2000, p. 63; Yim et al. 2014, pp. 17–18).

In an analysis of restaurant access in Brisbane (Australia), Yen et al. (2015, p. 10) found that most customers had short travel times (34% were under 10 min) and low costs. Only 18% of respondents travelled by car and the rest either used public transport or were active travellers (i.e. walked or cycled). About 20% of customers identified the active transport infrastructure (footpath, bicycle accessibility and pedestrian space) as incentives that encouraged them to visit a restaurant more often. More than 40% of customers said they would visit restaurants more often if public transport was improved and car traffic was reduced. These results show that, in some cases, locating cultural and tourist attractions close to public transport hubs can help to move the focus of owners from providing parking to the accessibility of public transport. The study also indicated that restaurateurs over estimated (by more than double) the numbers of customers who travelled by car and neglected the contribution of those who arrived by public transport. Moreover, it showed that clients and owners had different views on the importance of parking. Only 26% of customers thought they would find parking close to the restaurant. Restaurateurs ranked parking availability higher, implying that they believed that their customers often could not find somewhere to park (Yen et al. 2015, p. 12).

In the United Kingdom, customers consider several factors when deciding on the location of a restaurant: (1) driving distance, (2) convenience, (3) nearby entertainment, and (4) public transport (Longart 2015, pp. 193–195). The most important attribute was food quality (15.64%), followed by service (11.16%), atmosphere (9.91%), location (6.92%), food presentation (5.74%), menu options (3.58%) and appearance and cleanliness (3.43%) (Longart 2015, p. 232). In reference to location, 37% of respondents preferred parking facilities, while 25% chose public transport (Longart 2015, p. 215). The question of location seems to be more important for a night out with friends and family, possibly because people seek a convenient location that is accessible for everyone (from children to the elderly) (Longart 2015, p. 235).

The Brisbane study suggests found that restaurateurs should advocate for more public transport rather than parking areas, because these users spend more money. Customers who arrive on foot, by bike, or on public transport spend significantly more than others (Yen et al. 2015, p. 18), possibly because they consume more alcohol (whose price has a significant effect on the size of the final bill).

Simons (1992) suggests that the most important location-related factors for restaurants and other eating establishments are: (a) access, (b) visibility, (c) traffic count, (d) centre size, and (e) the presence of complementary businesses, including restaurants, bars, etc. (p. 522). Accessibility refers to the convenience of local transportation (including public transport) and parking. Visibility means the ability of potential customers to see the restaurant (from outside): the visibility of fast-food restaurants and bars is more important than their accessibility (Simons 1992, p. 527). Traffic count relates to the number of potential customers. The presence of other retail facilities means that a customer who visits a retail outlet is encouraged to visit a restaurant or ice cream bar. In other words, the presence of restaurants or bars in the same place is more attractive to customers than a specific facility. Potential customers frequent areas of the city where there are many restaurants, and only then decide which one to choose. These concentrations are referred to as gastronomic clusters (Simons 1992, p. 523) and are often called food courts, food streets or even food districts. There are good examples of food courts in the old part of Istanbul in the Çiçek, Hacopulo and Suriye arcades (located along the historical Istiklal Street), which are dominated by more than 40 eating outlets (Garip et al. 2013, p. 099:7). The notion that fast-food bars may benefit from the proximity of other quick-service restaurants (Simons 1992, p. 528) was initially decried. The appearance of food courts at shopping centres, however, offers support for this thesis.

Finally, we should note that there is a partial convergence between models describing the spatial behaviour of consumers and the theory of space and place. This is particularly true of the concept of 'lived space' (Lefebvre) and space (Tuan and Relph), which facilitate understanding of the behaviour of restaurant and bar customers.

2.4 Non-spatially Oriented Theories

As this book is about the relationship between gastronomy and urban space, concepts related to the non-spatial approach to culinary issues will only be briefly discussed.

First and foremost, it is important to note that food is not only one of the basic physiological needs, but also a very important element of culture and an activity that has an economic dimension. Although biological and medical science views of the role of food are important, they will not be considered here. The social and economic aspects of eating, however, cannot be ignored, because of their close links to geographical issues. Among the theories that are potentially useful in explaining the phenomenon of gastronomy are concepts from the social sciences (mainly cultural anthropology and sociology) and economics. In particular, theories that have emerged in the context of food—the concept of cultural heritage, the theory of cultural diffusion and the theory of assimilation—are all important.

Cultural heritage refers to the legacy of tangible (material) artefacts, and inherited intangible (value) attributes of a group or society. It is maintained in the present for the benefit of future generations. One of the important elements of cultural heritage is food. This is true of both food that is eaten and the ways in which it is prepared and served. Therefore, it can explain the persistence of regional and traditional gastronomy, especially given the recent renaissance of traditional cuisines.

The theory of cultural diffusion stands somewhat in contrast to the concept of cultural heritage, as it demonstrates that culture is constantly changing, which makes it increasingly difficult to identify original elements (Fig. 2.9). This penetration of cultural norms largely concerns food habits. Many dishes have been transferred—often as a result of migration. Italian pasta and pizza, for example, are now available worldwide. Pizzerias exist in places like North America and Asia, and in Turkish cities Turkish pizza is street food. At the same time, one of the most popular street foods in Poland is the Turkish döner kebab which, interestingly, came to Poland from Germany via Polish *Gastarbeiter* and tourists, rather than Turkey. Cultural diffusion can be party explained by the theory of assimilation. A common metaphor is the melting pot, which assumes that new conditions arising from an influx of immigrants lead to a society that integrates their cultural elements. Although the process of ethnic integration can be divided into the melting pot and true assimilation (i.e. immigrants adopt the cultural patterns of the society they have become part of) in the context of gastronomy this distinction seems unnecessary.

Economics suggests that economic base theory, ground rent theory and rational choice theory (also known as choice theory or rational action theory) are particularly useful in the context of gastronomy. Among more recent theories, the concept of the experience economy will also be considered.

In the context of gastronomy, economic base theory assumes, among other things, that food services are endogenous and exogenous. The former describes the situation in which restaurants and bars serve the local population and generate income by satisfying the needs of visitors and exporting food products. The latter mainly refers to cities with developed tourist functions. Ground rent theory can explain, to some extent, the location of eating establishments as well as the quality of catering services (restaurant categories), prices, etc. It explains, for example, why prices in downtown restaurants are higher than the same type of restaurant located on a city's outskirts. Rational choice theory (or rational action theory) can be used to explain the economic behaviour of individual actors. This theory focuses on the determinants of individual choices and assumes that people can choose among several available alternatives. A potential customer has access to relevant information, knows event probabilities and potential costs, and consistently chooses the

Fig. 2.9 Gastronomy from the perspectives of the theory of cultural diffusion and the theory of globalization. **a** The food court at Rehoboth Shopping Mall at Rehoboth (Namibia). The restaurant names suggest South African cuisine (biltong—on the left), as well as worldwide culinary trends (coffee shop and Steers—in the centre; pizza outlet on the right). The Steers restaurant chain is South Africa's leading burger brand. **b** McDonald's restaurant in the Defence Housing Authority district of Lahore (Punjab, Pakistan)

best action. Related to this group of theories is expected utility theory that explains how to make optimal decisions under risk. With reference to gastronomy, these theories help to understand the decisions and behaviour of owners, their employees, food suppliers, not to mention consumers' choices and behaviours.

The last concept addressed here is experience economy theory (Pine and Gilmore 1998). This suggests that businesses must create memorable events for their customers, and that memory itself becomes the product, i.e. the experience. The model includes four components that influence this experience: entertainment; education; aesthetics; and escapism (Pine and Gilmore 1998, p. 102). This idea has been applied in tourism research (Getz 2008, p. 413) and transferred to gastronomy research. The concept helps to understand the popularity of Mediterranean cuisine in the 1970s–1980s in Northern Europe. Similarly, the growing popularity of Thai cuisine in Europe and North America supports experience economy theory, rather than simply being attributed to the mass influx of people from Thailand.

There are five other concepts that can be used to explain both social and economic processes: the theory of modernisation; dependency theory; the theory of globalization; the theory of sustainable development; and the theory of postmodernity. Their common feature is also that they are general enough to be useful in a variety of situations. In the context of urban catering, the theory of modernisation can explain the evolution of culinary preferences and the provision of eating services. Dependency theory explains why, in Third World countries, local culinary traditions are being neglected in favour of bars offering hamburgers, fried potatoes or hotdogs. This appears to be not so much due to history (colonial past), but modern cultural neo-colonialism, partly introduced by overseas tourists. Both these theories are referenced in the concept of globalization. Examples include McDonald's restaurants in South Asia or Africa, Starbucks coffee shops in Vienna (the cafe capital of Central Europe since the seventeenth century) or Oktoberfest events in many North American and Canadian cities and towns. To some extent, globalization contrasts and conflicts with the concept of sustainable development. Although the latter can be applied to various problems and phenomena, in the case of gastronomy it is followed by the emergence and growing popularity of the Slow Food ideology (part of the so-called slow life). References to culinary heritage (reflected in a range of traditional dishes on a restaurant's menu, for example) can be regarded as instances resulting from the theory of sustainable development.

Postmodernism is another ideology that incorporates elements of all of the theories described. This philosophical position proposes that reality is ultimately inaccessible to human investigation, that knowledge is a social construct, and that the meaning of words is to be determined by readers not authors. In gastronomy, examples include recent trends such as new global cuisine, molecular gastronomy and fusion cuisine, or advertising food establishments as gay-friendly. The growing number of cities (e.g. Singapore, Copenhagen, Vancouver) or restaurants that refer to these trends demonstrates the close links between postmodernism ideology and contemporary gastronomy. A prime example comes from Singapore, where the government promotes so-called New Asian cuisine (Scarpato and Daniele 2003, pp. 310–312).

Finally, there is the concept of the 'city as an entertainment machine' (Lloyd and Clark 2001), which is derived from the concept of the city as a growth machine. Recent trends have seen a focus on consumption and growing importance placed on post-industrial services, including culture, tourism, etc. Lloyd and Clark note that, '[i]n a few square blocks of Chicago's Gold Coast, one encounters Thai, Japanese, Mexican, Indian, French, Cajun and Italian cuisine, or one may settle at Gibson's for the old Chicago standby, a Midwestern prime' (p. 6). They also observe that 'spaces of the industrialized past become mined for their aesthetic potential, like bars or theatres locating in former steel plants' (p. 9), as further support for the notion. On the one hand, this touches on the issue of globalization, encounters with different cultures, assimilation, etc. and, on the other hand, it is consistent with the trend of opening restaurants, bars and other entertainment establishments in former industrial areas. Such transformations are one of the symbols of the postmodern era (Fig. 2.10).

(a) (b)

Fig. 2.10 In Europe, perhaps the best example of the theory of the 'city as an entertainment machine' is Barcelona (Spain). **a** A food cluster at the Barcelona Port Olimpic. **b** Banners on the façade of a building in the La Ribera district. Night-time noise is a big problem, not only in Barcelona. Even in postmodern society, the concept of the 'city as an entertainment machine' has its opponents. Source: Anna Kowalczyk 2015

2.5 New Approaches to the City and City Planning in the Context of Urban Gastronomy

Previous sections may give the impression that everything that there was to be known about the role of gastronomy in urban space had already been discovered in the 1950s–1970s. What is more, the reader may think that since then, no new concepts have been put forward to help us to interpret the role of the restaurant sector in the contemporary city. This is, of course, not true, as many theories and models regarding cities and settlement systems have been formulated in recent decades. Most are drawn, however, from the social sciences. Although they concern urban planning and governance (including the adaptation of new information technologies), what they all have in common is the focus on different groups of city users[6] and the city's economy. In these theories, space, in the Euclidean sense, is often only understood as a background for social and economic processes.

---

[6]According to Martinotti (1996) contemporary cities (especially the largest) are becoming less and less dependent on residents and commuters, and more and more dependent on other groups of users: visitors (e.g. students, but especially tourists), and so-called metropolitan businessmen. These new groups have changed the social and physical shape of the city, as they force city authorities and investors to draw up urban policies that address their needs. One example of these new economic activities is the location of new and very expensive restaurants in zones visited by tourists and/ or businessmen. These restaurants, bar or cafés are often owned by foreign (independent) owners or global corporations. Although their pricing policy is not a problem for municipal authorities that collect taxes, they are often inaccessible to many residents. Kotus et al. (2015) carried out an interesting study of the spatial activity of city users (including students and tourists) in Poznań (Poland).

This section focuses on a few, selected concepts of the city. This sample was chosen to reflect a variety of questions related to the role of gastronomy. Since most of these ideas are relatively new, and have not yet been widely used in urban planning and policy, it is difficult to find empirical support. However, a few examples will help us to ascertain which of them have (or may have) an impact on the role of gastronomy in urban space. Specifically, we examine three contemporary theories: (a) the creative city; (b) the compact city; and (c) the smart city.

2.5.1 The Creative City

Richard Florida is the author of the creative city concept (2005, 2009). According to him, a creative city uses creativity to build a competitive advantage. Creativity is understood in the broadest sense, as an attitude, a mental process leading to the emergence of new ideas, concepts or associations, and links with existing ideas and concepts. In other words, a creative city is an environment in which values serve as a catalyst for development, where artists and other participants play a fundamental role. These places have developed living, working and entertainment conditions that are tolerant and welcome newcomers. The features that define a creative city include: (a) a political framework that encourages creative attitudes; (b) uniqueness and diversity; (c) openness and tolerance; (d) entrepreneurship and innovation; (e) strategic leadership and vision; (f) an environment favourable to learning and the development of talents (including appropriate infrastructure); (g) good transport, accessibility and networking; (h) an adequate quality of life (and high-quality public services); and (i) professionalism and efficiency. Another factor that influences whether a city can be considered as creative is the presence of so-called 'creative industries'. Generally speaking, these are areas of activity that are concerned with the generation or exploitation of knowledge and information. According to Florida, creative cities attract, and are inhabited by the creative class. He argues that these highly qualified employees are drawn by jobs in high-technology industries.

Among the features of a creative city are gastronomy facilities that reflect the high-quality of life required by the creative class. According to Florida's research, this class pays more attention to amenities that fit with their lifestyle than other, more traditional values (see Frenkel et al. 2013 for empirical evidence using the example of Tel Aviv, Israel). Focusing on the importance of amenities, Florida proposed a 'coolness' index, based on the percentage of the population aged 22–29, night life (number of bars, clubs, etc. *per capita*) and culture (number of art galleries and museums *per capita*). Finally, he stated that there is a significant correlation between the coolness index and a so-called 'talent' index (r = 0.469).

Since its formulation, the creative city idea has become very important in urban planning and policy. As Dimitrovski and Crespi Vallbona (2018, p. 399) write, *'the novelty in current urban planning is its dependence on cultural consumption. This cultural consumption in turn leads to the implicit tourism, entertainment and leisure model. Indeed, specific cities are attractive to the visitor due to the perception of the lifestyles led in them'*. Similarly, Richards (2014) argues that the concept of a creative city in the tourism domain can be divided into six aspects: (a) developing tourism products; (b) revitalisation of existing products; (c) enhancing cultural and creative assets; (d) economic spin-offs for creative development; (e) using creative techniques to enhance the tourism experiences, and (f) adding buzz and atmosphere.

However, other authors have criticized Florida's concept, arguing that he overestimates the role of the creative class in the city's development, and underestimates the role of innovative business (Moretti 2012). Others suggest that he sees the city from a neo-liberal perspective and fails to acknowledge that there are both enclaves of success and enclaves of poverty. This is confirmed by longitudinal research carried out in Barcelona and Madrid (Yáñez 2013). Left-leaning urban activists criticize the concept for leading to the development of programmes or development strategies that 'use' culture, art and creativity as an instrument (d'Ovidio and Rodríguez Morat 2017). These objections can also be applied to gastronomy. For example, Mattson (2015, p. 2) writes, *'Bars are beachheads of gentrification, featuring in almost every ethnography of gentrification as sites of desire and conflict where newcomers encounter oldtimers'*. The latter author takes the example of the Polk Gulch (now Polk Street) in San Francisco: following the implementation of a strategy inspired by the creative city concept, the number of LGBT bars fell from 13 (in 1999) to three (in 2010). The second-largest gay bar district transitioned to the city's top nightlife destination for young heterosexuals (Mattson 2015). Other authors have other objections. For instance, some question whether the experience from the United States is transferable to other countries (Richards 2014), while Baum argues that the 'creative class' does not exist in isolation from the employees who support it. In particular, the latter author notes that the hospitality industry includes the people it employs, which means that attention must be given to '*…the role that low skills personal service workers play in supporting the development of creative cities*' (Baum 2018, p. 23). However, this problem is very complicated; as Baum stated in an earlier article, it is difficult to reduce the question of the professional qualifications of people employed in the hospitality industry to a simple distinction between 'unskilled' and 'skilled' workers (Baum 1996, pp. 208–209). Taking the example of gastronomy, a chef can be an 'ordinary' chef who cooks 'ordinary' dishes, but (s)he can also be a 'celebrity' chef who appears on television, and who creates new patterns of cooking and eating.

Although the validity of the creative city idea in the context of urban gastronomy remains questionable, it does explain several issues. First, it confirms a phenomenon that has been established since at least the eighteenth century, namely that eating establishments are places where social (and cultural) life is concentrated. Second, it suggests that the social group that Florida calls the creative class (due to both their economic wealth and their intellectual needs), prefers to visit places where they meet people from the same class. Thirdly, Florida shows that dining establishments are very important for both the city's economy and its image. The latter observation is particularly important, because it explains the existence, since 2004, of the UNESCO Creative Cities Network, whose members promote, among other features, their gastronomy.

Lew (2017) examines the problem of adapting Florida's concept to tourism planning and policy. He considers the importance of the concept of *place making*, which he defines as '...*how people recognize, define, and create the places they often call home, whether intentionally or not*' (p. 450). He distinguishes it from *place-making* and *placemaking*.[7] Place-making refers to bottom-up local initiatives, often inspired by the needs of residents. On the other hand, placemaking refers to a top-down process that is inspired and implemented by planners, local authorities, etc. The impact of place-making/placemaking can be seen in empirical studies on gastronomy in large cities. An example is research carried out in Budapest (Lugosi and Lugosi 2008), which focused on bars in rundown downtown neighbourhoods that started to be visited by young people and tourists. In this book we present examples of both processes: placemaking is seen in actions taken by cities to promote their culinary heritage (Chap. 5), or the establishment of zones with a clearly marked gastronomic function (e.g. along the Vistula River in Warsaw, Poland, see Sect. 16.5). Also in Warsaw, the Breakfast Market (*Targ Śniadaniowy*) initiative is an example of place-making (Sect. 16.4). Such initiatives are sometimes called creative placemaking (rather than place-making), and are understood as cultural and artistic activities that reflect the physical and social character of a neighbourhood.

Another term that can be considered similar to place making is *cultural mapping*. This relatively new concept (also known as *cultural resource mapping* or *cultural landscape mapping*) is focused on the conservation/protection of cultural diversity. In the popular sense, it includes techniques and tools that are used to 'map' the tangible and intangible cultural assets of distinct groups within their landscape or environment. This understanding includes so-called 'gastronomic mapping' or 'culinary mapping' (du Randa et al. 2016) and can, for example, include an inventory of the most popular restaurants.

Similarly, *festivalisation* can be compared to place making (Cudny 2016).[8] The festival phenomenon is closely connected with the attractiveness of cities. Although the term is mentioned in Sect. 2.2, and numerous examples of culinary festivals are described in Sect. 3.6, here we devote attention to the concept itself, as it is one of the tools that is used most frequently to build a creative city. The terms *festivalisation* and *festivalscape* appear in the literature as corresponding to the concept of the *touristscape*. For Mason and Paggiaro (2010, pp. 1330–1331), the term means '...*the physical environment, putting together tangible factors and the event*

---

[7]The distinction between 'place-making' and 'placemaking' is not obvious. For instance, Salzman and Yerace (2018, p. 58) write, "*Placemaking is focused on the event, not the institution or the individual. Even though formal programming may be involved, a placemaking 'event' can be understood as whatever activates (or brings people to) the public space*".

[8]This concept corresponds to the concept of tourismification (Jansen-Verbeke 1998, 2007) or touristification.

atmosphere' and, in the context of a gastronomy event studied in Udine (Italy), it is '*...the general atmosphere experienced by festival visitors in food and wine festivals*'. This understanding comprises both material (tangible) and non-material (intangible) elements.

2.5.2 The Compact City

The concept of the compact city is a reference to the idea of so-called 'new urbanism' that was formulated in the United States in the early 1980s. It can be considered as a reaction to the phenomena of suburbanization and urban sprawl (the specific topic of gastronomy in the suburbs is described in Sects. 1.4.3 and 2.5) Therefore, the starting point for the compact city philosophy is the fact that was a reaction to the phenomenon of urban sprawl. The main features are: (a) a relatively small area; (b) a compact built-up area; and (c) land use that reflects geographical conditions. It therefore corresponds to high residential density with mixed land use. High residential density has the positive effect of lowering expenditure on technical infrastructure. Furthermore, compactness means that residents tend to use public transport, cycle or walk rather than drive their own car, which reduces energy consumption and pollution.

Although the compact city concept was first proposed in the 1970s (Dantzig and Saaty 1973), its popularity increased in the 1990s. There were several reasons for this, notably '*...higher density settlements are argued to be more socially sustainable because local facilities and services can be maintained, due to high population densities, and therefore accessibility to goods and services is more equitably distributed*' (Williams 1999, p. 168). In the context of gastronomy, the usefulness of the concept primarily concerns the organization of public space in the centres of large cities. Many feature a concentration of eating places along pedestrian routes and, in areas where there are restrictions on the movement of cars, residents can move around by bicycle.

One of the foundations of the compact city model is the concentration of economic activities (especially services) around a central point, typically the market square. This configuration of eating establishments was observed in European cities and towns as early as the Middle Ages. The trend gradually weakened in the nineteenth century, but still today many of Europe's cities and towns are characterized by a concentration of restaurants and bars in the core of the historical area. This can be contrasted with North America, where the pattern is less visible, especially in cities and towns that were founded later. At the same time, the process of so-called 'urban sprawl' has changed the distribution of eating places within the city. This phenomenon first appeared in the United States, and only later in Europe. One of its manifestations is a concentration of dining establishments in shopping centres, which are usually located outside the inner city.

Urban planners responded to the adverse effects of urban sprawl in North America; among others actions, the festival market place (FMP) concept was developed.[9] The essence of the FMP is the restoration of the market place function to downtown areas. Thus, while the idea is a novelty in American cities, this is not the case in Europe, where the market square has always been the central point. In conclusion, it is no coincidence that the compact city concept emerged in the United States, as European cities suffer less from the problem of urban sprawl. The situation is different again in cities in other parts of the world. The demographic and territorial expansion of cities in Asia, Africa and Latin America has been so dramatic that the compact city concept is largely an illusion.

2.5.3 The Smart City

The smart city concept is more recent than the others described in this section. There is no single, valid definition, for several reasons. First, it is still relatively new and it is difficult to date its foundation. Second, it covers a broad and dynamic spectrum of elements, and the definition is still evolving. According to Giffinger (2007), the smart city consists of six dimensions: (a) a smart economy (or competitiveness); (b) smart people (or social and human capital); (c) smart governance (or participation); (d) smart mobility (or transport and ICT technologies); (e) a smart environment (natural resources); and (f) smart living (quality of life). Giffinger describes each of these dimensions in broad terms.

Batty et al. (2012, p. 482) note that '*smart cities are often pictured as constellations of instruments across many scales that are connected through multiple networks which provide continuous data regarding the movements of people and materials in terms of the flow of decisions about the physical and social form of the city*'. A similar description is given by Hollands (2015, p. 64), who states that '*...a smart city is made up of IT devices, industry and business, governance and urban services, neighbourhoods, housing and people, education, buildings, lifestyle, transport and the environment*'. More rarely, other authors pay attention to new methods used in the design and planning of smart cities. Because the smart city idea is relatively recent, it is difficult to find empirical evidence in the literature. Songdo (South Korea) is one of the examples that is most frequently cited (Carvalho 2015).

However, in the context of gastronomy in urban space, the following are particularly important:

- a smart economy: innovative spirit, entrepreneurship and trademarks;
- smart people: level of qualification, ethnic plurality, creativity, open-mindedness;
- smart governance: transparency, political strategies and perspectives;
- smart mobility: local, interregional and international accessibility, availability of ICT infrastructure, transport systems;

---

[9]The concept of the festival market place is mentioned in several places in this book (first, in Sect. 2.2).

- a smart environment: attractive natural conditions, low pollution, environmental protection, sustainable resource management;
- smart living: cultural facilities, individual safety, attractive to tourists.

These factors merit closer attention.

A *smart economy* concerns both current issues (innovativeness) and the place's history and culinary traditions (trademarks). The gastronomic attractiveness rankings described later in this book are the result of these premises. Copenhagen (Danemark) and Stockholm (Sweden) are examples of cities that have focused on innovation, while Paris and Lyon (France) or Madrid (Spain) attract foodies because of their adherence to tradition. *Smart people* also have a big impact on which cities are perceived as global/world capitals of food. For example, the high ranking of New York, San Francisco or Singapore results not only from the creativity, open-mindedness and creativity of staff in restaurants but, above all, from their ethnic plurality. Among the elements that make up *smart governance*, the most important for the functioning of restaurants, bars, etc. (together with other services) is transparency. In particular, this concerns responsible fiscal policy and policy regarding renting premises for eating establishments. The importance of *smart mobility* applies to the entire catering industry as an economic sector as it can stimulate new forms of entrepreneurship. In the context of the gastronomic function in urban space, this concerns:

- mobile applications that make it easier for customers to find and visit a restaurant or bar, and stay in touch with it;
- home delivery of restaurant meals to the client's home or office catering, which may, in turn, limit visits by consumers to physical restaurants or bars.

Regarding a *smart environment*, some authors argue that the term 'smart city' is synonymous with the term 'self-sustainable city' (Romero-Borbón et al. 2015), suggesting that inhabitants are able to feed themselves from food products produced in the city itself. But is ordering food from organic farms, or shopping at farmers markets typical of a self-sustainable city? If this is not the case, then these activities are characteristic of the sixth factor: *smart living*. In addition to cultural facilities, individual safety and touristic attractiveness, smart living reflects better living conditions (e.g. taking care of health and a pro-ecological lifestyle), and has a significant impact on improving the quality of life.

Finally, other relevant concepts that can be used in the context of gastronomy include the *slow city* or the *green city* (previously often referred to as the garden city). Like the smart city, they can be considered as an element of so-called 'urban sustainable development'. According to Leźnicki and Lewandowska (2016), the concept of sustainable urban development includes the terms: the sustainable city, the eco-city, the green city, the ecological city, as well as the compact city and the smart city. In the context of gastronomy in urban space, differences between these concepts are irrelevant.

In conclusion, this section draws attention to two issues. First, in recent years, many authors have tried to revise earlier theories of urban development. These reformulated concepts can be classified as neo-liberal theories (e.g. the creative city), while others refer to the ideology of sustainable development (e.g. the compact city and the smart city). Second, new visions of the city mainly concern issues of living conditions, social justice and management (the implementation of pro-ecological policy or the introduction of new technologies). Spatial aspects are marginal and, to a large extent, refer to concepts and models formulated a few decades earlier (e.g. Howard's garden city). This applies to, notably, the question of the location of services in the urban space. Despite these new developments, it appears that the findings of earlier researchers such as Christaller, or Berry and Garrison (see Sect. 2.2) continue to be taken into account in decisions made by planners, entrepreneurs and consumers regarding the restaurant sector.

References

5 urban models, March 12, 2009. https://www.slideshare.net/ecumene/5-urban-models. Accessed 10 June 2017

Agnew JA (2011) Space and place. In: Agnew J, Livingstone D (eds) Handbook of geographical knowledge. Sage, London, Sage, pp 316–330

Alonso W (1964) Location and land use. Harvard University Press, Cambridge, MA

Augé M (1995) Non-places. Introduction to an anthropology of supermodernity. Verso, London—New York, English translation by J. Howe

Auty S (1992) Consumer choice and segmentation in the restaurant industry. Serv Ind J 12(3):324–339

Batty M, Axhausen K, Giannotti M, Pozdnoukhov A, Bazzani A, Wachowicz M, Ouzounis G, Portugali Y (2012) Creative cities of the future. Eur Phys J Spec Top 214:481–518

Baum T (1996) Unskilled work and the hospitality industry: myth or reality? Int J Hosp Manage 15(3):207–209

Baum T (2018) Changing employment dynamics within the creative city: exploring the role of 'ordinary people' within the changing city landscape. Econ Ind Democr 1–39. https://strathprints.strath.ac.uk/62181/1/Baum_EID2017_Changing_employment_dynamics_within_the_creative_city.pdf. Accessed 02 Oct 2018

Behera A (2017) Reimagining contemporary urban planning with placemaking, May 2017. MSc thesis, Georgia Institute of Technology, Atlanta, GA

Berry BJL (1959) Ribbon development in the urban business pattern. Ann Assoc Am Geogr 49 (2):145–155

Berry BJL (1963) Commercial structure and commercial blight. Retail patterns and processes in the city of Chicago. Research paper, 85, The University of Chicago, Department of Geography, Chicago

Berry BJL, Garrison WL (1958) The functional bases of the central place hierarchy. Econ Geogr 34(2)(April):145–154

Burgess EW (1925) The growth of the city: an introduction to a research project. In: Park RE, Burgess EW, Mc.Kenzie RD (eds) The city, The University of Chicago Press, Chicago and London, pp 47–62

Burke M, Brown AL (2007) Distances people walk for transport. Road Transp Res 16(3):16–29

Cadwallader M (1975) A behavioral model of consumer spatial decision making. Econ Geogr. 51 (4)(October):339–349

Cannarozzo-Tinoco MA, Duarte-Ribeiro L (2012) Main attributes of quality and price perception for a la carte restaurants. Management 2(2):40–48

Carvalho L (2015) Smart cities from scratch? A socio-technical perspective. Camb J Reg Econ Soc 8(1)(March):43–60
Chen JD (2015) The urban realms model by Vance, May 7, 2015. https://worldgeographyindonesia.wordpress.com/2015/05/07/the-urban-realms-model-by-vance. Accessed 10 June 2017
Cohen E (1979) A phenomenology of tourist experiences. Sociology 13(2)(May):179–201
Cresswell T (2004) Place: a short introduction. Blackwell Publishing, Malden, MA—Oxford—Carlton, Vic
Cresswell T (2008) Place: encountering geography as philosophy. Geography 93(3)(Autumn):132–139
Cresswell T (2009) Place. In: Thrift N., Kitchen R (eds) International encyclopedia of human geography, vol 8. Elsevier, Oxford, pp 169–177
Cresswell T (2013) Geographic thought: a critical introduction: critical introductions to geography. Wiley-Blackwell, Malden, MA—Oxford—Chichester
Cudny W (2016) Festivalisation of urban space. Factors, processes and effects. Springer Geography, Springer International Publishing AG, Cham
d'Ovidio M, Rodríguez Morat A (2017) Introduction to SI: against the creative city: activism in the creative city: when cultural workers fight against creative city policy. City, Cult Soc 8 (March):3–6
Dantzig GB, Saaty TL (1973) Compact city: plan for a liveable urban environment. W. H, Freeman, San Francisco, CA
de Villiers RET (2016) The V&A waterfront as workplace and leisure space for Capetonians, March 2016, Faculty of Arts and Social Sciences at Stellenbosch University, Stellenbosch, MA thesis
Dimitrovski D, Crespi Vallbona M (2018) Urban food markets in the context of a tourist attraction—La Boqueria market in Barcelona, Spain. Tour Geogr 20(3):397–417
Dine at Underground Atlanta (2017). http://www.underground-atlanta.com. Accessed 10 July 2017
Dine (2017). https://www.riverwalkneworleans.com/directory/dine.html. Accessed 10 July 2017
Dining Directory (2017). http://www.faneuilhallmarketplace.com/restaurants/all. Accessed 10 July 2017
du Randa GE, Booysenb I, Atkinson D (2016) Culinary mapping and tourism development in South Africa's Karoo region. Afr J Hosp Tour Leis 5(4). http://www.ajhtl.com/uploads/7/1/6/3/7163688/article_11_vol_5__4_.pdf. Accessed 10 Sept 2018
Farell K (1980) Insights into site selection. Restaur Bus 79(1 July):115–120
Ferreira S, Visser G (2007) Creating an African Riviera: revisiting the impact of the Victoria and Alfred Waterfront development in Cape Town. Urban Forum 18(3 September):227–246
Florida R (2005) Cities and the creative class, Routledge, New York—London
Florida R (2009) Who's your city?. Basic Books, New York
Frenkel A, Bendit E, Kaplan S (2013) Residential location choice of knowledge-workers: the role of amenities, workplace and lifestyle. Cities 35:33–41
Garip E, Salgamcıoglu ME, Kos FC, Menderes F (2013) A comparative analysis of urban arcades in "Istiklal Street" Istanbul and an evaluation of their potentials of use. In: Kim YO, Park HT, Seo KW (eds) Proceedings of the ninth international space syntax symposium, 31 Oct–3 Nov 2013, Sejon University, Seoul, 089, pp 099:1–099:12. http://www.sss9.or.kr/paperpdf/ussecp/SSS9_2013_REF099_P.pdf. Accessed 20 June 2017
Getz D (1993) Planning for tourism business districts. Ann Tour Res 20(3):583–600
Getz D (2008) Event tourism: definition, ecolution, and research. Tour Manage 29(3)(June):403–428
Giffinger R (2007) Smart cities ranking of European medium-sized cities. Centre of Regional Science, Vienna
Gravari-Barbas M (1998) La "festival market place" ou le tourisme sur le front d'eau. Un modèle urbain américain à exporter. Norois, 45(178)(April–June):261–278

Gregory D, Johnston R, Pratt G, Watts MJ, Whatmore S (eds) (2009) The dictionary of human geography, 5th edn, Blackwell Publishing Ltd, Chichester

Hall CM, Sharples L, Smith A (2003) The experience of consumption or the consumption of experiences? Challenges and issues in food tourism. In: Hall CM, Sharples L, Mitchell R, Macionis N, Cambourne B, (eds) Food tourism around the world. Development, management and markets. Butterworth-Heinemann, Oxford—Burlington, pp 314–335

Harris ChD, Ullman EL (1945) The nature of cities. In: Building the future city. Ann Am Acad Polit Soc Sci 242 (November):7–17

Harvey D (1973) The social justice and the city. Basil Blackwell Publishers, Oxford

Harvey D (1989) From managerialism to entrepreneurialism: the transformation in urban governance in late capitalism. Geogr Ann 71 (Series B):3–17

Harvey D (1990) Between space and time: reflections on the geographical imagination Ann Assoc Am Geogr 80(3)(September):418–434

Harvey D (2004) Space as a key word. In: Paper for Marx and philosophy conference, 29 May 2004, Institute of Education, London. http://frontdeskapparatus.com/files/harvey2004.pdf. Accessed 20 Dec 2018

Harvey D (2008) The right to the city. New Left Rev 53 (September–October):23–40

Harvey D (2012) Rebel cities. From the right to the city to the urban revolution. Verso, London—New York

Hollands RG (2015) Critical interventions into the corporate smart city. Camb J Reg, Econ Soc 8 (1):61–77

Hoyt H (1939) The structure and growth of residential neighborhoods in American cities. Federal Housing Administration, Washington

http://citeseerx.ist.psu.edu/viewdoc/download?doi=10.1.1.555.5915&rep=rep1&type=pdf. Accessed 17 Jan 2019

https://scholar.sun.ac.za/bitstream/handle/10019.1/98689/devilliers_waterfront_2016.pdf?sequence=2&isAllowed=y. Accessed 17 Jan 2019

https://smartech.gatech.edu/bitstream/handle/1853/58520/abhishek_behera_reimagining_contemporary_urban_planning.pdf. Accessed 15 Feb 2019

Huff DL (1963) A probabilistic analysis of shopping center trade areas. Land Econ 39(1) (February):81–90

Huff DL (1964) Defining and estimating a trading area. J Mark 28(3)(July):34–38

Huff DL (1966) A programmed solution for approximating an optimum retail location. Land Econ 42(3)(August):293–303

Jansen-Verbeke M (1998) Tourismification of historical cities. Ann Tour Res 25(3):739–742

Jansen-Verbeke M (2007) Cultural resources and the tourismification of territories. Acta Turistica Nova 1(1):21–41

Jaynes SL, Hoffman JO (1994) Discrete event simulation for quick service restaurant traffic analysis. In: Tew JD, Manivannan S, Sadowski DA, Seila AF (eds) Proceedings of the 1994 winter simulation conference, Orlando, Florida, USA, 11–14 Dec 1994, pp 1061–1066

Jeanne C (2016) Comparing the right to the city concepts of Henri Lefebvre and David Harvey, Working Paper, April 2016

Kearsley G (1983) Teaching urban geography: the Burgess model. NZ J Geogr 12(1) (October):10–13

Kincaid C, Baloglu S, Mao Z, Busser J (2010) What really brings them back? Int J Contemp Hosp Manage 22(2):209–220

King LA (2019) Henri Lefebvre and the right to the city. In: Meagher S, Noll S, Biehl JS (eds) Philosophy of the City Handbook, Routledge, London (in press). https://www.researchgate.net/publication/328491674_Henri_Lefebvre_and_the_Right_to_the_City. Accessed 15 Feb 2019

Kivela JJ (1997) Restaurant marketing: selection and segmentation in Hong Kong. Int J Contemp Hosp Manage 9(3):116–123

Kotus J, Rzeszewski M, Bajerski A (2015) Przyjezdni w strukturach miasta – miasto wobec przyjezdnych: studenci i turyści w mieście w kontekście koncepcji city users, Studia i Prace z Geografii, 52. Bogucki Wydawnictwo Naukowe, Poznań

Kowalczyk A (2014) The phenomenology of tourism space. Tourism 24(1):9–15

Law ChM (1996) Urban tourism. Attracting visitors to large cities. Tour Leis Recreat Ser. Mansell Publishing Limited, London—New York

Lefebvre H (1996a) The right to the city. In: Writings on cities/Henri Lefebvre. Selected, translated and introduced by Eleonore Kofman and Elizabeth Lebas. Blackwell Publishers Ltd./Blackwell Publishers, Inc., Oxford—Malden, MA, pp 147–159

Lefebvre H (1996b) Industralization and urbanization. In: Writings on cities/Henri Lefebvre. Selected, translated and introduced by Eleonore Kofman and Elizabeth Lebas. Blackwell Publishers Ltd./Blackwell Publishers, Inc., Oxford—Malden, MA, pp 65–85

Lefebvre H (1991) The production of space, Basil Blackwell Inc, Oxford—Cambridge, MA. English translation by D, Nicholson-Smith

Lew AA (2017) Tourism planning and place making: place-making or placemaking? Tour Geogr 19(3):448–466

Leźnicki M, Lewandowska A (2016) Contemporary concepts of a city in the context of sustainable development: perspective of humanities and natural sciences/Współczesne koncepcje miast w kontekście zrównoważonego rozwoju: perspektywa humanistyczno-przyrodnicza. Probl Ekorozwoju-Probl Sustain Dev 11(2):45–54. https://papers.ssrn.com/sol3/Delivery.cfm/SSRN_ID2858231_code1258602.pdf?abstractid=2858231&mirid=1. 10 Sept 2018

Lloyd R, Clark TN (2001) The city as an entertainment machine. In: Research Project no. 454, paper prepared for presentation at the annual meeting of the American Sociological Association, 2000. http://www.sociologia.unimib.it/DATA/Insegnamenti/4_2996/materiale/the/city/as/an/entertainment/machine.doc. Accessed 10 July 2017

Lo C-Y, Lin C-T, Tsai C-L (2011) Mobile restaurant information system integrating reservation navigating and parking management. Int J Eng Technol 3(2)(April–May):173–181

Longart P (2015) Consumer decision making in restaurant selection (volume I), April 2015, Faculty of Design, Media and Management, Buckinghamshire New University—Coventry University. Ph.D. dissertation. Accessed 01 June 2017

Lugosi P, Lugosi K (2008) The 'ruin' bars of Budapest: urban decay and the development of a genre of hospitality. In: Paper presented at the 17th CHME research conference, Strathclyde University, Glasgow, 14–16 May 2008. http://eprints.bournemouth.ac.uk/6892/2/P_Lugosi_and_K_Lugosi_Rom_hospitality.pdf. Accessed 05 August 2018

Martinotti G (1996) Four populations: human settlements and social morphology in the contemporary metropolis. Eur Rev 4(1)(January):3–23

Mason MC, Paggiaro A (2010) Investigating the role of festivalscape in culinary tourism: the case of food and wine events. Tour Manage 33(6):1329–1336

Massey D (1999) Power-geometries and the politics of space-time, Hettner-Lecture 1998, Hettner-Lectures 2, Department of Geography, University of Heidelberg, Heidelberg

Mattson G (2015) Bar districts as subcultural amenities. City, Cult Soc 6(1):1–8

Maye D (in press) 'Smart food city': conceptual relations between smart city planning, urban food systems and innovation theory. City Cult Soc. https://www.researchgate.net/publication/322269651_'Smart_food_city'_Conceptual_relations_between_smart_city_planning_urban_food_systems_and_innovation_theory. Accessed 10 Sept 2018

Moretti E (2012) The new geography of jobs. Houghton Mifflin Harcourt Publishing Company, Boston, MT—New York, NY

Nakanishi M, Cooper LG (1974) Parameter estimation for a multiplicative competitive interaction model: least squares approach. J Mark Res 11(3)(August):303–311

Narvaez L, Penn A, Griffiths S (2013) Spatial configuration and bid rent theory: how urban space shapes the urban economy. In: Kim YO, Park HT, Seo KW (eds) Proceedings of the ninth international space syntex symposium, 31 October–3 November 2013, Sejon University, Seoul, 089, pp 089:1–089:19, http://www.sss9.or.kr/paperpdf/ussecp/sss9_2013_ref089_p.pdf. Accessed 20 June 2017

O'Sullivan A (2003) Market areas and central place theory, on-line chapter from book Urban economics, 5th edn. McGraw Hill/Irvin, Boston. https://www.researchgate.net/publication/265030128_Market_Areas_and_Central_Place_Theory. Accessed 20 May 2017

Oosthuizen M The Victoria and Alfred Waterfront: evaluating the public space, Unpublished internal document, University of Stellenbosch, Department of Geography and Environmental studies, Matieland

Our history. https://www.waterfront.co.za/the-va/the-company/our-history. Accessed 17 Jan 2019

Overview. https://www.waterfront.co.za/business/tourism. Accessed 17 Jan 2019

Pacione M (2005) Urban geography. A Global Perspective, Routledge, London and New York

Park MH, Hong JH, Cho SB (2007) Location-based recommendation system using Bayesian user's preference model in mobile devices. In: Indulska J, Ma J, Yang LT, Ungerer T, Cao J (eds) Ubiquitous intelligence and computing, 4th international conference, UIC 2007, Hong Kong, China, 11–13 July 2007. Springer-Verlag, Berlin—Heidelberg, pp 1130–1139

Pillsbury R (1987) From hamburger alley to Hedgerose Heights: toward a model of restaurant location dynamics. Prof Geogr 39(3):326–344

Pine BJ, Gilmore J (1998) Welcome to the experience economy. Harv Bus Rev 76(4)(July–August):97–105

Pirie G (2007) Urban tourism in Cape Town. In: Rogerson ChM, Visser G (eds) Urban tourism in the developing world: the South African experience. Transaction Publishers, New Brunswick—London, pp 223–244

Pred A (1967) Behavior and location: foundations for a geographic and dynamic location theory, Tom 1. Lund studies in geography: human geography, No 27, Studies in Geography Series B (Human Geography), The Royal University of Lund, Department of Geography, C. W. K. Gleerup, Lund

Purcell M (2002) Excavating Lefebvre: the right to the city and its urban politics of the inhabitant. GeoJ 58(2/3):99–108

Purcell M (2013) Possible worlds: Henri Lefebvre and the right to the city. J Urban Affairs 36(1):141–154

Reimers V, Clulow V (2004) Retail concentration: a comparison of spatial convenience in shopping strips and shopping centres. J Retail Consum Serv 11(4):207–221

Relph E (1976) Place and placelessness. Pion Limited, London

Restaurants for whatever your taste. https://www.waterfront.co.za/leisure/food-drink/. Accessed 17 Jan 2019

Richards G (2014) Creativity and tourism in the city. Curr Issues Tour 17(2):119–144

Romero-Borbón DL, Larios VM, Romero LF (2015) Smart food: toward an integrated supply chain in a smart city. IEEE-GDL CCD Smart Cities white paper. https://smartcities.ieee.org/images/files/pdf/smart_food_towards_an_integrated_sc_in_a_smart_city.pdf. 10 Sept 2018

Sack RD (1997) Homo geographicus. Johns Hopkins University Press, Baltimore

Salzman R, Yerace M (2018) Toward understanding creative placemaking in a socio-political context. City Cult Soc 13 (June):57–63

Sauer CO (1925) The morphology of landscape. Univ Calif Publ Geogr 2(2):19–54

Scarpato R, Daniele R (2003) New global cuisine: tourism, authenticity and sense of place in postmodern gastronomy. In: Hall CM, Sharples L, Mitchell R, Macionis N, Cambourne B (eds) Food tourism around the world. Development, management and markets. Butterworth-Heinemann, Oxford—Burlington, pp 296–313

Schiff N (2015) Cities and product variety: evidence from restaurants. J Econ Geogr 15(6)(1 November), pp 1085–1123

Seamon D, Sowers J (2008) Place and placelessness, Edward Relph. In: Hubbard P, Kitchin R, Valentine G (eds) Key texts in human geography. Sage, London, pp 43–51

Simons RA (1992) Site attributes in retail leasing: an analysis of a fast-food restaurant market. Apprais J 60(4):521–531

Store Directory (2017a). http://www.harborplace.com/directory/restaurants-and-eateries/2138215890. Accessed 10 July 2017

Store Directory (2017b). http://www.stationsquare.com/directory/restaurants-full-service/2138210915. Accessed 10 July 2017

Susskind AM, Chan EK (2000) How restaurant features affect check averages: a study of the Toronto restaurant market. Cornell Hotel Restaur Adm Q 41(6):56–63

Timmermans HJP (1986) Locational choice behaviour of entrepreneurs: an experimental analysis. Urban Stud. 23(3):231–240

Timmermans H (1979) A spatial preference model of regional shopping behaviour. Tijdschrift voor economische en sociale geografie 70(1)(February):45–48

Tuan Y-F (1975) Place: an experiential perspective. Geogr Rev 65(2)(April):151–165

Tuan Y-F (1979) Space and place: humanistic perspective. In: Gale S, Olsson G (eds) Philosophy in geography, Theory and Decision Library, 20, D. Reidel Publishing Company, Dordrecht—Boston—London, pp 387–427

Tzeng G-H, Teng M-H, Chen J-J, Opricovic S (2002) Multicriteria selection for a restaurant location in Taipei. Int J Hosp Manage 21(2):171–187

Upadhyay Y., Singh SK, Thomas G (2007) Do people differ in their preferences regarding restaurants? An exploratory study. Vis J Bus Perspect 11(2)(April–June):7–22

V&A Waterfront Fact Sheet (2018) V&A Waterfront Tourism Marketing Department, Cape Town

Crespi-Vallbona MC, Pérez MD (2015) Tourism and food markets: a typology of food markets from case studies of Barcelona and Madrid. Regions Mag 299(1):15–17

van Zyl PS (2005) An African success story in the integration of water, working harbour, heritage, urban revitalisation and tourism development, CapeInfo. https://capeinfo.com/useful-links/history/115-waterfront-development.html. 21 Nov 2018

Williams K (1999) Urban intensification policies in England: problems and contradictions. Land Use Policy 16(3):167–178

Withers ChWJ (2009) Place and the "spatial turn" in geography and in history. J Hist Ideas 70(4) (October):637–658

Yáñez CJN (2013) Do creative cities have a dark side? Cultural scenes and socioeconomic status in Barcelona and Madrid (1991–2001). Cities 35:213–220

Yen B, Burke M, Tseng C, Ghafoor MMT, Mulley C, Moutou C (2015) Do restaurant precincts need more parking? Differences in business perceptions and customer travel behaviour in Brisbane, Queensland, Australia. In: Paper presented at the 37th Australasian Transport Research Forum (ATRF), Sydney, Australia, 29th Sept–2nd Oct 2015. http://atrf.info/papers/2015/files/ATRF2015_Resubmission_84.pdf. Accessed 01 June 2017

Yim ES, Lee S, Kim WG (2014) Determinants of a restaurant average meal price: an application of the hedonic pricing model. Int J Hosp Manage 39:11–20

Chapter 3
Dimensions of Gastronomy in Contemporary Cities

Andrzej Kowalczyk

Abstract Gastronomy can be described from several different perspectives. Not only are associated services part of the economy, but they are part of the culture. In the context of urban space the most important is its spatial dimension. This chapter focuses on the following issues: the importance of gastronomy in the economy and socio-demography of the city; urban planning; urban revitalisation and restructuring; and the use of gastronomy in territorial marketing. Examples which illustrate these phenomena include Lahore, Lisbon, Madrid, Warsaw, and many other cities.

Keywords Restaurant business · Urban policies and planning · Gentrification · Marketing

3.1 Introduction

In the modern era, gastronomy can be described from several different perspectives. Not only are associated services part of the economy, but they meet people's needs, are part of culture, and have a spatial dimension.

As Yen et al. (2015, pp. 2–3) noted, restaurants are a major contributor to night-time and weekend economies. They also attract office-based and other daytime trading businesses. The high flow of customers throughout the day contributes positively to the look and appeal of local centres (Yen et al. 2015, p. 4). Visitors to restaurants and bars have also been analysed. Dökmeci et al. (1997) suggested that the location of restaurants and other food facilities is determined by their function. Food facilities with weekday clients (usually during the working day) are not found in the same places as restaurants that are visited occasionally. The arrangement of the former, which usually offer dishes at lower prices, can be related to Christaller's central-place theory. Their location is often determined by the size and accessibility

---

A. Kowalczyk (✉)
Department of Tourism Geography and Recreation, Faculty of Geography and Regional Studies, University of Warsaw, ul. Krakowskie Przedmieście 30, 00-927 Warsaw, Poland
e-mail: akowalczyk@uw.edu.pl

© Springer Nature Switzerland AG 2020
A. Kowalczyk and M. Derek (eds.), *Gastronomy and Urban Space*,
The Urban Book Series, https://doi.org/10.1007/978-3-030-34492-4_3

of the service area, and adjacent residential and business areas. These factors are less important for facilities that are only visited sporadically (often in the evening and at weekends), where the dishes and beverages on offer are usually more expensive and which are usually in attractive locations. In this case, accessibility and the size of service area (usually much larger than those of the daytime establishments) are not considered particularly important. Ishizaki (1995), when analysing the fast-food network in Tokyo identified three facilities that focused on daytime clients (McDonald's, Lotteria, and Morinaga Love), while one (Mos Burger) specialised in offering services to clients who visited in the evening.

In this chapter, we focus on the following issues: the importance of gastronomy in the economy and socio-demography of the city; in urban planning; in urban revitalisation and restructuring (gentrification); and the use of gastronomy in marketing.

3.2 Gastronomy and Economic, Socio-Demographic and Socio-Cultural Processes

Determining the precise impact of the restaurant sector on a city's economy is difficult, both theoretically and practically, for two main reasons. Firstly, the role of the catering sector can be described by the number of people it employs. However, employees may live in other places, where they pay their personal income tax. Second, it can be described by levels of corporate income tax, which is usually paid where the company is registered. However, in the case of restaurant chains this means that it may be paid in another city, or even another country. It is therefore generally acceptable to use approximations. One way is to determine the share of registered eating establishments as a percentage of the total number of registered businesses in a specific area.

Table 3.1 shows Pearson's correlation coefficient between registered activity in gastronomy (Y) in Madrid (Spain) and Prague (the Czech Republic) and other companies. Independent variables (Xi) refer to:

$X1$—the total number of registered business units,
$X2$—the total number of accommodation establishments (including gastronomy),
$X3$—the number of registered units in the wholesale and retail trades (including motor vehicles and motorcycle repairs),
$X4$—the number of registered units in finance and insurance,
$X5$—the number of registered units in arts, entertainment and recreation,
$X6$—the number of registered units in manufacturing.

We can draw several conclusions from these figures. Firstly, in both cities, the presence of registered restaurant businesses is closely correlated with: (a) the total number of registered businesses, (b) the number of accommodation establishments (including gastronomy), (c) the number of registered units in the retail and

Table 3.1 Correlation coefficients between registered activity in gastronomy (*Y*) in Madrid and Prague (by district, n = 21 for Madrid, and n = 57 for Prague) and selected independent variables in 2015

Variables	Pearson correlation coefficient (r)		Coefficient of determination (R^2)		Percentage of variation in Y explained by X	
	Madrid	Prague	Madrid	Prague	Madrid	Prague
X1	0.918	0.994	0.843	0.988	84.3	98.8
X2	0.858	0.964	0.736	0.929	73.6	92.9
X3	0.908	0.991	0.824	0.982	82.4	98.2
X4	0.638	0.950	0.407	0.903	40.7	90.3
X5	0.950	0.966	0.903	0.933	90.3	93.3
X6	0.225	0.973	0.051	0.947	5.10	94.7

Note The *significance* level for *Pearson's correlation coefficients for X1, X2, X3 and X5 is 0.001; for X4 is it 0.002 is for X4. In Madrid, X6 is non-significant*
Source: Own calculations based on https://www.czso.cz/documents/10180/46014684/33012117q1s01.xlsx/95399516-d374-4b89-a5ca-accf9af11d6c?version=1.0 (Accessed 20 June 2017) and http://www.madrid.es/UnidadesDescentralizadas/UDCEstadistica/Nuevaweb/Economía/Empresasy-Locales/censo/D2110317.xls (Accessed 20 June 2017)

wholesale trades (including motor vehicles and motorcycle repairs) and (d) number of registered units in arts, entertainment and recreation. In Prague it is also correlated with the number of registered units in finance and insurance, and manufacturing. In all cases, Pearson's coefficient is significant (p = 0.001). Second, in Madrid, the number of gastronomy businesses has practically no correlation with the number of registered units in manufacturing (r = 0.225), whereas in Prague there is a strong correlation (r = 0.973). This could be because the economic processes in the two cities are somewhat different. In Madrid, industrial districts are isolated, whereas in contemporary Prague, each district can contain a mix of functions.

It is rare for the restaurant sector to play a leading economic role in terms of turnover, revenue and income. It does play a role, however, in smaller towns that rely on a developed tourist function. In terms of the labour market or local entrepreneurship a different picture emerges. As the number of firms engaged in the restaurant industry grows, they employ more people. This means that salaries and wages are relatively low (Graham 2016). Research in San Francisco has shown that even with obligatory minimum-wage laws, those employed in the restaurant sector are much less well off than the rest of the economy. Moreover, despite a 26–28% increase in the minimum wage, the average salary in fast-food restaurants only rose by 4–5% (Schmitt and Rosnick 2011, p. 9). One reason for this is that many restaurants and bars (not just those offering ethnic cuisine) in the cities of Europe, North America and Australia employ immigrants, sometimes working illegally. In Ireland, 35.5% of restaurant business employees in 2008 were not Irish. This is double the figure for any other employment sector (Exploitation in Ireland's restaurant industry 2008). Even low wages are spent, however, and the multiplier effect of the catering sector is considerable.

Low wages mean low labour costs. This means that during stagnation or recession, the food industry can continue to exist and even grow. In Madrid during the economic crisis (2005–2015), employment in the hospitality industry increased by 1.7%. It is worth noting that in the same period in Madrid there was a decrease in employment in manufacturing (−4.5%), the wholesale trade (−1.2%) and the retail trade (−1.3%) (*Situación y perspectivas de la Ciudad de Madrid* 2016). This does not mean, however, that activity in the restaurant business is predictable, economically safe and stable. According to some authors (Longart 2015, p. 1) as the gastronomy sector is one of the easiest to enter there is a constant threat of new competition. Moreover, there is the threat of substitution, if people choose to eat at home instead of eating out. Reports from the late 1990 s showed that the failure rate in the gastronomy business was very high: 26% of independent restaurants in the United States failed during their first year, 19% failed during the second year, and 14% in the third year (Parsa et al. 2005, p. 309). Of course, this depends on the area. In 2004 in the Los Angeles region the first-year failure rate was 24%, and the three-year cumulative rate was 51%. The picture is similar in other cities around the world. For instance, in India, seven of the leading 10 restaurants that opened in Mumbai in 2012 had closed by mid-2016 (India's food service industry: growth recipe 2016, p. 38).

It should be noted, however, that the restaurant sector is influenced by many factors, including the wealth of the population. Studies conducted in some of the largest cities in India showed that while in Ahmedabad only 0.8% of respondents reported that they left the house more than once a week, in Delhi and Greater Mumbai this percentage was 1.4%, in Pune 1.5%, in Hyderabad 2.2%, in Chennai 2.7%, in Kolkata 2.8% and in Bangalore—the country's IT centre and the richest city listed here—4.7% (Ahluwalia et al. 2004, p. 16).

Organisational and financial constraints make research on the impact of the eating sector on the urban economy relatively scarce. One example from Edinburgh showed that bars and pubs (grouped into one category) and restaurants and cafés (a separate category) had a greater direct and indirect multiplier effect on income and employment in the Old Town than hotels. Overall, the impact of restaurants/cafés over hotels was higher for employment, but lower for income (Parlett et al. 1995, p. 357, Tables 2 and 3; p. 358, Table 4). This can be explained by the fact that visitors spent £11.4 million on accommodation and £18.1 million on gastronomy, whereas day-trippers (who do not need accommodation) spent £17.9 million in restaurants, bars, pubs, etc. (Parlett et al. 1995, p. 359, Table 7).

As noted above, gastronomy is not only important for the economy, it also has a social dimension. This applies both to socio-demographic and socio-cultural processes. From the perspective of socio-demographic transformation, gastronomy is primarily associated with the age structure of the population. People who dine out are predominantly aged 20–60, for example. This does not mean, however, that for some segments of the dining industry other age groups are less desirable. Families with children and teenagers, for example, often frequent fast-food restaurants such as McDonald's and Pizza Hut. With regard to the urban space, this means that restaurants and bars are most often located in neighbourhoods where they are most likely to be visited by potential diners. Downtown areas, waterfronts, riversides and tourist districts, for example, are all visited more frequently than the outer zones of a city. This also applies to shopping centres, which are rarely visited by the elderly.

Another important social factor is socio-cultural processes. These relate to the mobility of the population and, above all, the influx of immigrants and tourists. The presence of immigrants fosters the emergence of new forms of gastronomy, often focused on specific parts of the city. This problem, noted earlier, is discussed elsewhere in the book. On the other hand, as visitors often arrive during the tourist season, some eating establishments are seasonal. Another problem associated with tourist-oriented gastronomy is the price of meals. Tourists are often richer than the average city resident; consequently, prices in tourist-oriented restaurants are higher, which limits the number of locals who can afford them. This can cause tension, especially in cities with developed tourist functions.

Another important socio-cultural factor is the education of the population. Generally speaking, people with a higher level of education are more likely to visit restaurants. Therefore, in neighbourhoods frequented by students or young professionals, the number of restaurants is higher than in neighbourhoods inhabited by the working class (blue-collar workers). This phenomenon is closely related to the processes described by Florida (2003) and the concept of the city as an entertainment machine (Lloyd and Clark 2001).

3.3 Gastronomy and Urban Policies

In urban spatial policy, the restaurant sector is rarely as important as transport, housing, green spaces or municipal services. Although in the nineteenth century local authorities organised canteens for the population of poor districts in some European cities, they treated other types of catering in the same manner as retail activities.

Things began to change in the twentieth century with the introduction of urban politics—initially in Europe and North America, followed by other countries. At first, this mainly covered the spheres of housing, public transport and communal services (water supply, sewage system) but, over time, also began to affect other areas of the city's economy and social life. Singapore was one of the first countries outside of Europe and North America to introduce legislation in the field of gastronomy. The new regulations targeted street food, a typical socio-economic phenomenon in many cities in Asia, Africa and Latin America. When Singapore gained independence in 1965, the government decided to regulate street vendors. Given the country's long history of selling cooked food in the street (Henderson 2016, p. 279) not everybody approved: 'Hawkers are an embarrassment for some officials whose aspiration for their city render selling food on the streets anachronistic' (te Lintelo 2009); 'Cooked food hawking provoked concerns about public health hazards emanating from poor hygiene and sanitation' (Henderson et al. 2012, pp. 850–851). The government moved all hawkers from the streets into newly built facilities in public housing estates. Relocation started in 1971 and ended in 1986 (Henderson 2016, p. 279). All street vendors were initially moved to 140 centres, although in 2010 this number had fallen to 109 (Henderson et al. 2012, p. 651). All are overseen by the Hawkers Department within the National Environment Agency (Henderson 2016, p. 279).

Regulating street food has also been studied in places such as India. Here, spatial development and restructuring are associated with the emergence of a growing middle class who are familiar with global trends (te Lintelo 2009, p. 65). The consequence was that 'the banning on cooking food in the streets of Delhi thus addressed the growing concern about food hygiene, but also revealed how the regulation of food hygiene was intricately interwoven with the management and design of urban space'. (te Lintelo 2009, p. 67). Mumbai has regulated hawking since 2003. Regulations prohibit cooking food on the street and hawking within 150 metres of railway stations, municipal markets, colleges, schools, hospitals and residential areas. It is also prohibited on streets that are less than eight metres wide, or from a table, stall or handcart (Anjaria 2006, pp. 2140–2141). Regulations on the provision of street food are also present in cities in other parts of the world, especially Africa. The problem of maintaining hygiene standards and, at the same time, the ability of vendors to earn a living, has been studied in Kumasi (Ghana) (Forkuor et al. 2017, pp. 4–5). Based on research in Kumba (Cameroon), Acho-Chi (2002) suggests that local authorities should pay specific attention to supporting and improving opportunities for job creation for poor families and should aim to develop street food enterprises into city food establishments.

In European countries with a centrally planned system (the former Soviet Union and other Soviet bloc countries), local authorities historically regulated the location of restaurants, bars, etc. For example, the number of inhabitants per restaurant, its surface area, the size of the kitchen and the number of meals (per day or per hour) were prescribed. Even nowadays, such regulations still apply in some countries. Nor do municipal authorities ignore the issue of the location of catering services. In some countries, authorities have designated special dining areas, which are often called food streets, food districts, food parks or food courts.

> One country where food streets have obtained official status and are treated as an element of city development is Pakistan. In that country food streets, which are pedestrianised, are formally planned and specifically designed. The first was established in the oldest part of Lahore and its cultural centre. It stretches over a few hundred metres and is surrounded by historical buildings (often from eighteenth to nineteenth centuries) designed in a style characteristic of Northern Pakistan, which is referred to as Persian–Kashmiri. In 2009, there were over 100 gastronomic establishments and shops selling food, employing over 2000 people (Raza 2009). Over time, another street appeared (in the Anarkali district, in 2002). At the same time food streets

appeared in Islamabad (Melody Food Street and in Blue Area, in 2002 and 2005), in Karachi (Burns Road, in 2005) and in Peshawar (Ghanta Ghar, also in 2005). Although the idea has faced resistance from small-scale merchants, current restaurateurs and the clergy (food streets are visited mainly by tourists and the Pakistani upper and middle classes), it is gradually being leveraged (Kowalczyk 2014, p. 151).

Food truck parks are another new phenomenon. These well-equipped car parks[1] are included in master plans (more often conceptual than detailed) and are becoming an integral part of urban regeneration programmes. To date, they are most popular in the United States but they also exist in other countries such as Australia.

3.4 Gastronomy and Urban Planning

Although the beginnings of gastronomy in urban planning can be found in the nineteenth century, the next decades have brought the development of this approach. According to Parham in modern cities '… food markets and cafes give local colour and vibrancy, food was to be added in once the important place design aspects, like parking' (Parham 2016, p. 16). However, in the twentieth century in cities in democratic countries, as well as in those with the liberal economy, the role of urban planning and design played a different role than in countries of so-called real socialism. In countries with a centrally planned economy, gastronomy was seen as a service to satisfy the needs of the population, for which the ruling Communist Party and government were responsible. Similarly to education, health care, culture or sport and recreation.

Until the political changes of 1989, gastronomy, like other spheres of public life, was subject to central planning in the Soviet Union and Soviet bloc countries. Regulations applied to, among other things, the number, capacity and distribution of eating facilities in urban areas. In 1965, technical rules for the design of dining establishments were introduced in Poland, which, depending on the type of facility, prescribed the number of covers and their daily use. These regulations divided eating establishments into basic and complementary categories (Table 3.2).

In 1972, planning standards were introduced that regulated the number and type of food infrastructure in urban areas and regions (Wilk 1984, p. 262). The regulations presented in Table 3.3 mainly relate to new housing estates, although their enforcement was relaxed over time. This was partly due to the housing crisis that hit Poland in the late 1970s and 1980s.

---

[1]Food trucks (2017).

Table 3.2 Technical standards applied to eating establishments in Poland in the 1960s–1970s

Eating establishments	Number of seats	Daily rotation per seat	Daily number of consumers
Basic infrastructure			
Casual restaurant	100–200	2–3	200–600
Restaurant	60–200	3–4	180–800
Canteen	100–200	5	500–1000
Quick-service bar	100–200	15	1500–3000
Tourist bar—independent	100–200	10	1000–2000
Tourist bar—dependent	70–150	10	700–1500
Drink bar	50–100	7–8	350–800
Milk bar	80–150	15	1200–2250
Complementary infrastructure			
Bar espresso	50–100	20	1000–2000
Coffee bar	40–80	20	800–1600
Cocktail bar	50–100	12	600–1200
Buffet (with a counter)	2–4	20	400–800
Café	50–200	7	350–1400
Confectionery	50–120	7	350–840
Wine bar	50–100	6	300–900
Beer bar	50–100	11	350–1100

Source: Adapted from Wilk (1984, p. 252, Table 2)

Other regulations remained in force until 1989. Planning recommendations stated that there should be 30 covers per 1,000 residents in housing estates, and 40 covers per 1,000 inhabitants in neighbourhoods consisting of at least several housing estates and a multifunctional civic centre (Wiśniewska 1984, pp. 370–371). They also limited eating establishments to restaurants, various types of bars (including milk bars and beer bars) and cafés or pastry shops. In central areas, the number of facilities was slightly higher, especially where the city served a tourist function. Higher-class restaurants were allowed in the downtown areas of (larger) cities, while some offered night-time entertainment. The types of eating establishments that could be located in communal buildings (residential, multifunctional) and in separate, standalone buildings was also regulated, as was the minimum size of dining outlets. For example, restaurants, milk bars, cafés and pastry shops in residential and multifunctional buildings had to have more than 100 seats (Wilk 1984, p. 260). Other regulations specified which eating establishments could be co-located. For example, restaurants could not be near canteens; canteens could not be close to quick-service bars; canteens and quick-service bars could not be near wine bars and beer bars; cafés, pastry shops and teahouses had to be separate from drink bars, etc. (Wilk 1984, p. 260, Table 6). The result was that in almost every

3 Dimensions of Gastronomy in Contemporary Cities

Table 3.3 Types of eating infrastructure at various levels of service for the urban population in Poland

Types of eating infrastructure	Levels of service to the population		
	Sub–local centre in the city	Local centre on the level of city borough	City or town
Casual restaurant	–	+	+
Restaurant	+	+	+
Canteen	+	+	+
Quick-service bar	–	+	+
Tourist bar (summer)/ Self-service bar	+	+	+
Drink bar/Milk bar	+	+	+
Bar espresso	–	+	+
Coffee bar	+	+	+
Cocktail bar and pastry shop	–	+	+
Buffet (with a counter)	–	+	+
Café	+	+	+
Pastry shop	–	+	+
Teahouse	–	–	+
Wine bar	–	+	+
Beer bar	–	+	+

Note '+' means YES, '–' NOT
Source: Adapted from Wilk (1984, p. 261, Table 2)

new housing estate eating facilities were standardised. It also meant that dining outlets and their interior decoration looked very similar.

Figure 3.1 shows a newly built housing estate in Bródno, north of Warsaw in 1970. There are four parts. Construction began in 1964 and, by the end of the 1970s, Bródno had about 100,000 inhabitants. Two-storey commercial and service pavilions were built at about the same as residential buildings, schools, kindergartens, medical clinics, libraries, etc. The master plan incorporated restaurants, milk bars, cafés and pubs. The task was to satisfy the catering and entertainment needs of the residents of nearby blocks (usually inhabited by 15,000–20,000 people). By the mid-1970s there was a restaurant, a self-service bar, a milk bar, two beer bars and three cafés. The map shows that these were fairly evenly distributed. There are two eating outlets in the same building (on Kondratowicza Street) because this street was planned to be the main street. This situation remained until the end of the 1980s. It was only after the transformation of the socialist economy into a liberal economy that gastronomy facilities were freed from central planning. The same situation was found in all of the cities and towns in the former Soviet bloc countries. It is common to find housing estates that were built between the 1950s and 1980s where there is a grocery store, clothing store, a restaurant or café, etc. From the outside and inside it is still the 1980s.

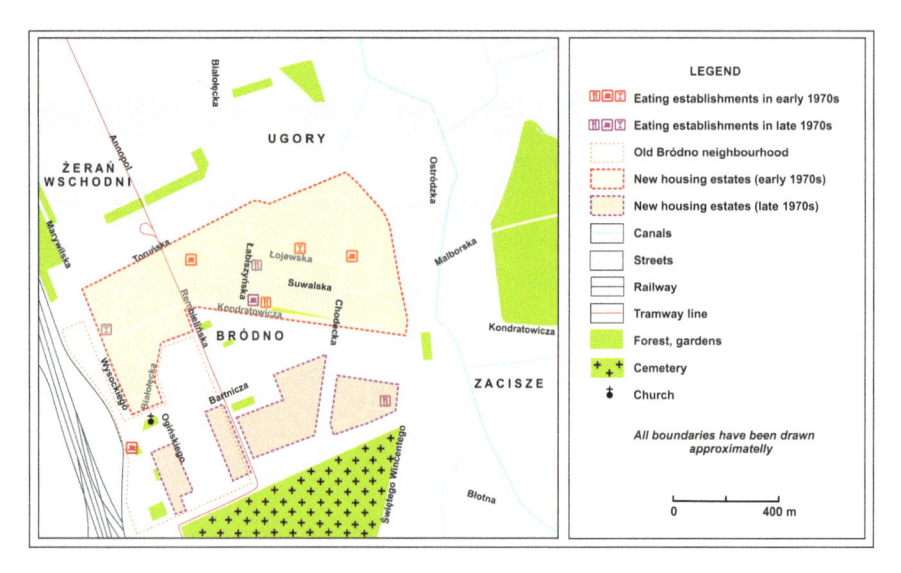

Fig. 3.1 Eating establishments at the Bródno housing project in Warsaw in the 1970s. Source: Information about facilities adapted from Bystrzanowski and Dutkowski (1975)

Referring to David Harvey's thesis (1973, pp. 101–108) about territorial distributive justice, one can assume that three criteria were sought in the policy of locating food services in the Bródno housing estate in 1970s: needs of inhabitants, contribution to common good, and merit. Of course, from today's neo-liberal perspective, this may not be so obvious, but in the conditions of real socialism, urban planners were guided in urban planning principles discussed in Harvey's publications.

In contemporary urban planning doctrine, as previously noted, the location of eating facilities is unimportant. This does not mean, however, that it is completely ignored. In recent years some cities have started including it in the context of environmental protection, particularly the protection of the acoustic environment. Noise is a big problem in cities where there are many restaurants and bars. It affects both large cities and towns. In Poland, it is an issue primarily in the old parts of Kraków, Warsaw, Wrocław and Poznań; in Spain, in Barcelona (especially at Barri Gòtic or Barrio Gótico) and Madrid; in Portugal in Lisbon (e.g. the Alfama district) and the old part of Porto; in the Czech Republic in the Staré Město and Malá Strana districts of Prague, etc. Noise can make the lives of residents (and some tourists) uncomfortable and some municipal authorities have attempted to address the problem. Figure 3.2, for example, shows how the presence of restaurants and bars affects noise levels in the northern part of Madrid's inner city. The maps shown in this figure highlight that noise intensity (shown in red on map B) is highest near nightclubs. Levels fall around restaurants and bars that do not play music (shown in

Fig. 3.2 The concentration of restaurants and bars in the area close to the Avenida de Brasil in Madrid (Spain) (map A) and associated noise levels (map B). Source: *Concentración de actividades de ocio nocturno* and *Mapa de los Niveles de Ruido Actividad de Ocio en la zona de AZCA —Avenida de Brasil* from the *Declaración de Zona de Protección Acústica Especial y Plan Zonal Específico de AZCA—Avenida de Brasil*, Boletín Oficial del Ayuntamiento de Madrid, Año CXIX, 26 de enero de 2015, Núm. 7.339, pp. 8–66, https://sede.madrid.es/csvfiles/UnidadesDescentralizadas/UDCBOAM/Contenidos/Boletin/2015/ENERO/Ficheros%20PDF/BOAM_7339_23012015134649429.pdf (Accessed 05 June 2017)

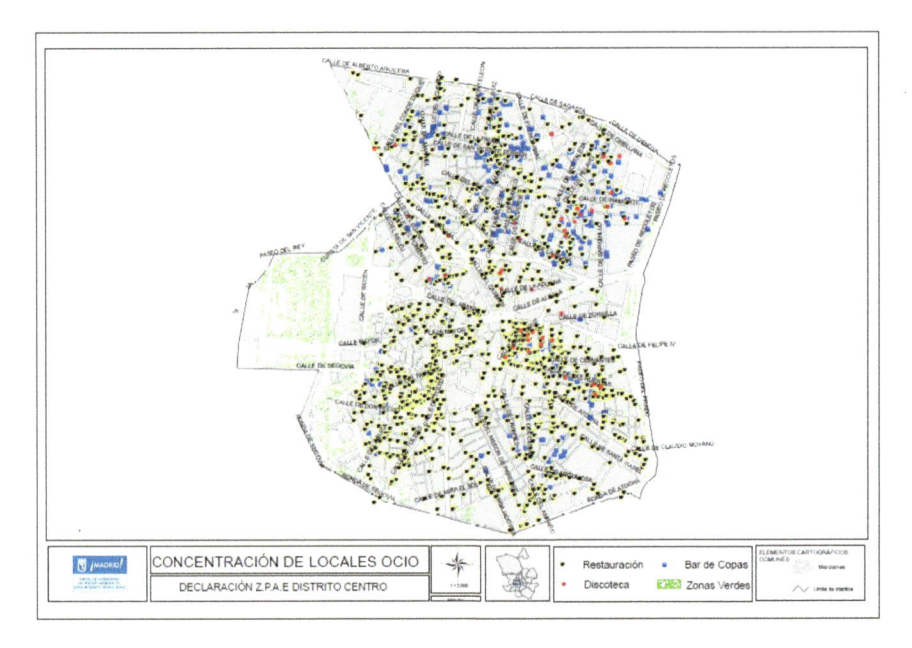

Fig. 3.3 Eating establishments in downtown Madrid (Spain). This map was annexed to the *Declaración de Zona de Protección Acústica Especial y Plan Zonal Específico del Distrito Centro* approved by the city council in September 2012. Source: *Declaración de Zona de Protección Acústica Especial y Plan Zonal Específico del Distrito Centro*, http://www.madrid.es/Unidad Web/Contenidos/Publicaciones/TemaMedioAmbiente/ZPAECentro/MemoriajustificativaZPAE Centrodefinit.pdf (Accessed 05 June 2017)

yellow). Importantly, most noise emanates from the courtyards between food establishments, rather than directly in the streets (Fig. 3.3).

The high concentration of restaurants and bars in downtown Madrid led city authorities to introduce measures in 2012 to reduce noise levels. These new regulations limit the times when food and drink can be served on restaurant and bar terraces. Acoustic protection zones have been introduced in some areas (*Zona de Protección Acústica Especial*), covering several neighbourhoods in the Centro district, part of Moncloa–Aravaca and the area surrounding the Avenida de Brasil. In 2016, they were extended to the La Lonja neighbourhood together with La Latina, Huertas and Malasaña. In these areas, restaurants and bars that do not play music can operate, but only if they are situated in buildings that are exclusively for non-residential use (Belver and Luque 2016) (Fig. 3.4).

Urban planners also have to deal with other sources of noise. In the United States, Parsa et al. (2005, pp. 306–307) noted that clusters of eating facilities increased both customer and vehicular traffic. The solution lies in the hands of local authorities, who are responsible for parking. The issue of parking zones is especially important in the busiest streets in the inner city, commercial city centres and suburban shopping centres. The challenge is to provide sufficient road space for different modes of transportation, and parking for drivers and cyclists (Fig. 3.5).

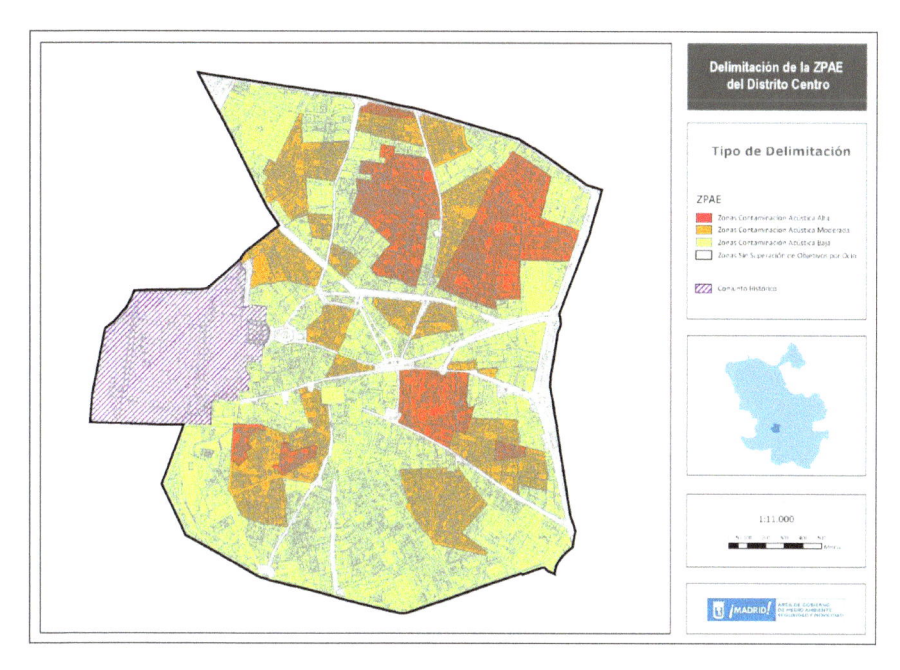

Fig. 3.4 Areas with high noise levels in downtown Madrid (Spain). Highest levels are shown in red. The shaded area on the left is the Royal Palace, the Cathedral and their surroundings. Source: *Normativa de Plan Zonal Específico de la Zona de Protección Acústica Especial del Distrito Centro*, https://sede.madrid.es/UnidadesDescentralizadas/UDCBOAM/UGMedioAmbientey Movilidad/ElementoBoletin/2012???/Anexo%20II%20Normativa%20ZPAE.pdf (Accessed 05 June 2017)

Fig. 3.5 Restaurant terraces. **a** The Old Town in Alanya (Turkey). Here, the terrace is an integral part of the restaurant. **b** A square in Venice (Italy). This is an example of the gastronomic invasion of public urban space. Source: Anna Kowalczyk 2010

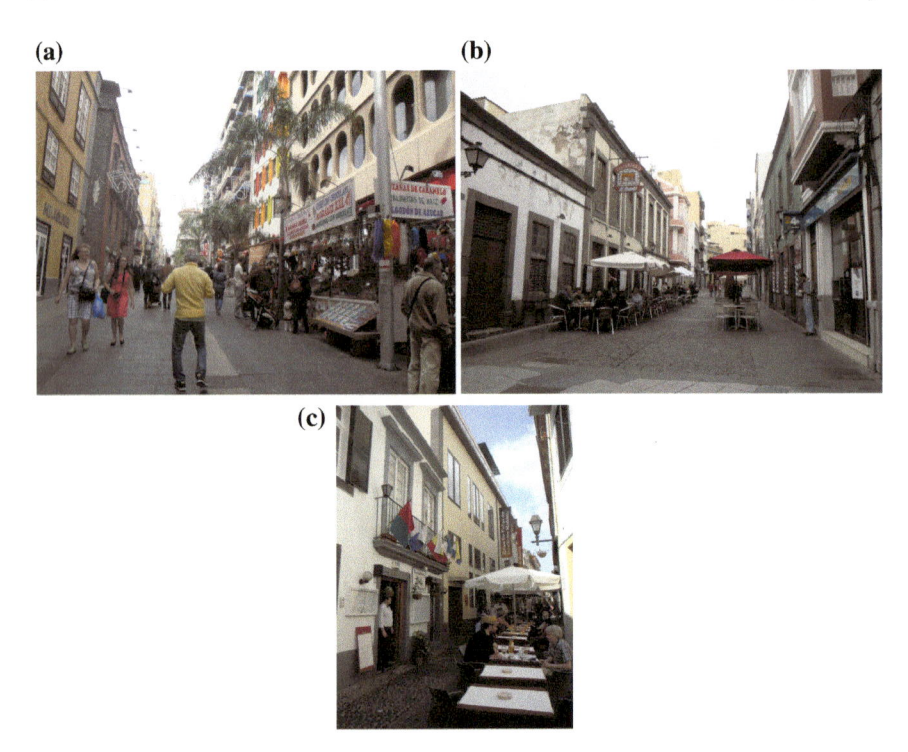

Fig. 3.6 Three examples of food streets in tourist cities. **a** The main street in Santa Cruz de Tenerifa, Tenerife (Spain). **b** A food alley in Las Palmas, Gran Canaria (Spian). **c** A food alley in Funchal, Madeira (Portugal). In all cases streets are pedestrianised. Source: Anna Kowalczyk 2012, 2014 and 2015

The need to consider to gastronomy is an issue worldwide. We have already mentioned Lahore, where gastronomy policies have been in place since 2002. In 2009, the Punjab Province government decided to open nine new food streets in the Lahore districts of Gulberg (a relatively modern trade and services centre), Shalimar Town, Nishtar Town, Samnabad, Allama Iqbal Town, Wahga (near the border with India), Aziz Bhatti Town, Ravi Town and Data Gunj Baksh Town (Habib 2011). Only one was ever implemented: the Fort Road Food Street project, which is discussed below. Similarly, in Karachi, in the spring of 2011, the Port Grand Food and Entertainment Complex was established. This included a footpath with 11 restaurants and bars (including Thai cuisine) partly located along the old port waterfront, the nineteenth century Native Jetty Bridge. This example of a city waterfront development also offers 30 shopping, service and cultural facilities, which can accommodate 4,000–5,000 visitors daily (Aqueel 2011; Ebrahim 2011) (Fig. 3.6).

Although the urban planning solutions used in Pakistan are not widely used in Europe, the presence of gastronomy in the public space is an important planning issue in many countries, as restaurants and bars not only adapt to the space they

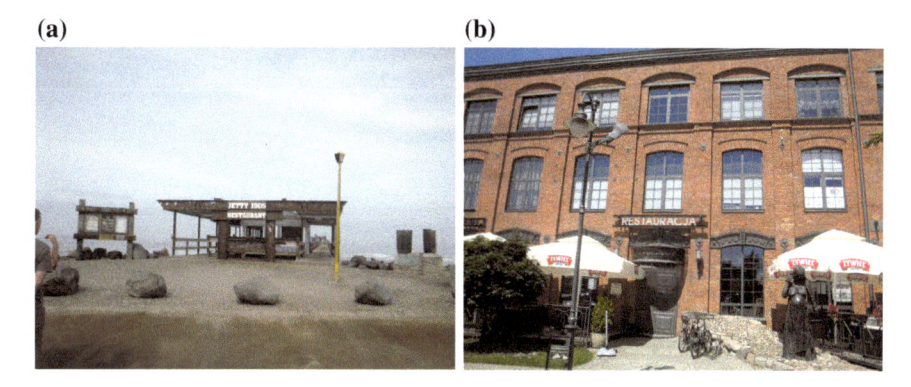

Fig. 3.7 a The entrance to the Jetty 1905 Restaurant in Swakopmund (Namibia). The remains of the old pier (from 1905) were converted into a fashionable restaurant. **b** The entrance to the most luxurious restaurant in Żyrardów (Poland). This restaurant was opened in a building in the former textile factory dating from about 1860. Its name, Szpularnia, suggests *szpulka* or spool, in English. On the right side is a sculpture of a pregnant woman designed to commemorate the women who worked in the nineteenth-century factory. Source: Anna Andrzej Kowalczyk 2016

occupy, but also expand into the public space. In general, this is legal and does not create problems, but sometimes this expansion can interfere with other urban space users. This is often the case in cities with particularly developed tourist functions: Rome, Venice and Naples in Italy; Barcelona in Spain; Lisbon in Portugal; Kraków in Poland; and Prague in the Czech Republic (Fig. 3.7).

In order to minimise spatial conflicts related to gastronomy in public spaces, many cities recommend that restaurants and bars should operate primarily in areas where cars are prohibited. The presence of dining facilities in pedestrian areas provides customers with greater security, which is particularly important in tourist cities. Creating urban pedestrian zones is one of the most important elements of urban politics in many European countries, as well as in North America and Australia, in some Asian cities (e.g. Singapore, Isfahan in Iran) and Africa (e.g. Johannesburg, Pretoria/Tshwane and Cape Town in South Africa, Windhoek in Namibia). Although their origins may be different, they often lead to the creation of clusters of restaurants and bars that resemble food streets and food districts (Figs. 3.8 and 3.9).

3.5 Gastronomy and Gentrification

Worldwide economic and socio-cultural changes mean that there are also deep functional and spatial changes in cities. This is largely due to a shift from industrial to service functions. The loss of a city's functions has always happened, to some extent, but has become more prevalent in the second half of the twentieth century. Initially, North American and Western European cities were most affected, followed

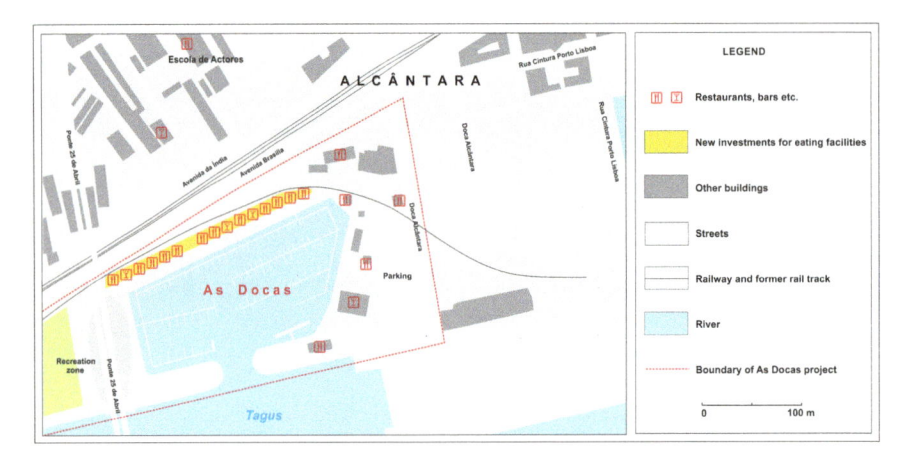

Fig. 3.8 The *As Docas* zone in the Alcântara district of Lisbon (Portugal). This is an example of a food district (or food court) planned and implemented in a post-industrial area. Source: Information about facilities adapted from *Bares*, http://lourencolx.wix.com/docas#!__bar-pt (Accessed 15 July 2017), *15 aniversario*, http://lourencolx.wix.com/docas#!__mapa-n (Accessed 15 July 2017), *Restaurantes*, http://lourencolx.wix.com/docas#!__rest-pt (Accessed 15 July 2017)

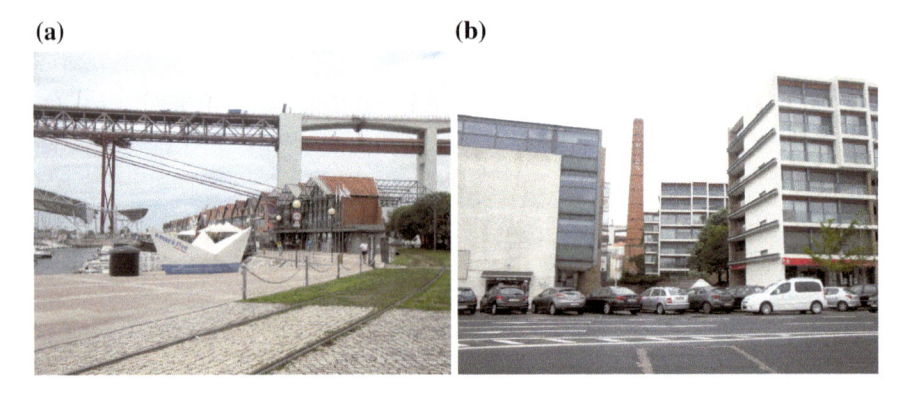

Fig. 3.9 The Alcântara district of Lisbon (Portugal). **a** The *As Docas* project showing restaurants and bars (centre), the marina (left), the railway line the old harbour (foreground) and the Ponte 25 de Abril. **b** Apartments and the chimney of the former factory, a reminder of the district's history. This is one example of the gentrification of post-industrial areas and neighbourhoods inhabited by working classes. Source: Anna Kowalczyk 2014

by other parts of the world. Without going into detail, the most important aspects are the decline in existing economic sectors, falling land prices, rising unemployment, rising poverty and reduced local government revenue. These processes provide a spatial dimension that manifests in the depopulation of neighbourhoods, the emergence of vacant dwellings and the redevelopment of disused factories, ports, railway stations, etc.

In order to halt the decay of the 1960s and 1970s, many cities in Europe (especially the United Kingdom, France and Germany) and the United States began to introduce revitalisation and restructuring programmes, notably in their downtown and post-industrial zones. These activities have expanded and continue to grow. The result is often gentrification, defined here as the process of renovation (re-generation or revitalisation) and restructuring of deteriorated urban neighbourhoods. The reasons for, and results of gentrification are an increase in interest in some parts of the city. The process is connected with increasing investments by real estate companies and local authorities, and leads to economic development. Although it makes the area more attractive to business, it also leads to population migration: existing inhabitants move out as housing prices rise; at the same time, new inhabitants move in. The newcomers are generally richer, younger tenants, many of whom are single urban professionals, or couples without children.

The gastronomic sector is often seen as a factor in the revitalisation of the urban space. This applies especially to central areas of the city, whose function has changed along with population outflows to the suburbs. Many examples indicate that food is a tool for all aspects of revitalisation—economic, social and cultural— and contributes to the regeneration of buildings and streets. Gastronomy plays a very important role in gentrification. Describing the changes that have occurred in the revitalised Kensington Market district in downtown Toronto, Dale and Newman observed that expensive restaurants, gourmet food shops and bakeries appeared first (Dale and Newman 2009, p. 675). In recent years, the gentrification process has started to happen more frequently in the cities of East Asia and South America, in places such as Singapore, Shanghai, Seoul and Kuala Lumpur or Mexico City, Bogota, Lima and São Paulo.

Many revitalisation programmes introduce entertainment and recreation, as well as eating functions to renovated districts. There are many examples of this, some of which were mentioned earlier when discussing the concept of the festival market place in American cities. Furthermore, changing the function of urban waterfronts facilitates their revitalisation and restructuring. Restaurants and bars have been introduced into former docks, warehouses and railway tracks alongside new public parks, bike paths, tennis courts and concert halls. However, as Norcliffe et al. (1996, p. 131) pointed out, although many of them are located near water, they often specialise in fish and seafood caught in other parts of the world, one effect of globalisation.

A good example of the practical application of the concept of the festival market place and the restoration and development of a waterfront is the Alcântara district in Lisbon, which lies on the Tagus River to the west of the city centre. In the mid-nineteenth century the former suburb began to transform into an industrial area (home to a textile factory, and chemical and metallurgy plants). This process accelerated when Alcântara was connected by rail to Lisbon. By the end of the century, docks and warehouses began to appear, and communal housing began to develop. The district lost its industrial and port functions in the second half of the twentieth century. In the early 1990s, Alcântara started to become a place for gastronomy (hosting restaurants, pubs, nightclubs). This was linked to its outer location which meant that night-time noise would not disturb residents. In 1995, dining establishments were added around the new marina, which in the past had a commercial function. Currently, there are about 20 trendy restaurants and bars in the former quay, together with fashionable boutiques, chandlers and recreation facilities. These new functions formed at the beginning of the gentrification process, and today, other post-industrial zones are being taken over by loft spaces and apartments (*Enquadramento histórico* 2008).

We return to another example of a city where the development of the eating function has been accompanied by activities aimed at revitalising the urban space: Lahore. In January 2012, at the initiative of Lahore's authorities, a new food street —Fort Road Food Street—opened in the neighbourhood of the historical Badashahi Mosque and the Lahore Fort. Its location was deliberately chosen to attract tourists visiting the city's most important attractions. It is also worth noting that Fort Road lies on the road connecting a major bus stop with the busy streets of the Lower Mall and Data Darbar Road. Restaurants located along the 1.4-km street were opened in 25 rebuilt and renovated buildings thanks to support from the City District Government (Tahir 2012). In the summer of 2012, 27 restaurants operated within the framework of the Fort Road Food Street project, making it one of the key regions offering food services in Lahore. Another key region is M. M. Alam Road, which was thoroughly modernised in 2011 and was intended to be the thoroughfare for the new part of Lahore. In the spring of 2014, M. M. Alam Road had 12 cafés and bars and 20 restaurants, including overseas chains such as McDonald's, Pizza Hut, Kentucky Fried Chicken, Subway, Domino's Pizza, Hardees and Nandos (Kowalczyk 2014, p. 152).

As mentioned earlier, urban regeneration through gentrification is often regarded as controversial. The unfavourable from the residents' point of view the effects of revitalization of city centres through gentrification are perceived by Harvey (2012, p. 78), writing that 'revitalization' in the opinion of residents in many cases means 'devitalization'.

3.6 Gastronomy and Marketing

3.6.1 Gastronomy and City Identity

A blogger recently wrote that: '…food is very important to city marketing, and vise-versa. Food is part of the culture of a city or a nation, it's part of its symbolic capital. Even so if the city doesn't have a strong gastronomic identity. It is important because food is not only about stomachs, but also about quality of life, about meeting people, sharing experiences, sharing a taste. Food comes with an imaginary and with rituals, it's a social construction. It can be understood as part of the symbolic capital of a city. (…) So, gastronomic city marketing and branding are not only about strategy, but also about meaning' (Maynadier 2014). Reality shows that there is a lot of truth in this. In recent years, gastronomy has been a key element in the marketing strategy of an increasing number of cities and towns. It is difficult to say with certainty whether these actions have been effective, as many initiatives are very recent. But some examples indicate that the use of food in marketing is effective.

Maynadier also wrote: 'Place branding is about meaning and strategy, it's about building consistency through values, stories and experiences. Gastronomy and food should not be ignored and should be taken seriously by political leaders and place marketing managers as a strategic asset, a symbolic capital for cities and places' Although Maynadier suggests that there are three basic terms in gastronomic marketing: values, stories and experiences, in the context of geographical research a fourth—location—should be added. Many cities have noted the benefits of using gastronomy to promote themselves, and this has led some local authorities and entrepreneurs to become complacent. Specifically, they assume that simply emphasising culinary traditions, supporting restaurateurs and organising festivals and culinary competitions will be profitable in all cases. However, very often this kind of action can lead to failure. That is why it is worthwhile to reflect once again on Maynadier's musings: 'This semiotics-inspired approach of a city brand and gastronomic marketing has to be completed by a strategic analysis: What positioning strategy to adopt? Does the city have a gastronomic icon? How to build one? With what other city to make alliances and co-branding operations? Is an event strategy relevant? Or storytelling strategy?' This statement suggests that city branding should promote and advertise culinary identity, as well as preserve its culinary sense of place. The gastronomic identity of a city or town has several distinguishable components:

- food:
    - basic food products (fruits, vegetables, fish, meat, etc.),
    - processed food products (meals, beverages),
- facilities:
    - restaurants famed for their quality,
    - restaurants known for the building they are in, or the interior design,
    - food and gastronomic museums,
- events:
    - food markets,
    - food festivals,
    - culinary contests,
    - cooking schools,
- technologies:
    - food manufacturing plants (wineries, breweries, bakeries, etc.),
    - kitchens in restaurants, bars, etc.

- people:
  - chefs,
  - winemakers,
  - authors of cookbooks,
  - culinary critics,
- others:
  - culinary routes (wine routes, beer routes, cheese routes, etc.),
  - literature and film.

3.6.2 Gastronomy and the Traditional Urban Space

The relationship between cities and gastronomy is sometimes reflected in language. Examples include *prosciutto di Parma*, *pizza napoletana*, *ragù Bolognese* ('Bolognese' sauce), *Debreceni kolbász* (sausage from Debrecen in Hungary), *cognac*, *jerez* (in English sherry, and in French *xérès*), *malaga* or port (*vinho do Porto* in Portuguese), hamburger, frankfurter or wiener (Vienna sausage, in other words: a hotdog), *Wiener Schnitzel* and many others. Some of these traditions have a long history, while others are more recent.

Beer is another tradition with a long history. Many breweries were founded in the Middle Ages, although their premises may have been renewed. One such brewery is the Augustiner brewery in Munich (now Augustiner-Bräu Wagner KG). Augustinian monks began brewing beer in their monastery (near the cathedral in Munich) in 1328. At the beginning of the nineteenth century, the monastery's own brewery was privatised and, in 1817, moved to Neuhauser Strasse. In 1885 a new brewery was built on Landsberger Strasse 19, which was, at that time, on the periphery of Munich (it is now about 1.5 km from the cathedral and Marienplatz, the main square in the old part of the city). Later a restaurant and beer hall, Zum Augustiner, opened in the old brewery building on Neuhauser Strasse 27. During World War II the brewery on Landsberger Strasse was partly destroyed, then rebuilt.[2]

The example of Augustiner beer in Munich is particularly illustrative. Munich is considered one of the world's beer capitals and Oktoberfest is undoubtedly one of the world's most important urban gastronomy events. It has been organised every year since 1810, apart from a few short interruptions. It started in the city centre, about 20 min on foot from the Marienplatz and the cathedral, but now takes place in the area called the Theresienwiese, a 42-ha zone in the Ludwigsvorstadt-Isarvorstadt

---

[2]*Our Augustiner-Bräu Wagner KG*, http://www.augustiner-braeu.de/en/home/about-us.html (Accessed 22 Aug 2017).

quarter (southwest of the historic centre).[3] The current location is close to the Augustiner-Bräu Wagner KG brewery mentioned above.

Other examples of the close relationship between a city and beer include Plzeň (Pilsen in German) and České Budějovice (Budweis in German) in the Czech Republic. The trademarks of *pilsner* and beers produced in České Budějovice are known worldwide and this has to do with their long traditions of brewing beer. České Budějovice, for example, started brewing beer in 1265, when the city was given special brewing rights. Despite its long history, the original brewery no longer exists and the company only moved to its current location in 1795. It was at this point that *Budweiser Bier* began to be produced.[4] Although for decades there have been international disputes over commercial rights to the name (Budweiser, Budvar, Bud) it is most closely associated with the city that was previously known as Budweis.

Gastronomy traditions are clearly visible in the space of cities such as Porto in Portugal and Jerez de la Frontera in Spain. In both places, there is a clear spatial concentration of wineries, wine wholesalers, etc., both in the city and restaurants and bars mainly used by tourists. In the case of Porto, part of the wine-related city is situated on the far bank of the Douro River, at Vila Nova de Gaia, where the most famous port wine cellars are located between the river (Cais de Gaia) and the southern railway line. In Jerez de la Frontera, most of the *bodegas* that tourists visit are located in the historic centre and can be reached using the dedicated Gonzalez Byass Bodega tourist train.[5]

Beverages are not the only factor supporting the gastronomic traditions of cities. In the case of Alkmaar, in The Netherlands, the cheese trade plays a key role. There has been a cheese market in the historic centre of town, Waagplein square, since 1593. Nowadays it is the city's main tourist attraction, although opening hours are restricted. In 2017, it was only held on Fridays, and from the end of March until the end of September from 10 am until 1 pm. It is not only the picturesque cheese market that attracts tourists. Next to it is the Dutch Cheese Museum (Het Kaasmuseum), which opened in 1983. There are also several restaurants and bars in the historical buildings along nearby streets.[6]

The fish market, however, is probably the best-known illustration of support for a city's culinary traditions. Nearly every city or town by the sea or adjacent to a fishing harbour has either an open-air fish market or specially built halls housing fish and seafood restaurants. Japan is home to the Namamugi Fish Market at Yokohama and the Tsukiji Fish Market at Tokyo; in Hong Kong autonomous

---

[3]Where is the Octoberfest?, https://www.oktoberfest.de/en/article/Oktoberfest+2017/About+the +Oktoberfest/Where+is+the+Octoberfest_-2-_/1992 (Accessed 28 Aug 2017).

[4]History of the brewery, http://www.budejovickybudvar.cz/en/o-spolecnosti/historie.html (Accessed 22 Aug 2017).

[5]Flores Watson F., *Sherry bodegas – winery tours*, http://www.andalucia.com/cities/jerez/sherrybodegas.htm (Accessed 30 July 2017).

[6]Alkmaar cheese trading started in 1365, https://www.kaasmarkt.nl/en/cheesemuseum-and-history/ins-and-outs-of-the-cheesemarket (Accessed 30 Aug 2017).

region, there is Sai Kung quay and the Aberdeen Fish Market, and in the United Kingdom there is the historic Billingsgate Fish Market in London.

Of course, other types of food (meat, fruits, vegetables, breads, pastas, etc.) can also support local culinary traditions; here we focus on the most important examples in the context of the urban space. These examples relate to gastronomy traditions in cities that are several centuries old. However, there are other food associations in towns and cities that do not have such a long history, such as the link between a particular type of sandwich and some American cities. Levine (2004), for example, listed Milwaukee and Sheboygan (Wisconsin) as bratwurst sandwich cities, while Des Moines, Kansas City, New York, Boston, Philadelphia, Miami, New Orleans, Los Angeles and other cities are places where visitors can eat different types of sandwich.

3.6.3 Gastronomy and Promotional Activities

Gastronomy is important for making a city more attractive. Culinary issues are often a component of both national and tourism marketing. In recent decades, such activities have often been included as part of a *re-imaging* campaign implemented by local authorities (Smith 2005), notably when trying to change a city's image from an industrial city to a city as an entertainment machine. Food tourism has become increasingly important in recent years. One important element of this is gastronomy, which can increase the attraction of particular cities and towns. The underlying element of culinary tourism is the geographical differentiation of gastronomic traditions. Key factors include: (a) the distinctiveness of a specific cuisine in relation to other cuisines; (b) its diversity; and (c) the use of traditional recipes (Kowalczyk 2005, p. 170). All of these characteristics contribute to the food image of a particular place, be it country, region, city or town. For example, the Guinness Storehouse at St James' Gate Brewery (founded in Dublin in 1759) says in its advertising slogan, 'Ireland's No 1 Visitor Attraction—The Home, Heart and Soul of Guinness®'.[7] It is worth noting here that although the brewery is located in Dublin, it is being promoted on a broader scale as *Ireland's* number one attraction.

In the context of using gastronomy to promote cities, urban marketing has relatively little to say. In most cases, the city is advertised as a whole, and there is little emphasis on individual attractions. Such publications therefore merit little discussion in the context of analysing the gastronomy of the urban space. Examples that illustrate this point include promotional materials produced for Barcelona, Tucson (Arizona), Florianópolis (Brazil)[8] and Whistler (Canada) (Fig. 3.10). Cities

---

[7]Guinness Storehouse, http://www.dublintourist.com/details/guinness-storehouse.shtml (Accessed 22 Aug 2017).
[8]Tucson, Arizona an international culinary destination, 2014, https://www.tucsonaz.gov/files/integrated-planning/Tucson_An_International_Culinary_Destination.pdf (Accessed 01 Sept 2017),

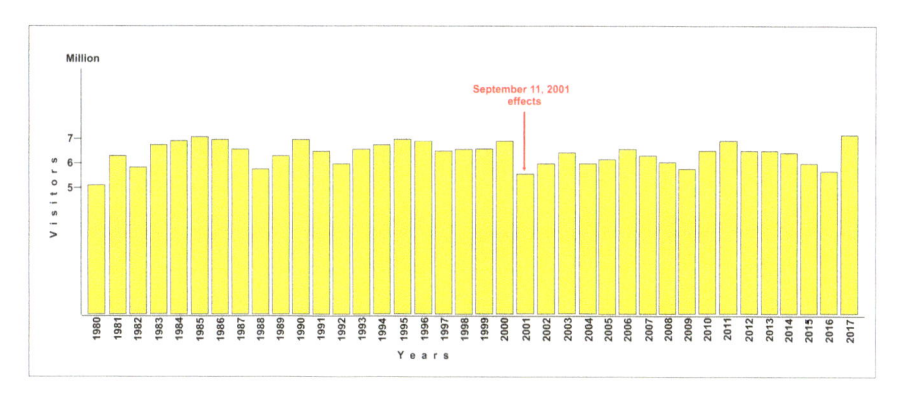

3 Dimensions of Gastronomy in Contemporary Cities   113

Fig. 3.10 Visitors to Oktoberfest in Munich (Germany) between 1980 and 2017. The fall in the number of visitors in 2001 was the consequence of the World Trade Center disaster in New York. In that year the festival was held between 22 September and 7 October. Source: *Anzahl der Besucher auf dem Oktoberfest in München von 1980 bis 2017 (in Millionen)*, https://de.statista.com/statistik/daten/studie/165511/umfrage/anzahl-der-besucher-auf-dem-oktoberfest-seit-1980 (Accessed 25 Aug 2017)

Fig. 3.11 An example of the promotion of gastronomy in a city. This webpage is from the official dining guide for Whistler, British Columbia (Canada). The website promotes the culinary attractions of Whistler in the context of nature and lifestyle rather than urban space. Source: *Whistler dining and restaurant guide*, https://www.whistler.com/dining (Accessed 05 Sept 2017)

often use television advertising, films, websites and mobile phone applications (Horng and Tsai 2010) (Fig. 3.11).

Despite misleading taglines and titles, some materials present information about the location of eating establishments. Examples include the booklets *Eating in*

Florianópolis. UNESCO City of Gastronomy, http://floripamanha.org/wp-content/uploads/2014/02/florianopolis-city-of-gastronomy-en.pdf (Accessed 30 July 2017).

Fig. 3.12 Booklets about gastronomy in relation to the urban space for Hong Kong Special Administrative Region of the People's Republic of China, the Neukölln district of Berlin (Germany), and Buenos Aires (Argentina). Source: *Hong Kong District Food Guide. January–March 2002*, Dimension Marketing Ltd., Victoria; *Neuköln kulinarisch. Das Beste von Pommes bis Schampus. Genießen und Feiern, Kultur und Kinos, Hotels und Pensionen, Bezirksamt Neuköln von Berlin*, Berlin; *Cafés y Bares Notables de Buenos Aires. Localización*, Secretaría de Cultura, Subsceretaría de Patrimonio Cultural, Direccíon General de Patrimonio – Secretaría de Desarrollo Económico, Subsceretaría de Turismo, Direccíon General de Promoción y Desarrollo Turístico, Buenos Aires

Guangzhou, the *Hong Kong district food guide*, *Monaco*. Die Restaurants 2000–2001, the *Barcelona restaurants map, 2011–2012* and the culinary guide to Iowa City (United States). Some booklets are shown in Figs. 3.12 and 3.13. Although sometimes these publications only list restaurants or bars, sometimes they also include maps of their location[9] along with general information about a part of the city, tourist attractions, etc.

The city that is perhaps most prominently promoted through gastronomy is Singapore, which for many years has been advertising itself as the capital of *New Asian* cuisine, a term that was first used in 1997. According to Henderson (2004, p. 72) food is an element of tourist promotion, and travel agencies organise

---

[9]Eating in Guangzhou 2003, Tourism Administration of Guangzhou Municipality, Guangzhou; *Hong Kong district food guide, January–March 2002*, Hong Kong; *Monaco. Die Restaurants 2000–2001. Spezialitäten, charakteristische Merkmale, angegebene Preise*, Direction du Tourisme et des Congrès de la Principauté de Monaco, Monte-Carlo; Barcelona restaurants map 2011–2012, Moritz Beer & Barcelona restaurants, Barcelona Turisme. Gastronomia, Barcelona; *The official dining guide of downtown Iowa City*, https://uiowa.edu/ismc/sites/uiowa.edu.ismc/files/wysiwyg_uploads/DiningGuide.pdf (Accessed 01 Sept 2017).

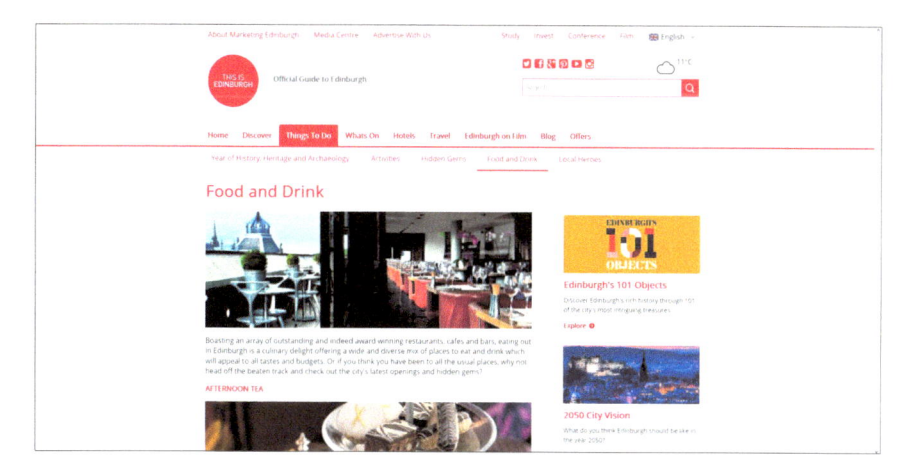

Fig. 3.13 An example of the promotion of gastronomy in the city using the web. Webpage of the official dining guide for Edinburgh, Scotland (United Kingdom). The first photo (top left) and the photo on the bottom right promote the culinary attractions of Edinburgh in the context of the urban space. Source: *Food and drink*, http://edinburgh.org/things-to-do/food-and-drink (Accessed 05 Sept 2017)

specialised tours, promising a 'gastronomic adventure'. Culinary attractions play a very important role in the 'Uniquely Singapore Shop and Eat Tours' project, which shows visitors the Singaporean way of life through local food. For example, the 'Ethnic Trail' covers the districts of Chinatown, Little India and Arab Street, while another 'Heartland Trail' covers the suburbs where most of the Singaporean population live. During these tours tourists can observe everyday life, and eat lunch or dinner with local residents. The case of Singapore shows that the promotion of culinary traditions is regarded as an important element of the city's development strategy. Another illustration of the dining function as a trigger for the economy of the city is Johannesburg (South Africa). The *Joburg 2030* programme includes the restaurant sector as one of the Sector Development Programmes that have been designed to support urban tourism (Rogerson 2006, p. 158).

References

15 aniversario. http://lourencolx.wix.com/docas#!__mapa-n. Accessed 15 July 2017
Acho-Chi C (2002) The mobile street food service practice in the urban economy of Kumba, Cameroon. Singap. J. Trop. Geogr. 23(2):131–148
Ahluwalia N, Singh D, Suri S, Pandey PCh, Sharma RR (2004) Restaurant industry in India—trends and opportunities. A research study, April 2004. Federation of Hotel & Restaurant Associations of India, New Delhi. https://www.hvs.com/content/1336.pdf. Accessed 01 June 2017
Alkmaar cheese trading started in 1365. https://www.kaasmarkt.nl/en/cheesemuseum-and-history/ins-and-outs-of-the-cheesemarket. Accessed 30 Aug 2017

Anjaria SA (2006) Street hawkers and public space in Mumbai. Economic and Political Weekly 41 (21):2140–2146

Anzahl der Besucher auf dem Oktoberfest in München von 1980 bis 2017 (in Millionen). https://de.statista.com/statistik/daten/studie/165511/umfrage/anzahl-der-besucher-auf-dem-oktoberfest-seit-1980. Accessed 25 Aug 2017

Aqueel N (2011) Good times: Port Grand finally makes a grand opening. The Express Tribune. http://tribune.com.pk/story/177989/good-times-port-grand-finally-makes-a-grand-opening. Accessed 02 July 2014

Barcelona, Catalonia's capital of gastronomy. Turismo de Barcelona, Barcelona

Barcelona restaurants map, 2011–2012. Moritz Beer & Barcelona restaurants, Barcelona Turisme. Gastronomia, Barcelona

Bares. http://lourencolx.wix.com/docas#!__bar-pt. Accessed 15 July 2017

Belver M, Luque I (2016) *Madrid tiene 15.000 bares y restaurantes, uno por cada 211 habitantes*. El Mundo, on-line version 22.08.2016. http://www.elmundo.es/madrid/2016/08/22/57bb446eca474137398b45c7.html. Accessed 05 June 2017

Bystrzanowski J, Dutkowski K (1975) *Wszystko o Warszawie. Informator*. Sport i Turystyka, Warszawa

Cafés y Bares Notables de Buenos Aires. Localización, Secretaría de Cultura, Subsceretaría de Patrimonio Cultural, Direccíon General de Patrimonio − Secretaría de Desarrollo Económico, Subsceretaría de Turismo, Direccíon General de Promoción y Desarrollo Turístico, Buenos Aires

Concentración de actividades de ocio nocturno and *Mapa de los Niveles de Ruido Actividad de Ocio en la zona de AZCA − Avenida de Brasil* from the *Declaración de Zona de Protección Acústica Especial y Plan Zonal Específico de AZCA − Avenida de Brasil*, Boletín Oficial del Ayuntamiento de Madrid, Año CXIX, 26 de enero de 2015, Núm. 7.339, pp 8–66, https://sede.madrid.es/csvfiles/UnidadesDescentralizadas/UDCBOAM/Contenidos/Boletin/2015/ENERO/Ficheros%20PDF/BOAM_7339_23012015134649429.pdf. Accessed 05 June 2017

Dale A, Newman LL (2009) Sustainable development for some: green urban development and affordability. Local Environ 14(7):669–681

Declaración de Zona de Protección Acústica Especial y Plan Zonal Específico del Distrito Centro. http://www.madrid.es/UnidadWeb/Contenidos/Publicaciones/TemaMedioAmbiente/ZPAECentro/MemoriajustificativaZPAECentrodefinit.pdf. Accessed 05 June 2017

Dökmeci V, Beygo C, Cakir F (1997) *Spatial analysis of restaurants in Istanbul*, European Regional Science Association Meeting, 37th European Congress, Rome, 26–29 August 1997, Rome, http://www-sre.wu.ac.at/ersa/ersaconfs/ersa97/sessions/paper-e/e11.htm. Accessed 02 July 2014

Eating in Guangzhou (2003) Tourism Administration of Guangzhou Municipality, Guangzhou

Ebrahim Z (2011) *Troubled Karachi finds an enclave of calm*. Al-Jazeera, 17th July 2011. http://www.aljazeera.com/indepth/features/2011/07/2011717132722871313.html. Accessed 02 July 2014

Enquadramento histórico, 2008, Junta Freguesia da Alcântara, Lisboa, https://web.archive.org/web/20110702134506/http://www.jf-alcantara.pt/Default.aspx?Module=ArtigoForm&ID=42. Accessed 20 Aug 2011

Flores Watson F. Sherry bodegas—winery tours. http://www.andalucia.com/cities/jerez/sherrybodegas.htm. Accessed 30 July 2017

Florianópolis. UNESCO City of Gastronomy. The UNESCO Creative Cities Network. http://floripamanha.org/wp-content/uploads/2014/02/florianopolis-city-of-gastronomy-en.pdf. Accessed 30 July 2017

Florida R (2003) Cities and the creative class. City Commun 2(1):3–19

Food and drink. http://edinburgh.org/things-to-do/food-and-drink. Accessed 05 Sept 2017

Food trucks (2017). http://atlantafoodtruckpark.com/foodtrucks. Accessed 15 July 2017

Forkuor JB, Akuoko KO, Yeboah EH (2017) Effective and inclusive regulation of street foods in Kumasi: promoting food safety, protecting consumers and enhancing the well-being of food vendors. Danish International Development Agency (DANIDA) Ghana Street Foods Project. http://dfcentre.com/wp-content/uploads/2017/03/2017-03-27_11-P21-GHA-Kwasi-Ohene-Yankyera.pdf. Accessed 20 July 2017

Graham J (2016) Minimum wages surged in 6 cities last year; then this happened, Investor's Bus Daily. Economy (on-line 02.04.2016) http://www.investors.com/news/economy/hiring-slowed-where-minimum-wage-surged-in-2015. Accessed 01 June 2017

Guinness Storehouse. http://www.dublintourist.com/details/guinness-storehouse.shtml. Accessed 22 Aug 2017

Habib Y (2011) What is Lahore with a food street. Pakistan Today. http://www.pakistantoday.com.pk/2011/11/20/city/lahore/what-is-lahore-with-a-food-street. Accessed 02 July 2014

Harvey D (1973) The social justice and the city. Basil Blackwell Publishers, Oxford.

Harvey D (2012) Rebel cities. From the right to the city to the urban revolution. Verso, London, New York

Henderson JC (2004) Food as a tourism resource: a view from Singapore. Tour Recreat Res 29(3):69–74

Henderson JC (2016) Foodservice in Singapore: Retaining a place for hawkers? J Foodservice Bus Res 19(3):272–286

Henderson JC, Yun OS, Poon P, Biwei X (2012) Hawker centres as tourist attractions: the case of Singapore. Int J Hospitality Management", 31, 3 (September), pp. 849–855

History of the brewery. http://www.budejovickybudvar.cz/en/o-spolecnosti/historie.html. Accessed 22 Aug 2017

Hong Kong district food guide, January–March 2002. Dimension Marketing Ltd., Victoria

Horng JS, Tsai ChT (S) (2010) Government websites for promoting East Asian culinary tourism: a cross-national analysis. Tourism Manag 31(1):74–85

India's food service industry: growth recipe. Consumer markets, Nov 2016, KPMG – FCCI. https://assets.kpmg.com/content/dam/kpmg/in/pdf/2016/11/Indias-food-service.pdf. Accessed 22 June 2017

Ishizaki K (1995) Spatial competition and marketing strategy of fast food chains in Tokyo. Geogr Rev Jpn 68(Ser. B)(1):86–93

Kowalczyk A (2005) *Turystyka kulinarna – ujęcie geograficzne*. Turyzm 15(1/2):163–186

Kowalczyk A (2014) From *street food to food districts – gastronomy dervices and culinary tourism in an urban space*. Turystyka Kulturowa 9(wrzesień):136–160. http://turystykakulturowa.org/ojs/index.php/tk/article/view/493/525. Accessed 22 June 2017

Levine E (2004) Sandwich pride. U.S. Soc Values 9(1):32–34. http://usinfo.state.gov/journals/itsv/0704/ijse/levine.htm. Accessed 25 July 2006

Lloyd R, Clark TN (2001) The city as an entertainment machine, "Research Project no. 454". Paper prepared for presentation at the annual meeting of the American Sociological Association, 2000. https://www.sociologia.unimib.it/DATA/Insegnamenti/4_2996/materiale/thecityasanentertainmentmachine.doc. Accessed 10 July 2017

Longart P (2015) Consumer decision making in restaurant selection (volume I), April 2015. Faculty of Design, Media and Management, Buckinghamshire New University – Coventry University. Ph.D. dissertation. Accessed 01 June 2017

Maynadier B (2014) The gastronomic sense of place, 7 April 2014, Branding the City. Les territoires & leurs marques par Boris Maynadier. http://www.brandingthecity.com/2014/04/the-gastronomic-sense-of-place.html. Accessed 25 May 2017

Monaco. Die Restaurants 2000–2001. *Spezialitäten, charakteristische Merkmale, angegebene Preise*, Direction du Tourisme et des Congrès de la Principauté de Monaco, Monte-Carlo

Neukölln kulinarisch. Das Beste von Pommes bis Schampus. Genießen und Feiern, Kultur und Kinos, Hotels und Pensionen, Bezirksamt Neukölln von Berlin, Berlin

Norcliffe G, Bassett K, Hoare T (1996) The emergence of postmodernism on the urban waterfront. J Transp Geogr 4(2):123–134

Normativa de Plan Zonal Específico de la Zona de Protección Acústica Especial del Distrito Centro. https://sede.madrid.es/UnidadesDescentralizadas/UDCBOAM/UGMedioAmbienteyMovilidad/ElementoBoletin/2012/Anexo%20II%20Normativa%20ZPAE.pdf. Accessed 05 June 2017

Our Augustiner-Bräu Wagner KG. http://www.augustiner-braeu.de/en/home/about-us.html. Accessed 22 Aug 2017

Parham S (2016) Food and urbanism: connecting urban design and food space. Food and the city. Urban Des Grp J 140(Autumn):16–18

Parlett G, Fletcher J, Cooper Ch (1995) The impact of tourism on the Old Town of Edinburgh. Tour Manag 16(5):355–360

Parsa HG, Self JT, Njite D, King T (2005) Why restaurants fail. Cornell Hotel and Restaurant Administration Quarterly 46(3):304–322

Raza A (2009) Food street is history now? Int News. http://www.thenews.com.pk/TodaysPrintDetail.aspx?ID=187474&Cat=5&dt=7/11/2009. Accessed 02 Sept 2009

Restaurantes. http://lourencolx.wix.com/docas#!__rest-pt. Accessed 15 July 2017

Rogerson CM (2006) Creative industries and urban tourism: South African perspectives. Urban Forum 17(2):149–166

Schmitt J, Rosnick D (2011) The wage and employment impact of minimum-wage laws in three cities, March 2011. Center for Economic and Policy Research, Washington, D.C.

Situación y perspectivas de la Ciudad de Madrid, 21, 2° semestre 2016, Análisis Socioeconómico, Área de Gobierno de Economía y Hacienda, Ayuntamiento de Madrid, Madrid. http://www.madrid.es/UnidadesDescentralizadas/UDCObservEconomico/SituacionCiudadMadrid/2016/Ficheros/21/Julio%202016_SituacionPerspectivas_Completo.pdf. Accessed 01 July 2017

Smith A (2005) Conceptualizing city image change: the 're-imaging' of Barcelona. Tour Geogr 7(4):398–423

Tahir A (2012) Fort road food street opens today. Pakistan Today. http://www.pakistantoday.com.pk/2012/01/21/city/lahore/fort-road-food-street-opens-today. Accessed 02 July 2014

te Lintelo DJ (2009) The spatial politics of food hygiene: regulating small-scale retail in Delhi. Eur J Dev Res 21(1):63–80

The official dining guide of downtown Iowa City. https://uiowa.edu/ismc/sites/uiowa.edu.ismc/files/wysiwyg_uploads/DiningGuide.pdf. Accessed 01 Sept 2017

Tucson, Arizona an international culinary destination (2014) City of Tucson. https://www.tucsonaz.gov/files/integrated-planning/Tucson_An_International_Culinary_Destination.pdf. Accessed 01 Sept 2017

Where is the Octoberfest? https://www.oktoberfest.de/en/article/Oktoberfest+2017/About+the+Oktoberfest/Where+is+the+Octoberfest_-2-_/1992. Accessed 28 Aug 2017

Whistler dining and restaurant guide. https://www.whistler.com/dining. Accessed 05 Sept 2017

Wilk J (1984) *Gastronomia*. In: Nowakowski M (ed) *Kształtowanie sieci usług*, Instytut Kształtowania Środowiska, Warszawa, pp 248–262

Wiśniewska H (1984) *Modelowy program obsługi ludności*. In: Nowakowski M (Ed) *Kształtowanie sieci usług*, Instytut Kształtowania Środowiska, Warszawa, pp 360–375

Yen B, Burke M, Tseng C, Ghafoor MMT, Mulley C, Moutou C (2015) Do restaurant precincts need more parking? Differences in business perceptions and customer travel behaviour in Brisbane, Queensland, Australia. Paper presented at the 37th Australasian transport research forum (ATRF), Sydney, Australia, 29 Sept–2 Oct 2015. http://atrf.info/papers/2015/files/ATRF2015_Resubmission_84.pdf

Part II
Changes

Chapter 4
Changes in Gastronomy and Urban Space—Introduction to Part II

Marta Derek

Abstract This short introduction to the second part of the book is focused on changes in gastronomy and urban space. It offers an overview of different dimensions that determine the development of gastronomy and shape the quantity, quality and location of bars, restaurants, cafés and other outlets in urban environments. It draws attention to the fact that gastronomy is a lens through which we can observe social and cultural changes, as well as their manifestation in urban space. It also introduces some of the major dimensions to be analysed in the following chapters.

Keywords Eating out · Urban gastronomy · Urban development · Urban lifestyle

In today's world, the consumption of goods and services is seen as playing a key role in who we are, how we construct our lives and how we relate to others (Williams 2012). Gastronomy, typified by the popular slogan, 'we are what we eat', which draws upon the words of Brillat-Savarin (1862) '*Dis-moi ce que tu manges, je te dirai ce que tu es*' (Tell me what you eat and I will tell you what you are), is an important element of this. Food has gone far beyond a necessity, and gastronomy has become a significant source of identity in postmodern societies (Richards 2002). With an increasing number of cookbooks, culinary programmes on television, food magazines, 'celebrity' chefs, food writers, food bloggers and even festivals and workshops for food bloggers, new patterns of cooking, eating and drinking have appeared, evolved and changed our patterns of consumption.

But for geographers, the main interest of gastronomy is that it is found in a particular PLACE. From this perspective we are (not only) WHAT we eat, but also WHERE we eat. Changing food habits, such as what we eat, how we prepare it and where we look for ingredients or inspirations, not only changes us and influences our ways of consumption, but also contributes to the ongoing evolution of the urban

M. Derek (✉)
Department of Tourism Geography and Recreation, Faculty of Geography and Regional Studies, University of Warsaw, ul. Krakowskie Przedmieście 30, 00-927 Warsaw, Poland
e-mail: m.derek@uw.edu.pl

© Springer Nature Switzerland AG 2020
A. Kowalczyk and M. Derek (eds.), *Gastronomy and Urban Space*,
The Urban Book Series, https://doi.org/10.1007/978-3-030-34492-4_4

space. As gastronomy responds to human needs, and as these needs differ in space, it is a reflection of the functional and structural changes that take place in the urban environment. This is all the truer given that gastronomy is a flexible, short-term activity. It is relatively easy to open a restaurant or bar almost anywhere. It is also relatively easy to change the menu to reflect global trends. In this highly competitive sector there is a constant need for new products and new solutions: the challenge is constantly being addressed in every single eating establishment. This makes gastronomy a perfect lens through which we can observe social and cultural changes, as well as their manifestation in the urban space.

Many dimensions determine the development of gastronomy and shape the quantity, quality and location of bars, restaurants, cafés and other outlets in urban environments. Most of these dimensions result from the ongoing process of urban development, understood as expanding city territories and population growth from one hand, and development of urban lifestyle from the other one. As suggested by Gregory et al. (2009), city life now encompasses most areas of society as 'urbanism as a way of life' becomes pervasive.

We will introduce some of the main points of the relationship between gastronomy and urban space starting with increasingly affluent societies. This can be particularly seen in both developing and post-communist countries. As an increasing number of people can afford to eat out, venues spring up, not only in the city centre, but also on the outskirts of urban areas. The food on offer is becoming more diverse, ranging from ethnic cuisines to traditional and regional food, as an increasing number of increasingly affluent people seek out new flavours.

But eating out is not only related to the wealth of society. It also reflects changing lifestyles and the working environment. An increase in freelancers in the labour market results increases the number of people working without permanent offices. These urban 'nomads', as the owner of a popular café chain in Warsaw calls them, spend their time with a computer and a latte in a busy café in the city centre, where he or she can meet collaborators or participate in urban life while working. More women in the labour market, as well as shrinking households, are another aspect of these lifestyle changes. As Aldrich and Cliff (2003) point out, taking the example of the United States of America, in the mid-twentieth century, 'family' usually meant a nuclear two-generational group with parents and children sharing the same household, with few women working outside the home. In the twenty-first century almost everything has changed. The proportion of single-person households has increased, divorce rates have risen, more women are in employment, and average household size has decreased (Aldrich and Cliff 2003). All of these changes regarding family roles and relationships have important implications for the development of eateries outside the home.

Mobility is another important influence on urban gastronomy, notably migration, employment and tourism. Rapid globalisation is a major driving force in making many national, regional and local dishes international, and some ethnic food has already been 'deethnicised' (van den Berghe 1984). But we should bear in mind that mobility has been responsible for the dissemination of cuisines across the world for a very long time. For example, many ethnic dishes were brought to Europe as a result of colonisation. The potato, which is now an intrinsic element of many regional dishes, was introduced to Europe following the Spanish conquest of the Inca Empire in the second half of the sixteenth century. This is only one of many possible examples.

Migration has always influenced gastronomy as migrants bring their traditional food habits with them. As Gabaccia (1998) notes, in the 1900s in the United States of America, consumers still bought much of their food from entrepreneurs 'of their own kind'. In Europe, migrants coming from overseas colonial territories opened eating establishments around ports, where they settled and worked (Leung 2010). This, however, has changed. Beginning in the second half of the twentieth century and particularly in the twenty-first century, migrants have shared their cooking traditions and experience with an international clientele—a clientele that is already familiar with a variety of tastes thanks to their trips abroad. The tourism and travel sector has itself created demand—upon returning home, tourists seek to extend their holiday by visiting a restaurant that serves the cuisine of the country or region they have just visited. Their consumption patterns change; what was once foreign is now local (Hall et al. 2003). As Richards (2002, p. 3) points out, 'tourists themselves are contributing to gastronomic mobility, by creating a demand in their own countries for foods they have encountered abroad'.

Trends and fashion also shape urban gastronomy. For example, the global trend for vegan food is reflected in a growing number of eateries that only serve this kind of food. Similarly, growing interest in the zero-waste philosophy has led to some cafés offering cheaper coffee if the client arrives with their own cup. The popularity of the Slow Food movement, criticism of mass consumption and under-consumption, together with greater interest in local and traditional food translate into more diverse restaurant menus, and an increasing number of cafés and bars in local neighbourhoods. City centre restaurants are becoming less popular, as residents prefer to visit their trendy neighbourhood restaurant instead. In a globalised world, the spread of new philosophies, fashions and ideas is extremely rapid. Not only is the global becoming localised, but the local is also becoming globalised (Richards 2002). We can meet our friends in a local Starbucks just around the corner, and savour Prosciutto di Parma in every Italian restaurant around the world.

Finally, changes in transport and technology have made the changes mentioned above possible. The industrial world has better access to a greater range of foods and produce than ever before (Hall and Mitchell 2002), while global distribution has made it possible to prepare and supply diverse menus.

These are just a few of many dimensions—others include habits, tradition, planning legislation and changes in food production. In Part 2 of this volume, we translate some of these issues into spatial aspects, by examining them from the perspective of the urban space.

References

Aldrich HE, Cliff JE (2003) The pervasive effects of family on entrepreneurship: toward a family embeddedness perspective. J Bus Ventur 18(5):573–596

Brillat-Savarin JA (1862) Physiologie du goût or méditations de gastronomie transcendante. Charpentier, Libraire-Éditeur, Paris, France

Gabaccia DR (1998) We are what we eat. Ethnic food and the making of Americans. Harvard University Press, Cambridge

Gregory D, Johnston R, Pratt G, Watts MJ, Whatmore S (eds) (2009) The dictionary of human geography, 5th edn. Blackwell Publishing Ltd., Chichester

Hall CM, Mitchell R (2002) Tourism as a force for gastronomic globalization and localization. In: Hjalager AM, Richards G (eds) Tourism and gastronomy. Routledge, London, pp 71–87

Hall CM, Sharples L, Smith A (2003) The experience of consumption or the consumption of experiences? Challenges and issues in food tourism. In: Hall CM, Sharples L, Mitchell R, Macionis N, Cambourne B (eds) Food tourism around the world. Development, management and markets. Butterworth-Heinemann, Oxford, UK, pp 314–335

Leung G (2010) Ethnic foods in the UK. Nutr Bull 35:226–234. https://doi.org/10.1111/j.1467-3010.2010.01840.x

Richards G (2002) Gastronomy: an essential ingredient in tourism production and consumption? In: Hjalager AM, Richards G (eds) Tourism and gastronomy. Routledge, London, pp 3–20

van den Berghe PL (1984) Ethnic cuisine: culture in nature. Ethn Racial Stud 7(3):387–397

Williams A (2012) Understanding the hospitality consumer. Butterworth-Heinemann, Oxford

Chapter 5
Culinary Attractiveness of a City—Old and New Destinations

Andrzej Kowalczyk

Abstract Gastronomic function of a city depends on satisfying the food needs of its inhabitants and visitors, but the concept of *culinary attractiveness* makes sense when we consider it primarily from the point of view of tourists. The underlying element of culinary tourism is the geographical differentiation of gastronomic traditions. This chapter lists different factors (internal and external) that influence the gastronomic attractiveness of a city or town, divided into key factors, additional factors and others. The first and most important group includes the distinctiveness of a specific cuisine in relation to other cuisines, its diversity, as well as the use of traditional recipes. Particular attention is paid to the Top best food cities in the world rankings.

Keywords Culinary tourism · City · Attractiveness · Rankings

5.1 What Is the Culinary Attractiveness of a City or Town?

Although the gastronomic function of a city or town depends on satisfying the food needs of its inhabitants and visitors, the concept of *culinary attractiveness* makes sense when we consider it primarily from the point of view of tourists, especially those who practise culinary tourism. The underlying element of culinary tourism is the geographical differentiation of gastronomic traditions. The key (or basic) factors which influence the culinary attractiveness of a country, region or city include the distinctiveness of a specific cuisine in relation to other cuisines, its diversity, as well as the use of the traditional recipes (Kowalczyk 2005, p. 170). For those who are engaged in culinary tourism, distinctiveness and diversity are of particular value, notably:

A. Kowalczyk (✉)
Department of Tourism Geography and Recreation, Faculty of Geography and Regional Studies, University of Warsaw, ul. Krakowskie Przedmieście 30, 00-927 Warsaw, Poland
e-mail: akowalczyk@uw.edu.pl

- the taste of meals,
- the basic products used to prepare meals,
- the other ingredients to prepare meals,
- the way in which meals are prepared (e.g. frying, boiling, baking, roasting),
- the way that meals are served,
- the typical way in which a meal is consumed for a given culture.

The gastronomic attractiveness of a particular city or town is determined by the co-existence of several factors or conditions (Fig. 5.1). They can be divided into internal (endogenous) and external (exogenous) factors. There are three internal factors: basic, additional, and others (Kowalczyk 2011). They differ from each other in the meaning of culinary attractiveness, and in some situations a given factor may

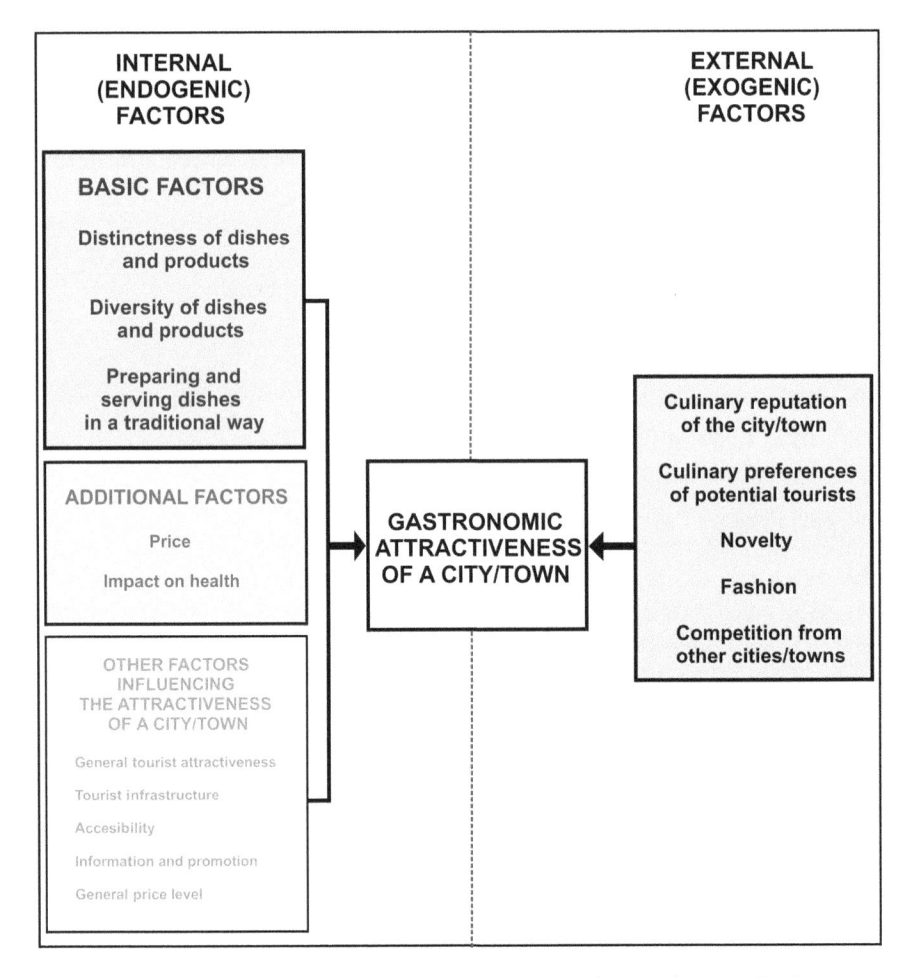

Fig. 5.1 Internal and external factors influencing the gastronomic attractiveness of a city or town

have more or less importance. As mentioned above, the basic internal factors for a city's gastronomic attractiveness are:

- distinctiveness of the dishes offered,
- diversity in the dishes that are typical of the local cuisine,
- processes used to prepare and serve dishes.

The first two factors are particularly important and are of interest to virtually all tourists (and other visitors). However, the ways of preparing meals can arouse the curiosity of people for whom cooking is a hobby or a profession. Apart from the basic factors, the reputation of the dishes offered is determined by tourists based on additional factors such as the price of the ingredients and ready-to-eat products, their health value, etc. Other external factors unrelated to the culinary art also contribute to the perception of gastronomic attractiveness, although they are part of the general attractiveness of the city and the development of tourist facilities, together with spatial accessibility.

Gastronomic attractiveness is also determined by external factors. The most important one is the city's reputation. For instance, in the seventeenth and nineteenth centuries, Paris was regarded by European gourmets as the capital of sophisticated gastronomy. Nowadays it is more difficult to single out just one city that would qualify for that title, but among food lovers, London, New York, Singapore and Tokyo are all very highly rated. The second important factor is the consumer's culinary preferences and associated behaviours. Although some cities may enjoy recognition from gourmets, culinary critics and chefs, the ordinary foodie may not appreciate the finer details of a particular local cuisine. Two further factors, novelty and fashion, are also associated with consumer preferences and behaviours. The role of the novelty (or newness) of a food or meal resonates with the willingness to learn something new. Many foodies like to experiment, looking for new flavours. Based on their own experiences they seek some imagined taste that prompts them to try new dishes. When searching for new flavours, however, they often are influenced by information from the mass media (e.g. food television, tabloids, the internet), and through literature, film, etc. In this process, the culinary stories of other people—mainly relatives and acquaintances—play an important role. Relying on such information, however, makes the boundary between searching for novelty and giving into fashion somewhat blurred. Examples of dishes that are treated by many people in Europe as novel, but also as fashion, are sushi, seaweed salads and other Japanese dishes. The final important external factor affecting a city's gastronomic attractiveness is competition from other cities or towns. It is relatively rare for culinary traditions to occur only in one, well-defined city. More often, they are connected with the region in which that city is situated. This means that the same food products and dishes are popular in several cities within a given area. Only the most discerning gourmets will be able to distinguish the specificity of products or meals in one city from those of a neighbouring one. Wine (in France, Italy, Spain, etc.), cheese (in The Netherlands or in France), kebabs (in Turkey or Iran) and pasta (in Italy) all illustrate this point. In such situations, local authorities organise outdoor culinary events, support cooking schools, organise gastronomic competitions, etc. in an attempt to emphasise their uniqueness. These sorts of activities are used by the Singapore government in campaigns to promote the city as the birthplace of *New Asian Cuisine* (Scarpato and Daniele 2003, pp. 310–312). In the same way, the city of Ayutthaya, near Bangkok, has long-standing and acclaimed traditions in Thailand that are linked with regional cuisine; it has recently been transformed into a centre of culinary tourism (Ladapha and Chiranut 2013). Another way to improve the attractiveness of gastronomy is by joining international organisations like the Creative City in Gastronomy.

5.2 'Top Best Food Cities in the World' Rankings

The gastronomic attractiveness of a particular city is affected not only by its actual resources (culinary offer), but also by (public) opinion. Usually, these opinions are expressed in newspapers and food television, culinary celebrities (e.g. Anthony Bourdain, Jamie Oliver, Nigella Lawson) or food travellers (e.g. James Cheah). On the web, the same role is played by the authors of gastronomic blogs. Although less prevalent nowadays, opinions about the gastronomy in a particular city also come from books or films. Of course, the opinions of chefs, culinary journalists, bloggers, etc., are subjective. Their views emerge from a combination of their own experience, and sponsorship from food corporations, travel companies, local authorities, etc. But an analysis of this kind of information shows that there is a group of cities in the world that regularly crop up in international 'best of' lists. As these rankings are determined using different criteria and are usually characterised by weak methodological underpinnings, they should not be considered as unbiased. Nevertheless, they should not be ignored either, because they can have a strong influence on tourists, who use the rankings to decide which cities to visit, which can turn those cities into important tourist destinations.

An interesting analysis of cities that are considered to have the biggest culinary attractiveness in the world was carried out by Cury (2009). The readers of American magazines (and their websites) *Food & Wine*, *USA Today* and *MSNBC* considered New York and Tokyo to be the most interesting cities, just ahead of London, Paris and Barcelona. Cury also added Sydney, Toronto, Los Angeles and Brussels to the list (Table 5.1). It is important to note that in these lists, the ranking of a particular city is less important than a particular city being incorporated in a group of cities that are considered as interesting for culinary reasons. Over time, the positions occupied by different cities can change. *Food & Wine* magazine in 2009, for example, classified Tokyo, Paris, New York, London and Barcelona as the five best food cities, but in one of its subsequent rankings the order has changed to: Tokyo, Barcelona, Copenhagen, London and New York (World's best food cities 2009).

The analysis of the rankings of the culinary attractiveness of international cities summarised below (Tables 5.2 and 5.3) includes the ratings published for the years 2014 and 2017.

Table 5.1 World's most highly evaluated cities in terms of their culinary attractiveness

Journal/website			Cury (2009)
Food & Wine	USA Today	MSNBC	
Tokyo	New York	Barcelona	New York
Paris	London	Brussels	Paris
New York	Paris	Hanoi	Barcelona
London	Tokyo	Las Vegas	Vancouver
Barcelona	Rome	Lyon	San Francisco
Sydney	Hong Kong	New York	Chicago
Madrid	San Francisco	San Francisco	Tokyo
Chicago	New Orleans	Rome	Hong Kong
Stockholm	Barcelona	Tokyo	Rome
Vancouver	Brussels	Vancouver	London

Source: Adapted from Cury (2009)

Table 5.2 World's most highly rated cities in terms of their gastronomy offer in 2014

Rankings		
'Top 15 best food cities in the world'	'Top 10 best food cities in the world'	'Top 10 world food cities'
New York	New York	San Francisco
Paris	Tokyo	Montréal
San Francisco	Lyon	New York
Rome	Barcelona	Melbourne
New Orleans	San Sebastian	Oaxaca
Tokyo	Paris	Singapore
Oaxaca	London	Lyon
Buenos Aires	Copenhagen	Cape Town
Mumbai	Bangkok	Bologna
London	São Paulo	Guangzhou

Source: Adapted from Kowalczyk (2014, p. 139)

The information contained in Table 5.2 shows that regardless of who compiled the lists in 2014, the same cities occupied the highest positions. In particular, New York, Paris, Tokyo and London and to a lesser degree, Lyon, San Francisco and Oaxaca (Mexico). In the case of New York, these rankings are not thought to be due to individual restaurants and bars, but to the wide range of gastronomy on offer, which reflects the cosmopolitan character of the city along with its diverse cultural and ethnic structure. A similar factor helped New Orleans to gain fifth place. In the case of Tokyo, its sixth place stemmed not only from the presence of restaurants famous for their magnificent cuisine, but also the Tsukiji Fish Market which is renowned as one of the city's most important tourist attractions. Similarly, Mumbai achieved ninth place because, among the people who enjoy culinary tourism, it too

Table 5.3 World's most highly rated cities in terms of their gastronomy offer in 2017

Rankings				
Nims (2018), Yagoda (2018)	'Ranked: the 50 best food cities of 2017'	'15 of the best "foodie" cities around the world…'	'The best food cities in the world'	'Top 10 best food cities in the world'
Rome	San Sebastian	Bologna	Los Angeles	New York
Florence	Tokyo	Tokyo	Florence	Tokyo
Paris	New York	Lima	London	Lyon
Barcelona	Barcelona	Melbourne	Toronto	Barcelona
New Orleans	Singapore	San Sebastian	New York	San Sebastian
New York	Paris	Zanzibar	Tokyo	Paris
Venice	Madrid	Hanoi	Paris	London
Madrid	Lima	Austin	Charleston	Copenhagen
Tokyo	London	Mumbai	Hong Kong	Bangkok
Bangkok	Munich	Mexico City	New Orleans	São Paulo

Source: Nims (2018), Yagoda (2018); The 50 best food cities of 2017 (2017), 15 of the best 'foodie' cities around the world (2018), The best food cities in the world (2017), Top 10 best food cities in the world (2018)

is renowned for its fish market at Sassoon Docks. Food markets helped Oaxaca into seventh place, because tourists who are interested in gastronomy recognise it for its confectionery, as well as fried grasshoppers (*chapulines*). Other cities considered as particularly attractive were Hong Kong (11), Istanbul (12), Brussels (13), Bangkok (14) and Sydney (15).

Table 5.3 shows the results for 2017. Although there are some differences, the leading culinary cities in the world remain Tokyo, New York and Paris, followed by Barcelona and San Sebastian and, to a lesser extent, Florence, New Orleans, Madrid, Bangkok, Lima and London.

The data presented in Table 5.3 requires some explanation. The list drawn from *15 of the best foodie cities around the world…* (published on the blog *God Save The Points*) is particularly interesting. It differs from the other rankings because it includes several cities that do not appear on other lists, such as Zanzibar (Tanzania) and Austin (Texas, United States). The list also includes Singapore, Buenos Aires, Chengdu (People's Republic of China), Vancouver, Nice and Cape Town (not shown). Some of these cities are rarely included in these kinds of rankings and cities from outside Europe predominate. In addition, there is no New York, London, Paris, Madrid or Barcelona, which are usually listed among the must visit cities.

Second, as previously mentioned, it is not only lifestyle and professional magazines that determine the culinary attractiveness of cities and regions. In recent years tourists are increasingly guided by the opinions of bloggers and information published on social networks. On the Mapping Megan blog, the small Dutch town of Edam is listed alongside Tokyo, Barcelona and New Orleans as a place to visit: "Cheese in Europe is a serious business, and for centuries the most popular cheese in the world has been sold in Edam (Jarred 2015)". During the summer, towns such as Alkmaar, Gouda, Edam and Hoorn host traditional cheese markets. Here, they enact the day-to-day scenes of cheese markets of bygone eras; watching farmers navigate the ancient canals and ferry taking rounded hunks of the cheese to the market is quite the scene!'. This quotation highlights the role of subjective feelings and experience in culinary rankings. Although the town of Edam can boast a long culinary tradition, for many years the town of Alkmaar has been recognised by the tourism industry as the main destination for Dutch cheese.

Among the rankings, the TripAdvisor list is of particular interest. In April 2018, their travel experts released a list of the world's best cities for foodies, based on ratings by site users (Nims 2018). However, understanding this ranking requires looking more deeply at the phenomenon it represents. According to Yagoda (2018) there are different reasons for the popularity of the top cities for foodies. This is particularly evident for Italian cities. In Rome, there are popular food tours at sunset in the Prati quarter, and in Florence they hold very popular cooking classes and market tours, based at farmhouses. The TripAdvisor data suggests that food tours are the fastest-growing experiences category and they increased by 61% between 2016 and 2017 (Yagoda 2018). Other examples of similar initiatives include the *Taste of Montmartre* project in Paris, food walking tours around the French Quarter in New Orleans and Brooklyn in New York, and interactive cooking experiences in Barcelona.

The fact that the gastronomic attractiveness of a city is strongly influenced by the diversity of its gastronomic offer has been investigated by the authors of the Bott and Co. blog. They analysed restaurant data for every city in the world with a population of over one million people. Data included restaurants dedicated to national cuisine listed in Google Maps (Which cities have the most diverse food scenes? 2018).

The data in Table 5.4 show that a significant proportion of the places in the world with the most diverse gastronomic offer are cities in the United States, Canada and Australia, as well as the cities of London, Paris, Berlin, Dubai, Tokyo, Madrid and Brussels. The occurrence of cuisines of different nations in the cities of North America and Australia can be explained by the history of these regions and their settlement by people from different parts of the world. The roots of this process date back to colonial times in seventeenth and eighteenth centuries. Another consequence of the colonial period is the high position of London, Paris and Madrid. But the high position of Berlin and Brussels and, above all, Dubai, results from processes that happened much later. We must also remember that Berlin is the capital of a country with great importance in the global economy, that Brussels is the seat of the European Commission and that Dubai has become a cosmopolitan city that attracts visitors (and workers) from virtually all over the world.

Table 5.4 Cities in the world with the largest number of national cuisines in 2018

City	Country	Number of national cuisines	City	Country	Number of national cuisines
New York	United States	94	Barcelona	Spain	48
London	United Kingdom	89	Frankfurt/Main	Germany	46
Toronto	Canada	73	Milan	Italy	45
Chicago	United States	66	Adelaide	Australia	44
Paris	France	65	Atlanta	United States	44
Berlin	Germany	64	Las Vegas	United States	44
Melbourne	Australia	64	Munich	Germany	44
Dubai	United Arab Emirates	62	San Diego	United States	44
Los Angeles	United States	61	Baltimore	United States	43
San Francisco	United States	60	Edmonton	Canada	43
Montréal	Canada	58	Houston	United States	43
Tokyo	Japan	58	Kuala Lumpur	Malaysia	43
Boston	United States	57	Manchester	United Kingdom	43
Sydney	Australia	56	Miami	United States	43
Philadelphia	United States	54	Orlando	United States	42
Madrid	Spain	53	Ottawa	Canada	42
Portland	United States	53	Perth	Australia	42
Seattle	United States	53	Singapore	Singapore	42
Vancouver	Canada	52	Brisbane	Australia	40
Brussels	Belgium	50	Charlotte	United States	40
Amsterdam	The Netherlands	49	Denver	United States	40
Calgary	Canada	49	Detroit	United States	40
Minneapolis	United States	49	Auckland	New Zealand	37
Vienna	Austria	49	Phoenix	United States	37
Bangkok	Thailand	48	Moscow	Russia	34

Source: Adapted from Which cities have the most diverse food scenes? (2018)

5.3 Conclusions

This analysis has shown that among cities which are considered as highly culinary attractive, there are cities with a long gastronomic tradition, as well as cities which have just appeared on the culinary map. The first category includes, among others, New York, Tokyo, Paris, London, Rome, Madrid, Barcelona and Lyon. The second category includes, for example, Copenhagen, Singapore, Dubai and Lima. Although the future always remains uncertain, we may assume that the high position of some cities will probably not change. At the same time, current evidence suggests that new cities will appear on the list of culinary attractions, although it is difficult to predict what they will be.

If we compare the lists of the top 'culinary cities'with the theories described in Chap. 2 you can see the relationship between practice and theory. Popularity of some culinary destinations among tourists and foodies may be a sign of their 'œuvre' (Lefebvre 1996), as well as a result of the justification of the principles of the creative class theory (Florida 2005).

References

15 of the best "foodie" cities around the world… (2018) God Save The Points, April 5, 2018. http://www.godsavethepoints.com/2018/04/05/best-foodie-cities-in-the-world. Accessed 30 May 2018

Cury JO (2009), The top 10 food cities in the world (or foodie cities? Or restaurant cities?). "Epicurious". http://www.epicurious.com/articlesguides/blogs/editor/2009/09/the-top-10-food-cities-in-the-world-or-foodie-cities-or-restaurant-cities.html. Accessed 02 July 2014

Florida R (2005) Cities and the creative class. Routledge, New York–London

Jarred M (2015) The best food cities in the world. Mapping Megan. https://www.mappingmegan.com/best-food-cities-in-the-world-foodie-destinations. Accessed 30 May 2018

Kowalczyk A (2005) Turystyka kulinarna – ujęcie geograficzne. "Turyzm":15(1/2):163–186

Kowalczyk A (2011) Turystyka kulinarna jako potencjalny czynnik rozwoju lokalnego i regionalnego. Turystyka i Rekreacja 7:13–26

Kowalczyk A (2014) From street food to food districts—gastronomy services and culinary tourism in an urban space. Turystyka Kulturowa 9(wrzesień):136–160. http://turystykakulturowa.org/ojs/index.php/tk/article/view/493/525. Accessed 30 May 2018

Ladapha P, Chiranut S (2013) Gastronomic tourism in Ayutthaya, Thailand. Int J Bus Appl Sci. http://www.ijbts-journal.com/images/main_1366796758/0043-Ladapha.pdf. Accessed 02 July 2014

Lefebvre H (1996) Industrialization and urbanization. In: Writings on cities/Henri Lefebvre. Selected, translated and introduced by Eleonore Kofman and Elizabeth Lebas. Blackwell Publishers Ltd./Blackwell Publishers, Inc., Oxford–Malden, MA, pp 65–85

Nims B (2018) The world's best food cities, according to TripAdvisor, will travel for food, "Huffpost Finds", 04/25/2018. https://www.huffingtonpost.com/entry/the-worlds-best-food-cities-2018-according-to-tripadvisor_us_5ae0b16ee4b055fd7fc72e42?guccounter=1. Accessed 30 May 2018

Ranked: the 50 best food cities of 2017, Love Exploring, 08.12.2017. https://www.loveexploring.com/gallerylist/69944/ranked-the-50-best-food-cities-of-2017. Accessed 30 May 2018

Scarpato R, Daniele R (2003) New global cuisine: tourism, authenticity and sense of place in postmodern gastronomy. In: Hall CM, Sharples L, Mitchell R, Macionis N, Cambourne B (eds) Food tourism around the world. Development, management and markets. Butterworth-Heinemann, Oxford-Burlington, pp 296–313

The best food cities in the world (2017). Food & Wine. https://www.foodandwine.com/travel/restaurants/best-food-cities-2017. Accessed 30 May 2018

Top 10 best food cities in the world (2018) The Ultimate Urban Destination Guides. https://www.ucityguides.com/cities/top-10-best-food-cities.html. Accessed 30 May 2018

Which cities have the most diverse food scenes? (2018) Bott and Co. https://www.bottonline.co.uk/blog/foodie-capitals-of-the-world. Accessed 30 May 2018

World's best food cities (2009) https://www.foodandwine.com/slideshows/worlds-5-best-food-cities. Accessed 30 May 2018

Yagoda M (2018) These are the 10 best food cities in the world, according to TripAdvisor number one should come as no surprise. Food & Wine, April 23, 2018. https://www.foodandwine.com/news/best-food-cities-world-tripadvisor. Accessed 30 May 2018

Chapter 6
Traditional and Regional Cuisine in Urban Space

Andrzej Kowalczyk and Magdalena Kubal-Czerwińska

Abstract An increase in interest in traditional and regional cuisine is one of the most important trends in contemporary gastronomy. Many restaurant owners and chefs have returned to traditional recipes that require local and regional products. This renaissance has been influenced by consumers who have become tired of the monotony of fast-food gastronomy. Another factor supporting these trends is the growing popularity of culinary tourism. The chapter illustrates the issue by analysing the distribution of restaurants and bars in Madrid offering regional Spanish cuisine; restaurants serving Polish cuisine in the Old Town of Warsaw; and stands selling Kraków's traditional *obwarzanki*.

Keywords Traditional and regional cuisine · Tourism · Madrid · Warsaw · Kraków

6.1 Where Did the Return to Traditional and Regional Cuisine Come from?

Observing trends in contemporary gastronomy, one can detect an increase in interest in traditional and regional cuisine. Although restaurant owners and chefs are still fascinated by fusion and molecular cuisine, in parallel there is a return to traditional recipes and local and regional products. This renaissance has been at least partly influenced by consumers becoming tired with the unified nature of

---

A. Kowalczyk (✉)
Department of Tourism Geography and Recreation, Faculty of Geography and Regional Studies, University of Warsaw, Ul. Krakowskie Przedmieście 30, 00-927 Warsaw, Poland
e-mail: akowalczyk@uw.edu.pl

M. Kubal-Czerwińska
Department of Tourism and Health Resort Management, Institute of Geography and Spatial Management in Kraków, Jagiellonian University, Ul. Gronostajowa 7, 30-387 Kraków, Poland
e-mail: magdalena.kubal@uj.edu.pl

© Springer Nature Switzerland AG 2020
A. Kowalczyk and M. Derek (eds.), *Gastronomy and Urban Space*,
The Urban Book Series, https://doi.org/10.1007/978-3-030-34492-4_6

fast-food gastronomy and has led to the establishment of the Slow Food movement. Developed primarily in Europe, the movement also gained support across North America. In Europe, a conscious policy from the European Union to support the agricultural sector through the introduction of a system of geographical indications and the protection of traditional specialties has played a major role in its resurgence. This policy has three pillars known as *protected designation of origin* (PDO), *protected geographical indication* (PGI), and *traditional specialities guaranteed* (TSG). The main goals are the promotion and protection of quality agricultural products and foodstuffs. Products registered under one of the three schemes may be marked with a logo.

> This policy began in 1992, when the European Community adopted legislation for agricultural products and foodstuffs. It 'was inspired by existing national systems, such as the French AOC (*Appellation d'Origine Contrôlée*) and the Italian DOC (*Denominazione d'Origine Controllata*)' (Dias and Mendes 2018, p. 492).
>
> One regulation protects the *designation of origin* (PDO), which is the name of a specific place (city, town, village), area (region, province, etc.) or (exceptionally) country, used as the designation for an agricultural product or foodstuff. The name is given to artefacts that come from such a place, whose quality or properties are determined by the geographical environment (natural and human factors) and whose production, processing and preparation take place within the determined geographical area.
>
> *Protected geographical indication* (PGI) is given to products that come from a particular place, which has qualities that are attributable to its geographical origin, and for which one of the stages of production, processing or preparation takes place in the area. This last issue distinguishes PDO from PGI.
>
> The last category, *traditional specialities guaranteed* (TSG), provides a protection regime for traditional food products. It does not certify that the food product has connections with the geographical area, but only guarantees that food must be of *specific character* and its raw materials, method of production or processing must be *traditional*.

The European Union system means that many food products are now considered to be closely related to a given locality. This is the case, for example, in the Czech Republic where, in the spring of 2018, 19 out of 34 food products were directly related to a particular city or town (or, in one case, to the rural commune) (Table 6.1).

Another international institution that has a policy to protect the culinary heritage of regions and cities is UNESCO (United Nations Educational, Scientific and Cultural Organisation). In 2004, this organisation created the UNESCO Creative Cities Network to promote cooperation within and among cities that have identified

Table 6.1 Food products in the Czech Republic protected by European Commission regulations in May 2018

Product	Type of protection	Year	Class	Connections with locality
Brněnské pivo/ Starobrněnské pivo	PGI	2009	Beer (Class 2.1.)	Brno
Budějovické pivo	PGI	2003	Beer (Class 2.1.)	*České Budějovice*
Budějovický měšťanský var	PGI	2003	Beer (Class 2.1.)	*České Budějovice*
Českobudějovické pivo	PGI	2003	Beer (Class 2.1.)	*České Budějovice*
Hořické trubičky	PGI	2007	Confectionary (Class 2.4.)	Hořice
Karlovarské oplatky	PGI	2011	Biscuits (Class 2.4.)	*Karlovy Vary*
Karlovarské trojhránky	PGI	2011	Rolls (Class 2.4.)	*Karlovy Vary*
Karlovarský suchar	PGI	2007	Biscuits (Class 2.4.)	*Karlovy Vary*
Lomnické suchary	PGI	2007	Biscuits (Class 2.4.)	*Lomnice nad Popelkou*
Mariánskolázeňské oplatky	PGI	2009	Biscuits (Class 2.4.)	*Mariánské Lázně*
Nošovické kysané zelí	PDO	2008	Processed cabbage (Class 1.6.)	Nošovice
Olomoucké tvarůžky	PGI	2010	Chesse (Class 1.3.)	Olomouc
Pardubický perník	PGI	2008	Cake (Class 2.4.)	Pardubice
Pražská šunka	TSG	2018	Ham (Class 1.2.)	Prague
Pohořelický kapr	PDO	2007	Carp (Class 1.7.)	Pohořelic
Štramberské uši	PGI	2007	Confectionary (Class 2.4.)	Štramberk
Třeboňský kapr	PGI	2007	Carp (Class 1.7.)	Třeboň
Znojemské pivo	PGI	2009	Beer (Class 2.1.)	Znojmo
Žatecký chmel	PDO	2007	Hop (Class 1.8.)	Žatec

Source: Adapted from European Commission, Agricultural and Rural Development DOOR, d'Origine Controllata. http://ec.europa.eu/agriculture/quality/door/list.html. Accessed 26 May 2018
Class 1.2. Meat products (cooked, salted, smoked, etc.); 1.3. Cheeses; 1.6. Fruit, vegetables and cereals fresh or processed; Class 1.7. Fresh fish, molluscs and crustaceans and products derived therefrom; Class 1.8. Other products of Annex I of the Treaty (spices, etc.); 2.1. Beers, 2.4. Bread, pastry, cakes, confectionery, biscuits and other baker's wares

creativity as a strategic factor for sustainable urban development. The network now has 180 members. These cities have a common goal: creativity and cultural industries are at the heart of their development plans at the local level, and they actively cooperate at the international level. In 26 of these cities, authorities and the local community have declared their interest as a result of their gastronomic tradition.

Table 6.2 Cities being that are members of the UNESCO Creative Cities network as a 'Creative City in Gastronomy' in 2018

Year	City (country)
2005	Popayán (Colombia)
2010	Chengdu (People's Republic of China)
	Östersund (Sweden)
2012	Jeonju (Republic od Korea)
2013	Zahlé (Lebanon)
2014	Florianopolis (Brazil)
	Shunde (People's Republic of China)
	Tsuruoka (Japan)
2015	Belém (Brazil)
	Bergen (Norway)
	Burgos (Spain)
	Dénia (Spain)
	Ensenada (Mexico)
	Gaziantep (Turkey)
	Parma (Italy)
	Phuket (Thailand)
	Rasht (Iran)
	Tucson (United States)
2017	Alba (Italy)
	Buenaventura (Colombia)
	Cochabamba (Bolivia)
	Hatay (Turkey)
	Macao (*Macao* Special Administrative Region of the People's Republic of China)
	Panama City (Panama)
	Paraty (Brazil)
	San Antonio (United States)

Source: Adapted from https://en.unesco.org/creative-cities/home. Accessed 26 May 2018

In 2005, the first declaration to promote culinary heritage was announced the city of Popayán in Colombia (Table 6.2). There were no more declarations until 2010 when Chengdu (People's Republic of China) and Östersund (Sweden) were added, followed by several others. The situation changed radically in 2015, when the declaration of the Creative City in Gastronomy network was announced by 10 cities. Among them was the first city from the United States, Tucson, Arizona. Then, in 2017, eight new cities followed, including San Antonio (Texas). This city is known not only for its Tex-Mex cuisine, but also for other culinary traditions (Spanish, French and German). San Antonio has more than 4,700 restaurants and bars, and houses the Culinary Institute of America, which is located in an old German brewery that was built in 1883. With 18 facilities, including a weekend farmers market and the city's first food hall, this project (called *Pearl*) has become a tourist attraction (San Antonio: A Creative City of Gastronomy 2017).

There are several similar organisations, such as the Délice Network. This was founded in 2007 in Lyon (France) and its founding members were the cities of Barcelona (Spain), Birmingham (United Kingdom), Brussels (Belgium), Guangzhou (People's Republic of China), Genoa (Italy), Gothenburg (Sweden), Lausanne (Switzerland), Leipzig (Germany), Lyon (France), Madrid (Spain), Milan (Italy), Montreal (Canada), Riga (Latvia) and Saint Louis (United States). Article 3 of the declaration that established the organisation stated that its purpose was 'to promote the appreciation of the culinary and gastronomic traditions, and of the regional specialties both from the land and from the sea, of its member cities' (Délice. Bylaws 2007, p. 3). In following years, several other cities joined: Aarhus (Denmark), Bordeaux (France), Helsinki (Finland), Lisbon (Portugal), Stavanger (Norway), Turin (Italy), Chicago (United States), Puebla and Mérida (Mexico), Hong Kong (Hong Kong Special Administrative Region of the People's Republic of China) and Izmir (Turkey) plus the Cape Winelands region (Republic of South Africa). In the same period, Genoa, Milan and Saint Louis all left. In spring 2018 it had 23 members: 15 European, four North American, three Asian and one African (Délice Member Cities 2018).

The promotion of culinary heritage in cities was also the main objective of the Gastronomic Cities Project initiative. This brought together five European cities, and was part of the URBACT II programme (2007–2013). Its members were Burgos and L'Hospitalet (Spain), Fermo (Italy), Korydallos (Greece) and Alba Iulia (Romania) (Sirše 2014). More information on the activities of this initiative in the context of sustainable development are provided elsewhere (Rinaldi 2016).

There have been many other initiatives of this kind in recent years, and their common goal is to strive to preserve the gastronomic heritage and use culinary traditions in the promotion of tourism. Many of these projects directly refer to the ideologies of slow food and sustainable development, which in the context of cities often comes down to the concepts of the slow city (*cittàslow*) and the creative city (described earlier). Cities and towns that have an active tourism policy based on their culinary heritage adopt different strategies. These consist of a number of activities that can be divided into several main categories. The most commonly used solutions that are reflected in the urban space include:

- organising outdoor gastronomic events (food festivals, culinary contests, etc.);
- planning and marking culinary trails;
- opening museums related to food and gastronomy (sometimes in historical buildings);
- planning and designating zones where gastronomic events take place (e.g. food truck festivals, farmers markets).

Municipal authorities, social organisations and local entrepreneurs participate in these activities. While some are carried out on a non-profit basis, others have a commercial character (Fig. 6.1).

Fig. 6.1 Slow Food in practice. Farmers market in the vicinity of the cathedral in Barcelona (Spain). The inscription above the stalls rads *Productes tradicionals catalans* (**a**). Stall with homemade cheese at the Boxhagener Platz in Berlin-Friedrichshain district (Germany) (**b**). Source: Anna Kowalczyk 2015; 2014

In recent years, perhaps the most frequently used strategy is to organise open-air gastronomic events. The success of these events is closely related to the season and weather. Therefore, a better solution is to use covered facilities, e.g. market halls or museums. Where museums are used to popularise the history of gastronomy, the best tend to be those that have a particular theme related to culinary heritage (Fig. 6.2). There are many such museums throughout the world. Many European cities have established facilities at breweries and wineries, Italy has pasta museums, Spain has ham (*jamón*) museums, Japan has noodle museums and China has tea museums. Some are highly original and very popular. The Cupnoodles Museum—instant noodles—in Yokohama (Japan), which opened in 2011, for example, attracts more than one million annual visitors. Cupnoodles Museum Welcomed its Five Millionth Visitor (2016). Likewise, many people visit a similar museum in Ikeda in the greater Osaka area. Its official name is 'Momofuku Ando Instant Ramen Museum' because Ikeda is the birthplace of Momofuku Ando, one of the inventors of instant noodles. Both museums belong to the Nissin Foods Product Co. corporation, and the museum in Ikeda advertises itself with the slogan 'The birthplace of instant noodles'.

One example of a museum that maintains a tradition related to gastronomy is the Madeira wine museum in Funchal on the island of Madeira (Portugal). Known as Blandy's Wine Lodge, it is located in a complex of baroque buildings. It includes the oldest Madeira wine cellars, dating from the seventeenth and eighteenth centuries. These cellars are called the São Francisco Cellars and are the part of the former São Francisco Convent, which was demolished in the nineteenth century. The main buildings, which now house the São Francisco Wineries, date mainly from the seventeenth and eighteenth centuries and some are old warehouses. The museum is owned by the Madeira Wine Company (founded in 1913) and includes an exhibition area, warehouses, tasting rooms and a shop.

Fig. 6.2 Museums related to gastronomic heritage. The Madeira wine museum Blandy's Wine Lodge in Funchal, Madeira (Portugal) (**a**). The chocolate museum, Heindl SchokoMuseum in Vienna (Austria) (**b**). Teaware museum—'Flagstaff House—Museum of Tea Ware' in Victoria (Hong Kong Special Administrative Region of the People's Republic of China) (**c**). Source: Anna Kowalczyk 2009

Another example is the chocolate museum in Vienna. This is located in the southern suburbs of Vienna, one kilometre east of the Kaufpark and Wohnpark Alterlaa residential and commercial complex. The Heindl SchokoMuseum (which opened in 2001) is part of a factory owned by Confiserie Heindl. Established in 1953, it moved to its current site in 1967. In addition to the history of chocolate, displays cover the history of Heindl and Pischinger (founded in 1849).

A third example is the teaware museum in Victoria on the island of Hong Kong (Hong Kong Special Administrative Region of the People's Republic of China). It is situated in Hong Kong Park (located in the city's downtown) and is housed in the historical Flagstaff House building that was constructed in 1846. This building originally served as the office and residence of the Commander of the British Forces in Hong Kong. It was converted into the Museum of Tea Ware in 1984 and, in 1995, a new wing was added. It attracts over 200,000 visitors a year (in 2014/2015 there were 214,000, in 2015/2016, 215,000 and in 2016/2017, 202,000). This makes it the fifth most visited museum in Hong Kong (Statistics Report 2018). Its popularity is at least partly related to its location in the city centre, as well as to the interest of visitors in the cultivation of tea and the art of brewing and serving it.

Gastronomic museums are especially important for supporting culinary traditions and regional cuisine. They are an element of the city's material structure and therefore not only is their role in the gastronomy of the urban space particularly significant, but they also play a very important educational role.

6.2 Regional Cuisine in the Gastronomic Space—An Example from Madrid (Spain)

The evidence above indicates that culinary heritage is being sustained not only in smaller cities, but also large cities. The Délice Network, which is almost exclusively made up of big cities, offers a good example. Among the cities that are members of network is Madrid, the capital of Spain. Located almost in the middle of Spain, Madrid has long been a destination for migrants. Originally they came from other regions of Spain, but nowadays they also come from other countries of the world. The migration from other parts of Spain, which began in the sixteenth century, has contributed to the varied gastronomic landscape. Foodies can eat dishes from virtually every region of Spain that are often prepared by chefs using traditional recipes. In the following, we look in more detail at the geographical distribution of the restaurants and bars in Madrid that advertise regional cuisine.

The study of the spatial distribution of restaurants and bars in Madrid follows a study of Atlanta, Georgia (United States) in the 1980s (Pillsbury 1987, p. 327). The list of dining establishments was derived from the telephone directory's yellow pages (*páginas amarillas* in Spanish) for 2013. The results of these studies can be used to draw many conclusions, some of which may also apply to other large cities.

Figure 6.3 shows the location of restaurants and bars in Madrid listed in the telephone directory as regional restaurants. The analysis covers only the most popular regional Spanish cuisines. The first regularity that appears is that most restaurants are in the downtown. An anomaly is that the number of restaurants and bars offering Castilian cuisine is relatively high in the Fuencarral-El Pardo district, which lies in the northern periphery. To a lesser extent, another anomaly relates to Galician restaurants, which are slightly larger in number in the Latina district in the south-western part of the city. Galician cuisine can also be found in the northern part of the inner city district of Tetuán. The second conclusion is that most restaurants serving regional dishes are in the Centro district. Andalusian cuisine is slightly more prevalent in the Salamanca district, which is part of the inner city. Although Fig. 6.3 suggests that the spatial dispersion of eating establishments differs according to the cuisine, it is difficult to see its extent. However, it can be estimated using statistical measures of standard deviation and coefficient of variation. Table 6.3 shows that Andalusian cuisine has the highest spatial variation

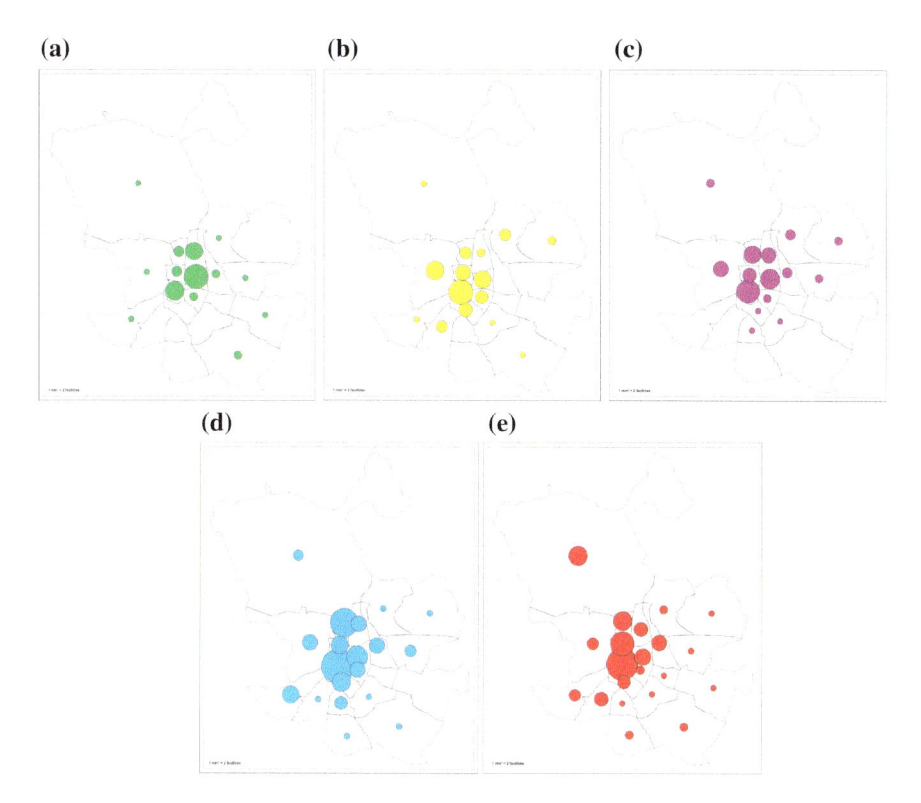

6 Traditional and Regional Cuisine in Urban Space   143

Fig. 6.3 Distribution of restaurants and bars in Madrid (Spain) with selected regional Spanish cuisine in 2013. **a** Andalusian. **b** Asturian. **c** Basque. **d** Galician. **e** Castilian

($V = 163.9$), indicating that Andalusian restaurants are most concentrated. This is confirmed by analysing Fig. 6.3a. In 2013, 67.3% of the eating establishments in the Salamanca, Centro and Chamartín districts offered Andalusian cuisine. For the others, the first three districts with the highest number of regional restaurants accounted for less than 50% of the total number of establishments (Table 6.4).

In summarising these results, it should be stated that the spatial distribution of eating establishments for all analysed regional cuisines was similar. The convergence of restaurants with Andalusian, Asturian and Basque cuisine was particularly noticeable, while the location of Galician and Castilian establishments was slightly different. Figure 6.3 suggests that restaurants offering Galician food in the central part of the city are concentrated in a stretch along a north–south axis. Furthermore, downtown Castilian and Galician restaurants and bars were similarly concentrated and formed a cluster in the northern suburbs (Fig. 6.4).

Table 6.3 Distribution in Madrid (Spain) of restaurants and bars offering regional cuisine in 2013 (by district)

Districts	Regional cuisines							
	Andalusian (*andaluza*)		Asturian (*asturiana*)		Basque (*vasca*)		Galician (*gallega*)		Castilian (*castellana*)	
	Number	%	Number	%	Number	%	Number	%	Number	%
Arganzuela	–	–	6	8.0	1	1.3	10	6.8	5	4.2
Barajas	–	–	2	2.7	2	2.6	1	0.7	1	0.8
Carabanchel	–	–	4	5.3	–	–	1	0.7	6	5.0
Centro	11	20.0	18	24.1	16	21.2	34	23.2	30	24.7
Chamartín	9	16.4	2	2.7	9	11.8	7	4.8	6	5.0
Chamberí	3	5.5	7	9.3	6	7.9	9	6.2	16	13.1
Ciudad Lineal	2	3.6	–	–	3	3.9	7	4.8	7	5.8
Fuencarral-El Pardo	3	5.5	1	1.3	2	2.6	3	2.1	11	9.1
Hortaleza	1	1.8	4	5.3	3	3.9	1	0.7	2	1.7
Latina	1	1.8	1	1.3	–	–	9	6.2	4	3.3
Moncloa-Aravaca	1	1.8	10	13.4	7	9.3	7	4.8	4	3.3
Moratalaz	–	–	–	–	–	–	–	–	1	0.8
Puente de Vallecas	–	–	1	1.3	1	1.3	1	0.7	1	0.8
Retiro	2	3.6	5	6.7	2	2.6	7	4.8	2	1.7
Salamanca	17	30.9	9	12.0	12	15.9	14	9.7	8	6.6
San Blas	1	1.8	–	–	2	2.6	4	2.7	1	0.8
Tetuán	3	5.5	4	5.3	9	11.8	24	16.3	10	8.3
Usera	–	–	–	–	1	1.3	5	3.4	1	0.8
Vicálvaro	1	1.8	–	–	–	–	–	–	1	0.8
Villa de Vallecas	–	–	1	1.3	–	–	1	0.7	2	1.7
Villaverde	–	–	–	–	–	–	1	0.7	2	1.7
MADRID	55	100.0	75	100.0	76	100.0	146	100.0	121	100.0

Source: Adapted from information published in *Páginas Amarillas/Páginas Blancas 2014. Madrid Capital/Zona Centro, Consulta la guía de restaurantes más completa*, 1/11/2013, Connecti S.A.U. Impeso por Ediciones Informatizadas, S.A., Madrid, pp. 433–435, 437, 439–440

Table 6.4 Statistical measures showing spatial variation in the location in Madrid (Spain) of restaurants and bars serving regional cuisine in 2013

Measure	Regional cuisine				
	Andalusian	Asturian	Basque	Galician	Castilian
Standard deviation (δ)	4.293	4.403	4.413	8.243	6.704
Arithmetic mean (μ)	2.619	3.571	3.619	6.952	5.762
Coefficient of variation (V)	163.9	123.3	121.9	118.6	116.3

Source: Based on information published in *Páginas Amarillas/Páginas Blancas 2014. Madrid Capital/Zona Centro, Consulta la guía de restaurantes más completa*, 1//11/2013, Connecti S.A.U. Impeso por Ediciones Informatizadas, S.A., Madrid, pp. 433–435, 437, 439–440

6.3 Traditional Polish Cuisine in Warsaw (Poland)

In a rapidly changing world, the importance of traditional dishes has decreased. The development of the international food trade, the expansion of large food corporations, international migration, tourism and many other factors have resulted in changes in traditional nutrition. Nowadays, it is difficult to say which dishes are original and stem from the culinary heritage, and which are the result of cultural diffusion. This problem is reflected in gastronomy. The popularity of typical fast-food cuisine is a threat to traditional cooking, as are the casual restaurants that serve fusion cuisine.

Here, we take central Warsaw as an example. Research conducted in 2013 in the Śródmieście borough showed that there were 64 restaurants that offered traditional Polish cuisine, which were mostly situated in the northern and central part of the district (Derek 2013, p. 89 and Fig. 2 on p. 95). Interestingly, in surveys conducted in the same area in 2003, only a few of the restaurants clearly referred to the traditions of Polish cuisine and their distribution did not show any statistical tendency (Brożek 2007). At that time, a significant proportion of the restaurants and bars advertised themselves as offering international or ethnic cuisine. What does the increase in the number of restaurants offering traditional Polish cuisine between 2003 and 2013 mean? It could be taken as evidence that business owners noted the advantages associated with referring to their culinary heritage. This is based on the fact that 56% of foreign tourists and 53% of tourists from the Polish diaspora visited the Old Town and its surroundings (Derek 2013, p. 89). The data indicates that the main cluster of restaurants in Warsaw offering traditional Polish dishes is in the Old Town (the official name of this area is Stare i Nowe Miasto), and this is the part of the inner city that we will focus on here.

Since Derek's (2013) study, the overall number of dining establishments in the Old Town has not changed much. This is also true of restaurants that explicitly refer to their Polish heritage.

Figure 6.5 shows that eating outlets are distributed unevenly in the neighbourhoods of the Old Town and the New Town (Nowe Miasto). First, in Nowe Miasto, the western part is largely made up of housing estates from the 1950s, the Supreme Court, a High School of Theater Arts and the field cathedral of the Polish Army

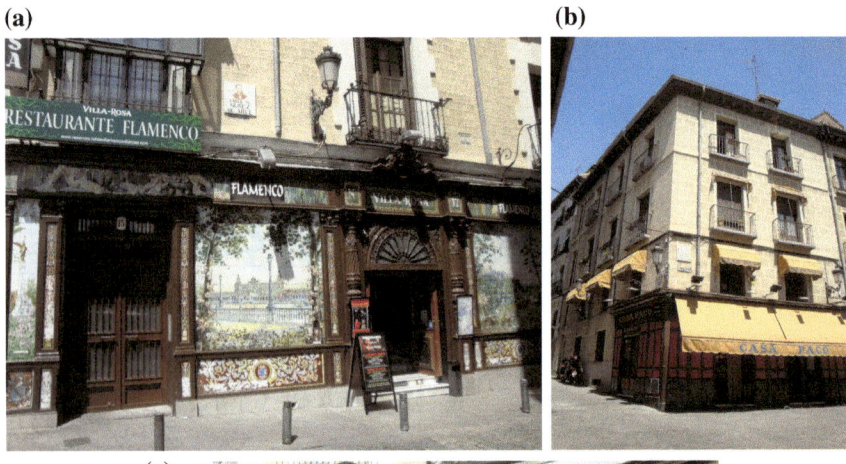

Fig. 6.4 Restaurants offering Spanish regional cuisine in central Madrid (Spain). **a** Established in 1911, the Andalusian restaurant Restaurante Tablao Villa-Rosa at Plaza Santa Ana 15. **b** Established in 1933, the Castilian (*cocina castellana*) restaurant the Casa Paco at Puerta Cerrada 11. **c** The Casa Labra restaurant is famous for *soldaditos de Pavía* (battered cod strips wrapped in red pepper) commemorating the political events in Madrid in 1874 and dissolving of Parliament by hussars under the command of General Pavía. The other example of restaurant offered *cocina castellana* in the centre of Madrid. Source: Anna Kowalczyk 2015

(known as the Church of Our Lady Queen of the Polish Crown), none of which are dining establishments. Second, traditional restaurants in this part of Warsaw form a large cluster around the Old Town Square (Rynek Starego Miasta) and along two streets that cross the Old Town from the south-east to north-west: Świętojańska Street and Freta. These streets, which date from the sixteenth and seventeenth centuries are the main communication axes traversing this part of Warsaw and thousands of tourists pass through them every day.

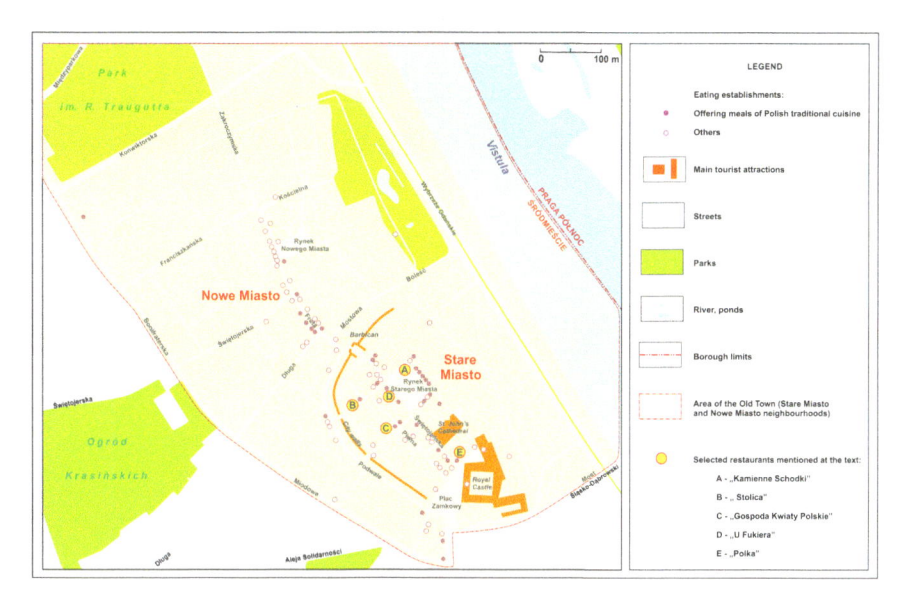

6 Traditional and Regional Cuisine in Urban Space 147

Fig. 6.5 Eating establishments in the Old Town of Warsaw (Poland) in 2018

Several restaurants in the Old Town referred to their tradition of Polish cuisine in the 1970s and 1980s. This is probably because, even then, this part of the city was visited by foreign tourists and represented a kind of 'postcard' for Warsaw and Poland. This thesis is supported by the fact that in 1975, seven out of 25 restaurants and bars offered traditional Polish cuisine (Bystrzanowski and Dutkowski 1975, pp. 73–82). The trend has continued and strengthened in the new political and socio-economic conditions. Later, in 2013, research found that there were 28 restaurants and bars in the Old Town offering traditional Polish cuisine (Derek 2013). They can be divided into three general categories:

- those that offered traditional Polish cuisine before 1989 (during the Communist period);
- those with Polish cuisine that opened later, but are located in outlets that previously served other cuisines;
- those that opened recently and have always specialised in Polish cuisine.

Another way of categorising dining establishments with Polish traditional cuisine is to divide them into independent restaurants and bars and those that are part of a chain. The first group includes only four restaurants. All are high quality (reflected in their sophisticated dishes and high prices) and were already advertised in the 1970s as specialising in traditional Polish dishes. The high prices mean that they are visited by foreign tourists, and used for corporate meetings and family celebrations. All are located in the central square of the Old Town (Rynek Starego Miasta), which is visited by nearly every tourist who comes to Warsaw. One of

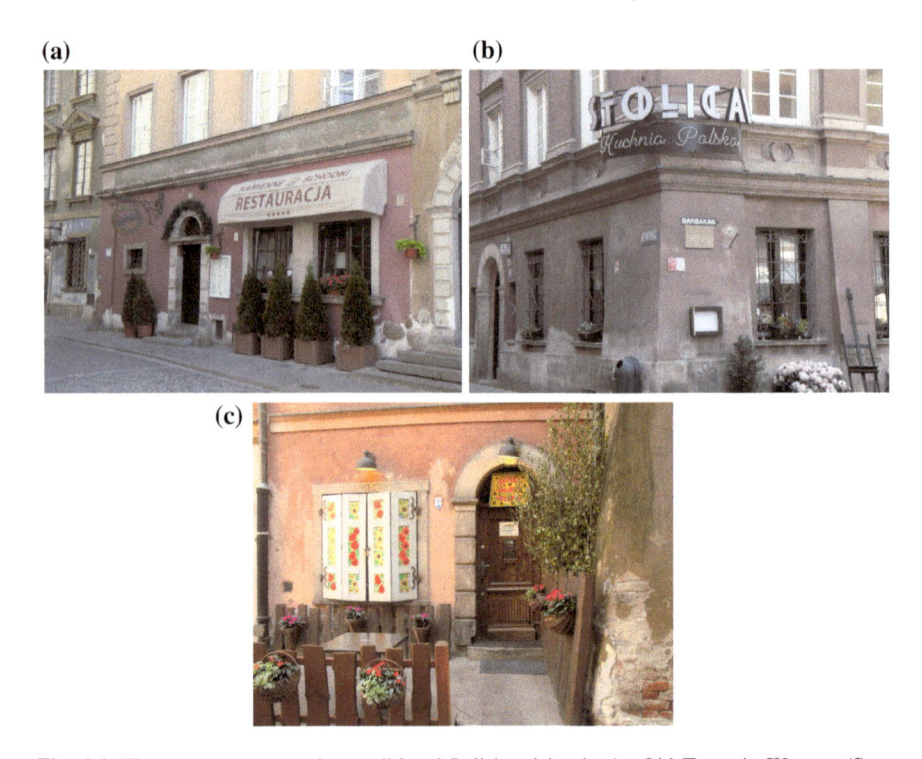

Fig. 6.6 Three restaurants serving traditional Polish cuisine in the Old Town in Warsaw (Stare Miasto) in 2018. Kamienne Schodki, which existed during the Communist period (**a**). Stolica restaurant (former Senator), which changed the name in 2017. Under the name of the restaurant is added Polish cuisine (Kuchnia polska in Polish) (**b**). New established restaurant Gospoda Kwiaty Polskie (**c**)

these restaurants, Kamienne Schodki is shown in Fig. 6.6. The second group consists of restaurants that previously had a similar function, but changed ownership (up until 1989 most were owned by either the state or municipality) and now have a different name. One such example is Stolica (Capital City). This was previously called Senator, and it also used to serve Polish cuisine. Although it is not situated in the Old Town square, it is only about 100 m away. The third category are restaurants and bars that have only opened relatively recently. They are situated in premises that previously served other functions. Most were shops, bookstores or service establishments (e.g. shoemaking, hairdresser, tailor). This type of dining outlet predominates and this is due to social and functional changes in this part of Śródmieście. In recent years, many residents have moved out and rented apartment blocks (many of them part of the Airbnb system) and hostels have been created. These changes caused a drop in demand for existing services and an increase in demand for food and beverage outlets. These new restaurants and bars serve traditional cuisine. An example is the restaurant Gospoda Kwiaty Polskie (Polish Flowers Inn). Many have fascinating inscriptions on their façades, originally

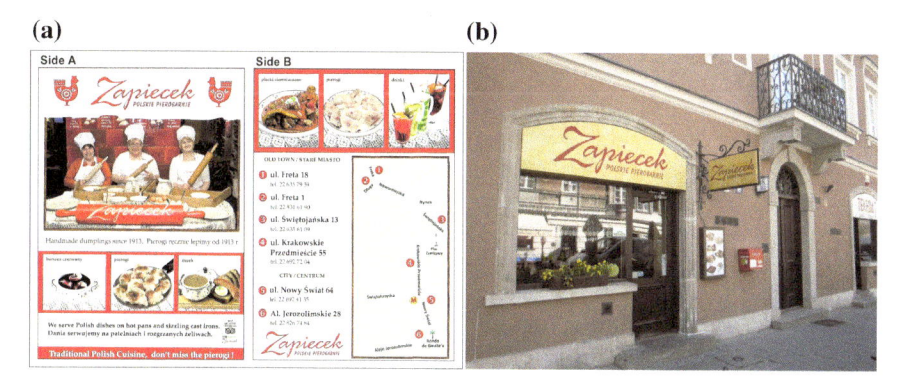

Fig. 6.7 The Zapiecek-Polskie Pierogarnie restaurant chain. Leaflet about restaurants offering traditional Polish cuisine in a part of Warsaw (Poland) visited by domestic and foreign tourists in 2018. The information on the leaflet is somewhat misleading, as it can be understood that restaurants have been operating since 1913. The map is also fraught with a mistake, because in reality the top part of the map is on a scale of about 1:10,000, and the bottom part is roughly from 1:25,000 to 1:30,000 (**a**). The Zapiecek-Polskie Pierogarnie outlet at Freta Street in New Town area of Warsaw (**b**). Source: Leaflet from Andrzej Kowalczyk collection, photo Andrzej Kowalczyk 2018

decorated entrances and interiors referring to Polish folklore. Their dishes are designed to attract tourists and they very often serve dumplings, which are both traditional and fashionable. In order to attract attention, waiters often wear clothes associated with Polish folklore.

Many of the Polish restaurants in the Old Town are independent, but some are part of chains. One, Zapiecek-Polskie Pierogarnie (Fig. 6.7), has restaurants that specialise in dumplings, and the word *pierogarnie* means a place where dumplings are made and served (*pieróg* means dumpling, *pierogi* means dumplings). The chain was created when the first (20 seat) restaurant was opened in January 2003 in the Old Town of Kraków. From the very beginning, it was very popular, serving 10,000 customers every month. By 2005 its popularity had increased, and it was handling 20,000 visitors a month (O nas 2018). The first restaurant in Warsaw was opened in 2005 (in the Old Town, opposite the cathedral). In 2006, a second restaurant followed. This was situated on Nowy Świat Street, one of the most expensive streets in the city. In ensuing years, a further six outlets were opened, three in the Old Town (Lokale 2018). By the spring of 2018 there were eight chain restaurants in Warsaw: four in the Old Town, and two more in the part of the downtown frequently visited by domestic and foreign tourists. Two other restaurants were located in zones that are less popular with tourists.

The Gessler family, which owns and manages three restaurants, is one of the most important players in the gastronomy of the Old Town. Two of their outlets clearly advertise themselves as serving traditional Polish dishes. They belong to Ms. Magda Gessler, who is known in Poland as a leader in the gastronomy sector. At the beginning of 2018 she had seven restaurants. Only two were outside Warsaw (in Łódź and Zakopane) (Restauracje 2018). One, U Fukiera, is one of the oldest and most famous in Warsaw. It is located in the Old Town Square, close to the

(a)   **(b)**

Fig. 6.8 Two restaurants owned by Ms. Magda Gessler (a food celebrity in Poland) in the Old City of Warsaw (Poland). Tourists in front of the Polka restaurant (**a**). Total renovation of the restaurant U Fukiera at the Old Town Square (**b**)

Museum of Warsaw and the most prestigious part of the Old Town. Gessler has owned this restaurant since 1997. In 2012, U Fukiera received its first Michelin recommendation, which has been retained to 2019. Between 2013 and 2017 U Fukiera received two "covers", which indicate comfort and quality of a restaurant (in particular interior decor, table setting, and service quality) (Fig. 6.8).

> The history of U Fukiera dates back to the beginning of the sixteenth century. At that time, the wine merchant, Grzegorz Korab, opened a wine cellar. In the early eighteenth century the Fukier family became the owners of this cellar. By the nineteenth century, it held a collection of the oldest wines in the world. This situation continued until the outbreak of World War II. The Old Town was rebuilt after 90% of it was destroyed during the war. The wine bar reopened in the old cellars, which quickly became a cult place for Warsaw intellectual elites. This period marked the beginning of the transformation of the political system in Poland in 1989. Following the new political situation, Princess Anne, the British Princess Royal, became one of the world-famous restaurant's first guests in 1991. In 1993, the Prime Minister of Poland Ms. Hanna Suchocka met here with the Prime Minister of Spain, Mr. Felipe González. The other guests have included Her Majesty Margrethe II, the Queen of Denmark, Mr. Yitzhak Rabin, the Prime Minister of Israel, Ms. Madeleine Albright (United States Secretary of State), Queen Sophia of Spain, and many actors (including Catherine Deneuve), politicians (e.g. Javier Solana, Henry Kissinger) and other famous people (e.g. the model, actress and singer Naomi Campbell) (Discover Fukier 2018).

The Gesslers' second restaurant in the Old Town is Polka (the word *polka* means dance in Polish, and *Polka* (capitalised) is a woman of Polish nationality).

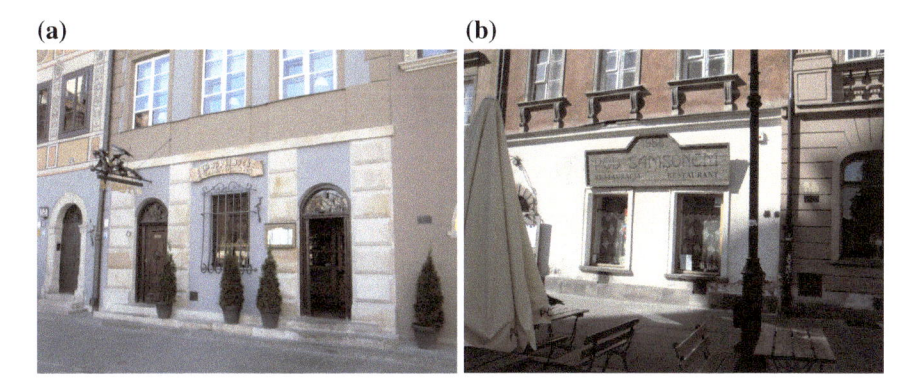

Fig. 6.9 Restaurants in the Old Town in Warsaw (Poland) attract tourists by referring to their historical heritage. The Bazyliszek in the Old Town Square is marked out by its legendary dragon (**a**). The Pod Samsonem restaurant at Freta Street offering dishes of Polish Jews (**b**)

This restaurant is very conveniently situated on Świętojańska Street, which connects the Royal Castle with the Old Town Square. This is the main street in the Old Town and is traversed by over 90% of foreign tourists visiting this part of Warsaw. The restaurant is located at the end of the street, only 50 m from the Royal Castle and 100 m from the cathedral. This means that it is the first restaurant a tourist sees on entering the Old Town. Gessler opened this restaurant in 2008 and traditional Polish cuisine has been on the menu since its opening (Restauracja Polka 2008).

Other places visited by tourists to the Old Town are Bazyliszek and Pod Samsonem (Fig. 6.9). The first is in the Old Town Square and draws attention because of the iron dragon that sits above its entrance. The restaurant has existed since the 1950s and still has the same name, because the dragon, called Bazyliszek, is a Warsaw legend. Pod Samsonem, which opened in 1958, is also noteworthy because it offers Polish Jewish cuisine.

> Some of the restaurants, cafés, pastry shops, ice cream parlours, etc. in the Old Town make explicit reference to their history, whereas those in the other parts of Warsaw do not. For example, the Strzałkowski confectionary shop displays information explaining that it is the successor to Camargo (a state-owned enterprise, established in 1975), while the tiny ice cream shop tells customers that it has been offering ice cream since 1968 (Fig. 6.10a). Another example is Restauracja Zapiecek, which informs guests in several places that it was opened in 1960 (Fig. 6.10b).

The main conclusion from this research in the Old Town area in Warsaw is that the distribution of catering establishments is very diverse. The second conclusion, confirmed by the presence of numerous restaurants offering traditional Polish

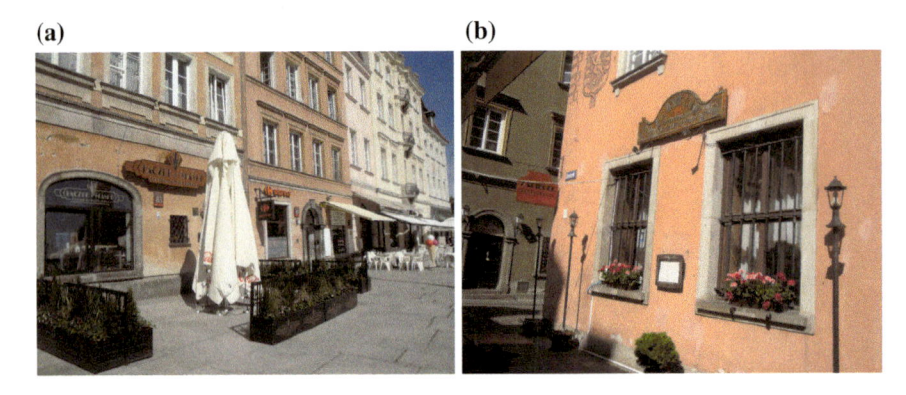

Fig. 6.10 Eating establishments in the Old Town in Warsaw (Poland) that attract customers by referring to their historical heritage in 2018. The signboard with name 'Pączek w maśle. Cukiernia warszawska' is similar to signs at stores in Warsaw before World War II. Further to the right (behind the grocery store chain Carrefour Express) is an ice cream shop with traditions dating back to 1968 (**a**). A similar signboard to the previously described is the inscription above the Restauracja Zapiecek, which has been operating since 1960 (**b**)

cuisine, is that gastronomy in this part of Warsaw is primarily oriented towards foreign tourists. This applies equally to the geographical location of restaurants, their design and decor and menus.

6.4 *Obwarzanki* in the Urban Space of Kraków (Poland)

Kraków's *obwarzanek*, for both residents and tourists, is regarded as a product of the city and one of the world's few fast foods with several hundred years of tradition. It has been the symbol of the city since its founding and stands where it is sold are a permanent fixture in the urban landscape. It is commonly believed that the best obwarzanek come from Kraków, 'The loaded-ranking national symbol, crunchy pastry with a shape similar to the Kraków garland. Derived from the vicinity of the Wawel Castle, a popular delicacy is reigning in the streets of closer and further cities (…) The world triumphs do not change, however, the well-established opinion that the real value of this sprinkled pastry with poppy seed, salt or sesame cake is best known not elsewhere, but there, where the Vistula River goes to Rudawa River, Wilga River or Białucha River'. (Makłowicz and Mancewicz 2001, p. 90).

The history of obwarzanki begins in the thirteenth century. It is mentioned in the founding document of Kraków issued by Duke Bolesław V the Chaste in 1257. This document contains information about Kraków's bakers and their associated privileges to sell their products at the market.[1] It is also mentioned in a resolution by

---

[1] Obwarzanek krakowski (2006).

Kraków City Council on baking bread and remuneration in 1529. On 26 May 1496, King Jan Olbracht gave Kraków bakers the right to exclusively make and sell *obwarzanki*.[2] It was only after 1849 that all bakers in Kraków could bake obwarzanek. The manufacturing process (*obwarzania*) means scouring dough in water. Other names are used to describe this pastry: *precel* (within the boundaries of the former Prussian partition) or *bagel* (Makłowicz and Mancewicz 2001, p. 90). Evidence of the unbroken tradition of producing obwarzanek is provided by a range of other documents, decisions, recipes, statements and photographs as well as the memories of generations of Krakovians. At present, producers and sellers of Kraków's obwarzanek are members of a guild, whose establishments are situated in the city and in neighbouring counties.[3] On 28 November 2006, *obwarzanek krakowski* was included in the list of traditional products from Poland and, on 30 October 2010, the European Union registered it in the Protected Geographical Indications Registry.

Up until the 1950s, Kraków's obwarzanek was sold directly from wicker baskets.[4] Nowadays, about 99% of products are sold on the street, from mobile stands (trolleys). There are about 170–180 of these trolleys, and nearly 150 thousand units are sold every day.[5] The main outlets are Market Square and its immediate vicinity, in the vicinity of the main railway station, and in the vicinity of some hospitals and health clinics (Stuch 2001). In 2016, in the Old Town, within the limits of the Old Town Cultural Park, there were 24 stands. The most popular points are Main Square, at the corner of Main Square and Floriańska Street and Grodzka Street. There are over 100 outdoor stands, governed by the Municipal and Transport Infrastructure Board, and 24 in communal areas (Gdzie w centrum kupimy obwarzanka? 2016). The location of sellers in the Old Town is governed by the regulations of the Cultural Park, whose borders coincide with the central district. These regulations prohibit the advertising of baked goods (Chojnowska 2012). The Department of Administrative Affairs administers tenders for its sale.[6]

Stall locations either have heavy tourist traffic or are public transport nodes that serve both residents and tourists. The main locations lie within the boundaries of the first ring road (Fig. 6.11). Eight stands are located in the immediate vicinity of the

---

[2]Obwarzanek krakowski (2006).

[3]W sprawie okreslenia zasad udzielania zgody na czasowe zajecie nieruchomosci lub przestrzeni bedacych wlasnoscia Gminy Miejskiej Kraków oraz nieruchomosci lub przestrzeni bedacych wlasnoscia Skarbu Panstwa a polozonych w obrebie Gminy Miejskiej (2011).

[4]Obwarzanek krakowski (2006).

[5]Rozporządzenie Rady (WE) nr 510/2006, *Obwarzanek krakowski" nr WE: PL-PGI-005-0674 ChOG (X), ChNP ().*

[6]W sprawie określenia zasad udzielania zgody na czasowe zajęcie nieruchomości lub przestrzeni będących własnością Gminy Miejskiej Kraków oraz nieruchomości lub przestrzeni będących własnością Skarbu Państwa a położonych w obrębie Gminy Miejskiej Kraków, 2011.

http://www.bip.krakow.pl/zarzadzenie/2011/1712/w_sprawie_okreslenia_zasad_udzielania_zgody_na_czasowe_zajecie_nieruchomosci_lub_przestrzeni_bedacych_wlasnoscia_Gminy_Miejskiej_Krakow_oraz_nieruchomosci_lub_p.html. Accessed 13 May 2017.

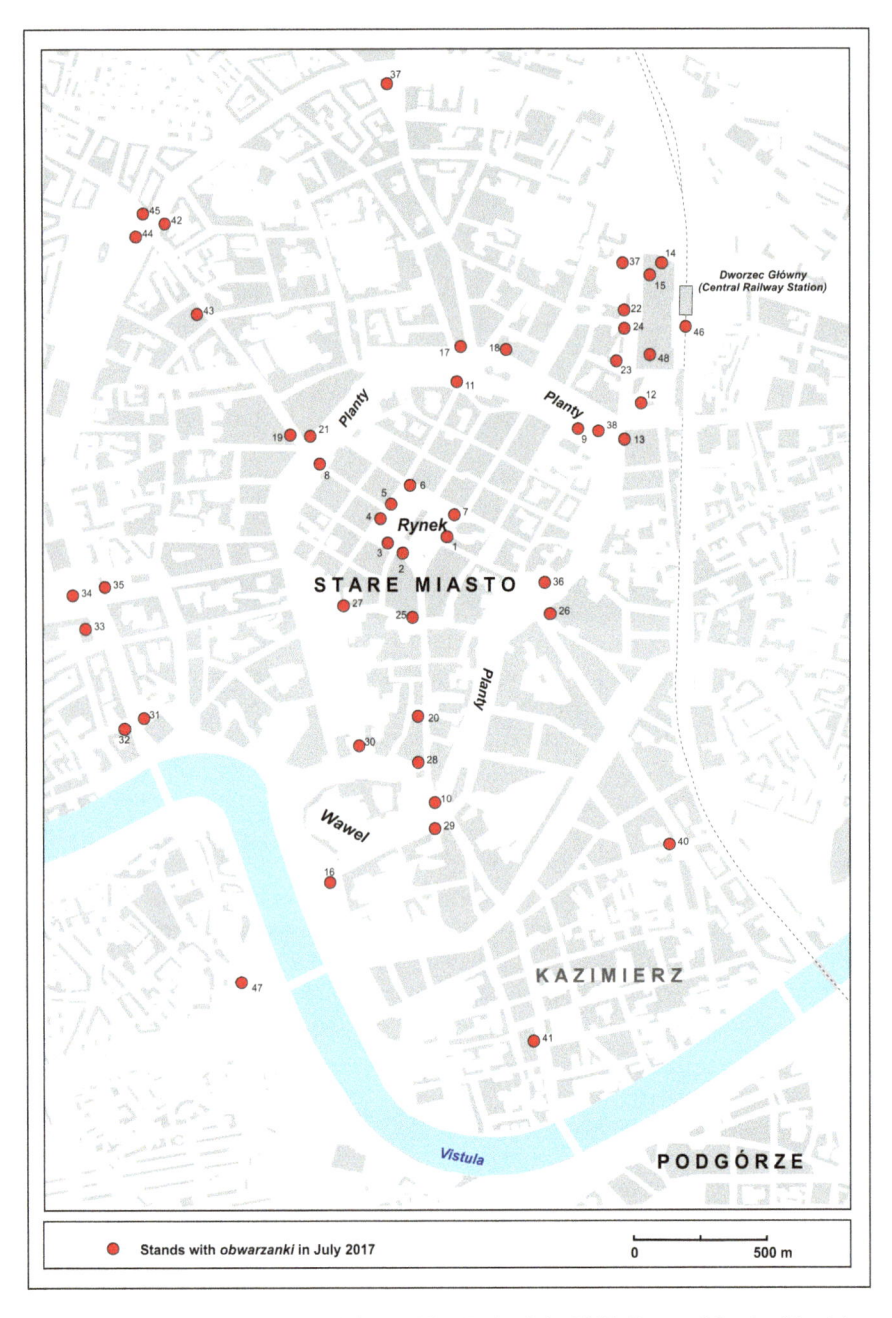

Fig. 6.11 Stands selling obwarzanki in Kraków (Poland) in 2017. Source: Map by Magdalena Kubal-Czerwińska

main railway station: two are in the immediate vicinity of the Galeria Krakowska shopping centre, and six are near the railway station, the surrounding streets and the Galeria Krakowska. These places are very advantageous because small passenger buses that bring people into town for work stop there. There are a further 16 stalls in the green belt surrounding the Old Town. This area, known as the *Planty* is a popular walking spot for tourists and locals. Seven stalls are located in Main Square. Many stalls are located along the streets leading from Main Square to other tourist attractions including Sławkowska Street, Sienna Street, Szewska Street and Grodzka Street, which are part of the Royal Route that leads from Main Square to the Royal Castle on Wawel Hill where there are three more stands, which are used mostly by tourists. Other stands in the city space are usually adjacent to public transport nodes and interchanges that are used by both residents and tourists.

Although the district of Kazimierz is just as popular as the Old Town, there are fewer stands. Here, trolleys have been supplanted by trucks that offer street food. Kazimierz is best known for its casseroles, which are sold in stands located in old Okrąglak in Nowy Square.

6.5 Conclusions

As demonstrated by the example of Madrid, the distribution of restaurants and bars with regional cuisine is very unequal in urban space. Next to the districts with a high concentration of gastronomic establishments offering dishes of one particular region, there are parts of the city where all regions are represented to a similar degree. However, examples of Warsaw and Kraków have shown that traditional cuisine plays an important role in gastronomic services concentrated in historical parts of cities.

To conclude this chapter, we must recognise the close relationship between traditional cuisine, and what is called a 'place' in human geography. This statement applies in particular to dishes specific to a given city, which are not only a part of its cultural heritage, but also attractions for tourists. In other words, they are part of the Lefebvre's (1991) 'lived space', and 'place' defined by theoreticians of human geography (Relph, Tuan, Seamon and others).

References

Brożek M (2007) Zmiany w przestrzennym zróżnicowaniu bazy gastronomicznej w dzielnicy Warszawa-Śródmieście na przestrzeni lat 1994–2003. Wydział Geografii i Studiów Regionalnych, Uniwersytet Warszawski, Warszawa. MA thesis

Bystrzanowski J, Dutkowski K (1975) Wszystko o Warszawie. Informator. Sport i Turystyka, Warszawa. Maps drew by Ostrowski J, Ostrowski W

Chojnowska A (2012) Stoisko z obwarzankami droższe o 2700 procent. Komunikaty.pl, 17.03.2012. http://www.komunikaty.pl/komunikaty/1,80849,11361152,Stoisko_z_obwarzankami_drozsze_o_2700_procent.html. Accessed 22 July 2017

Cupnoodles Museum Welcomed a Five Millionth Visitor. News release, July 2016, Nissin Group. https://cdn.nissin.com/gr-documents/attachments/news_posts/5325/bc0e03961274db25/original/20160715CNM-eng.pdf?1468560129&_ga=2.238174976.1663843244.1527400222-950200569.1527400222. Accessed 25 May 2018

Délice Member Cities (2018) http://www.delice-network.com/cities. Accessed 26 May 2018

Délice, Bylaws (2007) The network of worldwide good food cities. http://www.hel.fi/static/helsinki/paatosasiakirjat/Kh2009/Esityslista10/Liitteet/DELICE_n_saannot.pdf?Action=sd&id={82BE113C-E35F-451A-9365-D13DA1F08879}. Accessed 10 July 2017

Derek M (2013) Kierunki rozwoju usług gastronomicznych w warszawskiej dzielnicy Śródmieście. Prace i Studia Geograficzne 52:85–100

Dias C, Mendes L (2018) Protected Designation of Origin (PDO), Protected Geographical Indication (PGI) and Traditional Speciality Guaranteed (TSG): a bibiliometric analysis. Food Res Int 103:492–508

Discover Fukier (2018). http://ufukiera.pl/discover-fukier/?lang=en. Accessed 28 May 2018

Es war eimal...ein Wiener Konditormeister und der hieß Walter Heindl (2018). https://www.heindl.co.at/geschichte. Accessed 25 May 2018

Flagstaff House Museum of Tea Ware (2018). http://www.lcsd.gov.hk/CE/Museum/Arts/en_US/web/ma/tea-ware.html. Accessed 24 May 2018

Gdzie w centrum kupimy obwarzanka? (2016) "Magiczny Kraków". 05 Sept 2016. http://krakow.pl/aktualnosci/203265,26,komunikat,gdzie_w_centrum_kupimy_obwarzanka_.html. Accessed 13 May 2017

Lefebvre H (1991) The production of space. Basil Blackwell, Inc., Oxford—Cambridge, MA, English translation by D. Nicholson-Smith

Lokale (2018). http://www.zapiecek.eu/lokale.html. Accessed 26 May 2018

Madeira Wine Museum (2018). http://www.visitmadeira.pt/en-gb/explore/detalhe/madeira-wine-museum. Accessed 25 May 2018

Makłowicz R, Mancewicz S (2001) Zjeść Kraków. Przewodnik subiektywny, Znak, Kraków

Obwarzanek krakowski (2006). http://www.minrol.gov.pl/Jakosc-zywnosci/Produkty-regionalne-i-tradycyjne/Lista-produktow-tradycyjnych/woj.-malopolskie/Obwarzanek-krakowski. Accessed 22 July 2017

O nas (2018). http://www.zapiecek.eu/onas.html. Accessed 26 May 2018

Páginas Amarillas/Páginas Blancas (2014) Madrid Capital/Zona Centro, Consulta la guía de restaurantes más completa, 1//11/2013. Connecti S.A.U, Impeso por Ediciones Informatizadas, S.A., Madrid

Pillsbury R (1987) From hamburger alley to Hedgerose Heights: toward a model of restaurant location dynamics. Professional Geographer 39(3):326–344

Restauracje (2018). http://www.magdagessler.pl/restauracje.html. Accessed 25 May 2018

Restauracja Polka (2008) BCMFD Media Communications. http://www.bcmfd.pl/339-restauracja-polka. Accessed 28 May 2018

Rinaldi C (2016) Addressing sustainable development through food and gastronomy: the Gastronomic Cities case. In: Laven D, Skoglund W (eds) Valuing and evaluating creativity for sustainable regional development, VEC Conference, 11–14 Sept 2016. Mid Sweden University, Östersund, Sweden, pp 29–34

Obwarzanek krakowski (2006) Rozporządzenie Rady (WE) nr 510/2006. nr WE: PL-PGI-005-0674 ChOG (X), ChNP (), "Dziennik Urzędowy Unii Europejskiej", C 38/8, 16 Feb 2010. www.infor.pl/download/site/pl/oj/2010/c_038/c_03820100216pl00080012.pdf. Accessed 22 July 2017

San Antonio: A Creative City of Gastronomy (2017) Just About Travel, 07 Nov 2017. http://www.justabouttravel.net/2017/11/07/san-antonio-a-creative-city-of-gastronomy. Accessed 20 Mar 2018

Sirše J (2014) Gastronomic cities. Baseline study, March 2014, European Union-URBACT, Burgos. http://www.urbact.eu/sites/default/files/media/gastronomic_cities_baseline_study_final. pdf. Accessed 12 Sept 2017

Statistics Report (2018). http://www.lcsd.gov.hk/en/aboutlcsd/ppr/statistics/cultural.html. Accessed 19 May 2018

Stuch M (2001) Czy preclami rządzi mafia?, naszemiasto.pl Kraków, 31 Aug 2001. http://krakow. naszemiasto.pl/archiwum/czy-preclami-rzadzi-mafia,178105,art,t,id,tm.html. Accessed 10 July 2017

W sprawie określenia zasad udzielania zgody na czasowe zajęcie nieruchomości lub przestrzeni będących własnością Gminy Miejskiej Kraków oraz nieruchomości lub przestrzeni będących własnością Skarbu Państwa a położonych w obrębie Gminy Miejskiej Kraków (2011). http:// www.bip.krakow.pl/zarzadzenie/2011/1712/w_sprawie_okreslenia_zasad_udzielania_zgody_ na_czasowe_zajecie_nieruchomosci_lub_przestrzeni_bedacych_wlasnoscia_Gminy_Miejskiej_ Krakow_oraz_nieruchomosci_lub_p.html. Accessed 13 May 2017

Chapter 7
Changes in the Distribution of Gastronomic Services in the City Centre

Marta Derek, Andrzej Kowalczyk, Konstantin A. Kholodilin,
Leonid Limonov, Magdalena Kubal-Czerwińska and Dana Fialová

Abstract This chapter underlines the key role of a city centre in urban space gastronomy. It offers a four-step perspective, ranging from urban to local. First, the example of Saint Petersburg (Russia) shows that gastronomy reflects the major phases of urban growth. Here, eating establishments are used as a proxy for the city centre. Second, the example of Warsaw's Śródmieście district in Poland indicates the constant growth in catering services in this central borough since 1994. Using

---

M. Derek (✉) · A. Kowalczyk
Department of Tourism Geography and Recreation, Faculty of Geography and Regional Studies, University of Warsaw, ul. Krakowskie Przedmieście 30, 00-927 Warsaw, Poland
e-mail: m.derek@uw.edu.pl

A. Kowalczyk
e-mail: akowalczyk@uw.edu.pl

K. A. Kholodilin
Deutsches Institut für Wirtschaftsforschung (DIW Berlin), Mohrenstr. 58, 10117 Berlin, Germany
e-mail: kkholodilin@diw.de

K. A. Kholodilin · L. Limonov
National Research University "Higher School of Economics" in St. Petersburg, Kantemirovskaya ul., 3, 194100 St. Petersburg, Russia
e-mail: limonov@leontief.ru

L. Limonov
ANO International Centre for Social and Economic Research "Leontief Centre", 25, the 7th Krasnoarmeyskaya, 190005 St Petersburg, Russia

M. Kubal-Czerwińska
Department of Tourism and Health Resort Management, Institute of Geography and Spatial Management in Kraków, Jagiellonian University, ul. Gronostajowa 7, 30-387 Kraków, Poland
e-mail: magdalena.kubal@uj.edu.pl

D. Fialová
Faculty of Science, Department of Social Geography and Regional Development, The Geography of Leisure Research Centre, Charles University, Albertov 6, CZ 128 43 Prague 2, Czechia
e-mail: dana.fialova@natur.cuni.cz

© Springer Nature Switzerland AG 2020
A. Kowalczyk and M. Derek (eds.), *Gastronomy and Urban Space*,
The Urban Book Series, https://doi.org/10.1007/978-3-030-34492-4_7

density analysis, it shows gastronomy hotspots in the centre of the city. Next, the case of Kraków (Poland) focuses on the centre of a historical tourist city, where there has been both quantitative growth in the number of eating establishments and a change in their distribution. The last examples offer a local perspective, specifically they concern the district of *Żoliborz* in Warsaw, Poland, and the neighbourhood of *Podskalí* in Prague, the Czech Republic, which are near the city centre.

Keywords City centre · Density analysis · Eating establishments · Saint Petersburg · Warsaw · Kraków

7.1 Introduction

Traditionally, every city has a central zone where eating establishments predominate—helped by a large number of customers and ready access. Nowadays, however, they do not only serve to feed people, but they also constitute an important group of urban amenities. Charismatic restaurants, the type of cuisine, its social function, etc. make such outlets a place to be seen and a place to meet others. Moreover, a concentration of these establishments makes a place vibrant and liveable and contributes to a city's image. Therefore, a city centre—understood as both a central business district and a historical centre—constitutes a natural cluster for eating establishments.

A study of the role of the city centre in the distribution of restaurants was carried out in Hamilton (New Zealand) by Prayag, Landré and Ryan (2012). The authors used the Yellow Pages telephone directory to identify addresses of outlets, geocoded them, and then performed spatial analyses of their distribution. The results showed that between 1996 and 2008 around 74 to 84% of the restaurants located in the city were concentrated in the CBD. Concerning the core of the CBD the figures were lower and fluctuated from 39% (minimum) to 51% (maximum) (Prayag, Landré and Ryan 2012, p. 445). One of the reasons for the lower concentration of restaurants in the CBD was the emergence of a residential area on the outskirts of the city, and the development of restaurants in that area. The spatial distribution of restaurants revealed sensitivities to the changing population and land usage patterns in the city. Despite the emergence of new residential nodes and thus increased demand for restaurants, most restaurants were found in the city centre throughout the research period (Prayag, Landré and Ryan 2012).

Data on registered economic activities also show that gastronomy concentrates in the central districts of cities. In Madrid (Spain), for example, more than a quarter of all business units registered in the Centro district belongs to this sector (Table 7.1). In Prague (Czech Republic), 9% of all the eateries are registered in the central district of Praha 1—an area constituting just 1% of the total area of the city with 2.3% of its population. This means that there are 63 gastronomic enterprises per 1000 inhabitants of the central district, compared to an average of 15 for Prague as a whole (Table 7.2).

7 Changes in the Distribution of Gastronomic Services … 161

Table 7.1 Business sectors in districts of Madrid on 1 January 2017

District	Percent of total registered business units				
	Gastronomy	Manufacturing	Trade	Finance	Culture
Arganuzela	16.7	1.5	35.3	3.4	3.0
Barajas	23.1	0.9	32.0	2.7	2.3
Carabanchel	14.3	4.2	41.4	2.5	2.2
Centro	26.5	1.6	43.4	2.1	2.8
Chamartín	17.7	1.2	34.8	4.4	2.7
Chamberí	19.2	1.4	39.1	3.5	2.2
Ciudad Lineal	14.5	3.0	39.2	2.9	2.5
Fuencarral-El Pardo	17.5	2.1	34.5	3.6	3.2
Hortaleza	17.7	2.4	33.1	3.1	3.3
Latina	15.1	2.9	41.3	2.6	2.6
Moncloa-Aravaca	20.1	1.8	32.4	2.8	3.5
Moratalaz	15.2	1.2	37.9	2.9	2.8
Puente de Vallecas	15.6	3.9	40.6	2.0	2.2
Retiro	17.2	1.0	40.0	4.1	2.4
Salamanca	16.9	1.2	43.3	3.7	2.3
San Blas	16.9	3.9	37.1	3.1	2.5
Tetuán	16.4	3.9	37.7	3.7	2.5
Usera	16.2	4.5	39.5	2.2	2.2
Vicálvaro	16.0	5.6	37.5	2.3	2.9
Villa de Vallecas	11.4	7.8	46.7	2.3	2.1
Villaverde	12.8	6.3	40.2	2.0	2.0
MADRID	17.4	2.8	39.4	2.9	2.5

Source: *Censo de Locales y Actividades*, Madrid 2018

Table 7.2 Business sectors in districts of Prague on 31 December 2016

District	Gastronomy	Population	Area (hectare)	Registered activity in gastronomy per	
				1000 persons	10 ha
Praha 1	1855	29,587	554	63	33.5
Praha 2	1418	49,335	419	29	33.8
Praha 3	1679	73,095	648	23	25.9
Praha 4	2010	128,301	2420	16	8.3
Praha-Kunratice	100	9599	810	10	1.2
Praha 5	1415	84,165	2750	17	5.2
Praha-Slivenec	49	3534	759	14	0.7
Praha 6	1298	102,858	4152	13	3.1
Praha-Lysolaje	15	1433	248	11	0.6
Praha-Nebušice	21	3276	368	6	0.6

(continued)

Table 7.2 (continued)

District	Gastronomy	Population	Area (hectare)	Registered activity in gastronomy per 1000 persons	10 ha
Praha-Přední Kopanina	4	692	327	6	0.1
Praha-Suchdol	56	7099	514	8	1.1
Praha 7	731	43,362	709	17	10.3
Praha-Troja	22	1292	337	17	0.7
Praha 8	1428	104,224	2179	14	6.6
Praha-Březiněves	22	1539	338	14	0.7
Praha-Ďáblice	47	3612	738	13	0.6
Praha-Dolní Chabry	65	4383	499	15	1.3
Praha 9	717	57,048	1331	13	5.4
Praha 10	1582	109,336	1861	15	8.5
Praha 11	853	77,522	979	11	8.7
Praha-Křeslice	7	1034	344	7	0.2
Praha-Šeberov	53	3124	500	17	1.1
Praha-Újezd	29	3064	370	10	0.8
Praha 12	676	55,522	523	12	12.9
Praha-Libuš	175	10,235	2333	17	0.8
Praha 13	511	61,945	1320	8	3.9
Praha-Řeporyje	41	4493	990	9	0.4
Praha 14	560	46,577	1353	12	4.1
Praha-Dolní Počernice	40	2424	576	17	0.7
Praha 15	525	33,286	1025	16	5.1
Praha-Dolní Měcholupy	43	2802	466	15	0.9
Praha-Dubeč	43	3710	860	12	0.5
Praha-Petrovice	63	5997	179	11	3.5
Praha-Štěrboholy	34	2203	297	15	1.1
Praha 16	92	8401	930	11	1.0
Praha-Lipence	33	2692	825	12	0.4
Praha-Lochkov	5	765	272	7	0.2
Praha-Velká Chuchle	29	2423	603	12	0.5
Praha-Zbraslav	131	10,010	985	13	1.3
Praha 17	247	24,485	325	10	7.6
Praha-Zličín	70	6566	717	11	1.0
Praha 18	189	19,291	561	10	3.4
Praha-Čakovice	112	10,781	1018	10	1.1
Praha 19	82	7027	600	12	1.4
Praha-Satalice	30	2526	380	12	0.8
Praha-Vinoř	31	4251	600	7	0.5
Praha 20	169	15,304	1694	11	0.1

(continued)

Table 7.2 (continued)

District	Gastronomy	Population	Area (hectare)	Registered activity in gastronomy per 1000 persons	10 ha
Praha 21	140	10,807	1015	13	1.4
Praha-Běchovice	35	2623	683	13	0.5
Praha-Klánovice	42	3438	590	12	0.7
Praha-Koloděje	16	1500	376	11	0.4
Praha 22	139	10,882	1562	13	0.9
Praha-Benice	10	668	277	15	0.4
Praha-Kolovraty	30	3686	650	8	0.5
Praha-Královice	6	371	496	16	0.1
Praha-Nedvězí	3	303	381	10	0.1
PRAHA	19,823	1,280,508	49,616	15	4.0

7.2 Eating Establishments as a Proxy for the City Centre. The Example of Saint Petersburg (Russia)

Here we take the example of Saint Petersburg (Russia) to verify whether the centre of a city can be identified by studying the distribution of eating establishments. Saint Petersburg is a very good example, as there have been several important turning points in its development, which may have influenced a shift in the city centre during the three hundred years of its history. Two major turning points were the October Revolution of 1917, which ended the Imperial Era, and 1991 when a centrally planned economy was replaced by a market economy. These three main periods can be observed by studying the density of eating establishments: pre-1917 the number of eateries per 100,000 people was at its highest level (73–78 outlets). One reason was the numerical dominance of men in the population, due to the concentration of civil servants and military personnel in the then capital of the Russian Empire, which resulted in a high demand for food services outside of the home.

After the October Revolution, the number of establishments fell significantly (despite a slight rebound in 1935), reaching 14 outlets per 100,000 inhabitants in 1973. This may be due to several reasons: difficult post-war conditions; the reintroduction of a system of food stamps in 1928–1935 and then again in 1941–1947; low-quality food and service; the destruction of the elites, who could afford to eat out; and changing lifestyles, influenced by a new political and social system. In the Soviet period, people usually ate lunch at the workplace canteen; these places were not registered as separate establishments and consequently not reflected in statistics. Visiting a restaurant for dinner was a luxury, even for relatively highly ranked people and families could only afford to visit a restaurant several times per year, for special occasions (birthdays, weddings, etc.). Later, the number of eating

Table 7.3 Eating establishments in Saint Petersburg 1894–2017

Year	Coefficient of spatial variation	Number of establishments	Population (in thousands)	Establishments per 100,000 persons
1894	1.87	836	1098[a]	76
1905	1.43	1283	1635	78
1915	1.23	1784	2315	77
1917	1.30	1671	2300	73
1923	1.98	506	1093	46
1935	1.25	970	2716	36
1973	1.47	595	4220	14
1982	1.18	1155	4711	25
2005	1.22	2050	4581	45
2017	1.83	2992	5226[b]	57

Sources: Kostantin A. Kholodilin, and Leonid E. Limonov based on address directories, Petrostat, and own calculations
Legend: [a] Due to a lack of data, the population in 1895 is used; [b] Population for 2015 is used

establishments started to grow rapidly. A huge change occurred between 1982 and 2005, when the number of eateries almost doubled due to *Perestroika* (launched in 1985), and between 2005 and 2017, when 942 new establishments were opened in 12 years (an increase of 31%; Table 7.3). Nevertheless, per capita numbers continue to remain below the pre-1917 level.

Not only did the number of eateries in Saint Petersburg change, so did their location. The spatial concentration of eating establishments was computed as a coefficient of variation for the distribution of outlets within the city's boundaries (Table 7.3). The concentration was highest in 1894, meaning that the vast majority of establishments were clustered in the centre. Later, they developed on the outskirts of the centre. By 1923, the spatial concentration had increased significantly. This can be explained by the large reduction in the population, the displaced workers from the periphery to its central districts, and the fact that then-Petrograd had become a city at the border with the new states of Finland, Estonia and Latvia (which had broken away from the former Russian Empire). Therefore, many establishments located on roads leading to these formerly Russian provinces, now independent states, closed (Fig. 7.1). After this time, it began to fall. In 1935, it reached a minimum, highlighting a significant decentralisation of public catering, in the form of canteens at factories, schools and offices. At this time, factories in the city were located on the fringe of the historical centre making up a 'grey stripe' that separated the CBD from residential areas, and eating establishments were less centralised. Beginning in 1973, the coefficient of variation began to increase, with a major change between 2005 and 2017. By 2017 it had regained its 1894 level (Table 7.3) indicating that in the past 12 years, a huge number of eateries have been opened in the centre. These changes were a consequence of the end of the Soviet system of decentralised job distribution, a surge in the services sector, a huge rise in privately owned real estate, and a rapid increase in the motorisation of the population.

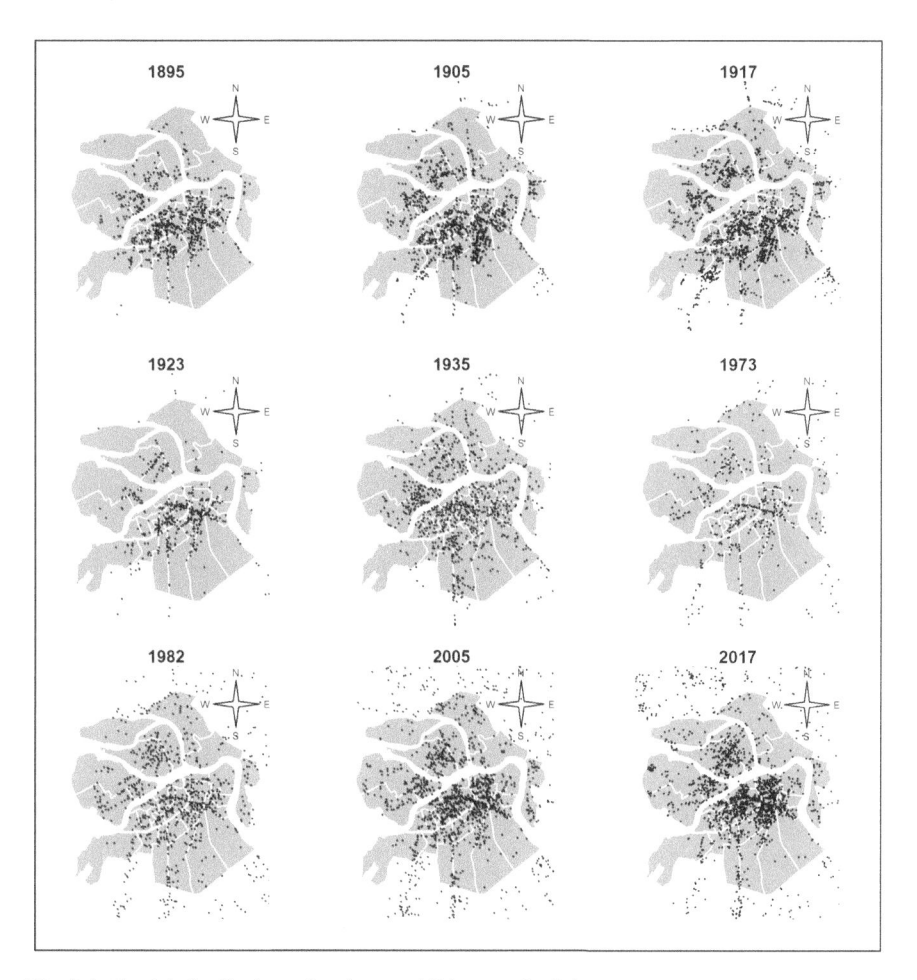

Fig. 7.1 Spatial distribution of eating establishments in Saint Petersburg 1894–2017. Source: Kostantin A. Kholodilin and Leonid E. Limonov 2018

Based on data showing the exact localisation of eating establishments through the years, a 2-D kernel density estimation was used in order to identify the central point. This was compared to the statistically measured city centre estimated using population and employment density. The spatial analysis revealed that eating establishments may be a very useful proxy to identify the centre of a city.

As Fig. 7.2 shows, the results of the three estimates place the city centre in a very similar location. The maximum distance between the centre indicated by eating establishments and population is 0.6 km, compared to 1.3 km for eating establishments and employment. In most years, the centre is around Sennaya Square, which used to be an important marketplace; this is slightly east of employment and population centres, on the opposite (left) bank of the Fontanka river. As both Sennaya, with its famous market, and significant parts of Sadovaya

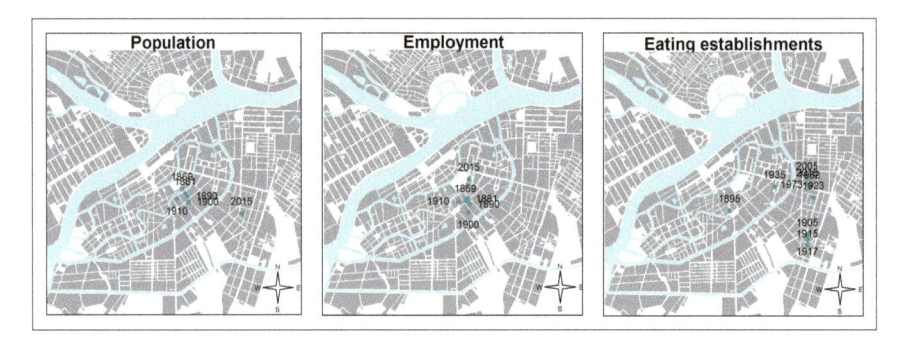

Fig. 7.2 Shift in the city centre of Saint Petersburg, 1869–2017. Source: Kostantin A. Kholodilin and Leonid E. Limonov 2018

Street, Ekaterininsky Canal Embankment and several other streets between Nevsky Prospect and Sennaya were occupied by trade and retail businesses (Gostinyj Gvor, Apraksin Dvor, Perinnye Riady, Mebel'nye Riady, etc.), the density of the daytime population was much higher than in other parts of the city. It should be noted, however, that the centre of the city estimated by population also shifted to the left bank of the river in 2015.

7.3 Gastronomy Hotspots in the Central Borough. The Example of Warsaw (Poland)

Like Saint Petersburg, in Warsaw, gastronomy is a service that is concentrated in the centre of the city. As many as 19% of all gastronomic enterprises are registered in the central borough (Śródmieście). This gives a density of 15 entities per 1000 inhabitants—three times more than the average for the city. Śródmieście is one of 18 boroughs of Warsaw, comprising 16 km^2 (3% of the total area) and 117,005 inhabitants (6.6% of the total population of Warsaw). It is home to a railway station, metro, the seat of the Polish government and the main office of the Polish Parliament, universities, as well as a concentration of offices, ministries, embassies, theatres and museums. It is also the main tourist centre. Fieldwork carried out in the autumn of 2017 in the district found a total number of 1,172 eating establishments, indicating that catering services have constantly grown in number since 1994. Average annual growth is, however, decreasing (Fig. 7.3).

After more than 50 years of communism, when the vast majority of Poles ate at home, eating establishments in the urban space of central Warsaw boomed in the 1990s. Between 1994 and 2003 the number of catering outlets grew by 68%, which translates into an annual average rate of 7.6% (Table 7.4). As Fig. 7.3 highlights, although the number of outlets is growing, growth is slowing. That could indicate that the market in central Warsaw is reaching a saturation point. After a huge

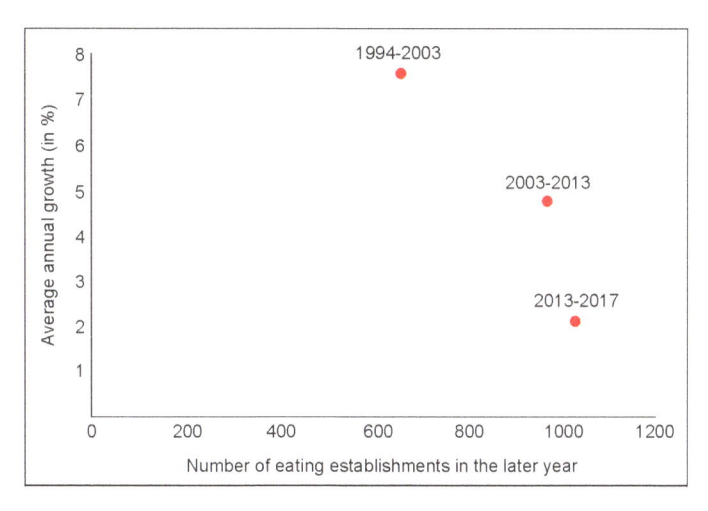

Fig. 7.3 Growth in the number of eating establishments in Warsaw's central borough of Śródmieście between 1994 and 2017. Source: Rokicka-Donica (1995), Kaczorek (2002), Derek (2013) and Marta Derek's own research

proliferation in gastronomy at the end of the 1990s and beginning of the twenty-first century, in recent years growth has fallen to an annual average rate of 2.1%.

This trend differs, however, if we compare different types of catering outlets. Between 2013 and 2017, the number of restaurants in Śródmieście grew three times faster than the average for all eating establishments (4.6% per year on average). In 2017, up to 40% of all eateries in central Warsaw were restaurants. This may be explained by the gradual enrichment of Polish society and, consequently, the growing trend of eating out. Other types of eateries whose numbers grew significantly between 2013 and 2017 are takeaways (understood here as all outlets that do not provide meals on the premises), pastry shops, pizzerias and bars.

On the other hand, some outlets decreased in number between 2013 and 2017. The greatest fall was seen for pubs, bars, wine bars and music clubs. These places of entertainment focus on serving beverages rather than food. Between 1994 and 2013, they made up 10–12% of the total number of eating establishments; this had dropped to 8.2% in 2017. The other group that saw a major fall was cafés and coffee bars. This fall, however, was preceded by a huge increase between 2003 and 2013, when the number of outlets specialised in serving coffee almost doubled. Paradoxically, this may be a result of the economic crisis. Adam Ringer, president of a very popular chain of coffee bars in Warsaw argues that cafés are not only less fragile in a crisis than many other establishments, but they may also substitute for them. 'The crisis has caused new customers to appear in coffee bars and cafés. Those who used to go for business lunches, today meet with customers in a café with coffee and a sandwich. They save on expenses. This business is resistant to the crisis compared to many others' (Naszkowska 2013). This could explain the huge increase in cafés between 2003 and 2013 (during an economic crisis) followed by

Table 7.4 Changes in number and type of eating establishments in Warsaw's borough of Śródmieście 1994–2017

Type of catering facility	1994 Number	1994 Share (%)	2003 Number	2003 Share (%)	2003/1994 growth (1994 = 100)	2013 Number	2013 Share (%)	2013/2003 growth (2003 = 100)	2017 Number	2017 Share (%)	2017/2013 growth (2013 = 100)
Restaurants	114	29.2	210	32.0	184	346	35.7	165	410	39.8	118
Bars	93	23.8	222	33.8	239	266	27.5	120	313	30.4	118
Cafés and coffee bars	109	27.9	107	16.3	98	202	20.8	189	179	17.4	89
Pubs, drink bars, wine bars and music clubs	40	10.2	73	11.1	183	117	12.1	160	84	8.2	72
Pizzerias	25	6.4	23	3.5	92	17	1.8	74	21	2.0	124
Milk bars	10	2.6	14	2.1	140	14	1.4	100	16	1.6	114
Tea houses	0	0.0	8	1.2		7	0.7	88	6	0.6	86
Wine bars	4	1.0	6	0.9	150	6	0.6	100	7	0.7	117
Takeaways	.		.		.	54		.	69		128
Pastry shops	.		.		.	58		.	74		128
TOTAL	391	100.0	657	100.0[a]	168	1081	100.0[a]	147	1172	100.0[a]	108

Source: Adapted from Rokicka-Donica (1995), Kaczorek (2003), Derek (2013) and Marta Derek's own research
Legend: [a]Excluding pastry shops and food outlets due to the lack of data in 1994 and 2003

their decrease after 2013, especially as the number of restaurants increased significantly at the same time (Table 7.4).

The main eating establishment zones were identified using the Kernel density function. A radius of 100 m was used and the quantile method of division into classes was applied. This found that the distribution of eating establishments in the central district of Warsaw is uneven. As Fig. 7.4 shows, there are several main areas:

- The core of the central district. This compact area, populated with offices, shops, and cultural and entertainment venues, holds the biggest cluster of eateries in the whole district. The Centrum (centre) metro station is nearby, and pedestrianised shopping street Chmielna crosses it. It corresponds to types B and D described in Part 1 of this book (a commercial centre; theatre district and entertainment zone), although there are also many offices and hotels.
- The old town. This is a typical historical city zone (inner city type A, described in Part 1). It is listed as a UNESCO World Heritage Site as 'a unique European experience [which] contributed to the verification of conservation doctrines and practices' (Derek 2018). The agglomeration of eating establishments in this part of the city is mainly due to appeal for tourists: it is the tourist precinct of Warsaw, with a high concentration of leisure activities. Fieldwork carried out in April 2016 showed that 38% of all listed services in the core area of the Old Town were eating establishments (Derek 2018).
- The northern part of the Royal Trail (Krakowskie Przedmieście Street), which is the main tourist route in Warsaw, runs south from the Old Town. Although this area is part of the historical city, unlike the Old Town, there are more residents, students and other visitors.
- Powiśle is a former industrial district that was transformed into a residential area in the twentieth century. After the transition to a market economy in 1990, these attractive riverside areas have begun to take their rightful place in the structure of the city. Major new investments have been made, including the construction of a new, modern building for the University of Warsaw Library (1999), building a road tunnel under the river (2002), the Copernicus Science Centre (2010), and redeveloping the waterfront as an attractive, public space (the first part of the investment was opened in 2015). New offices and apartments soon followed. As these changes to the urban space unfolded, gastronomy was one of the drivers that attracted people. The development of this district reflects several types of inner city eating establishments defined in the first part of this book: mostly type K (riverside or waterfront zone), but also H (post-industrial revitalised area zone), J (public park zone) and L (recreation zone).

The inner city also includes a central railway station zone (type I). In Warsaw, this area is enlarged by the Arkadia shopping mall, the biggest in Warsaw.

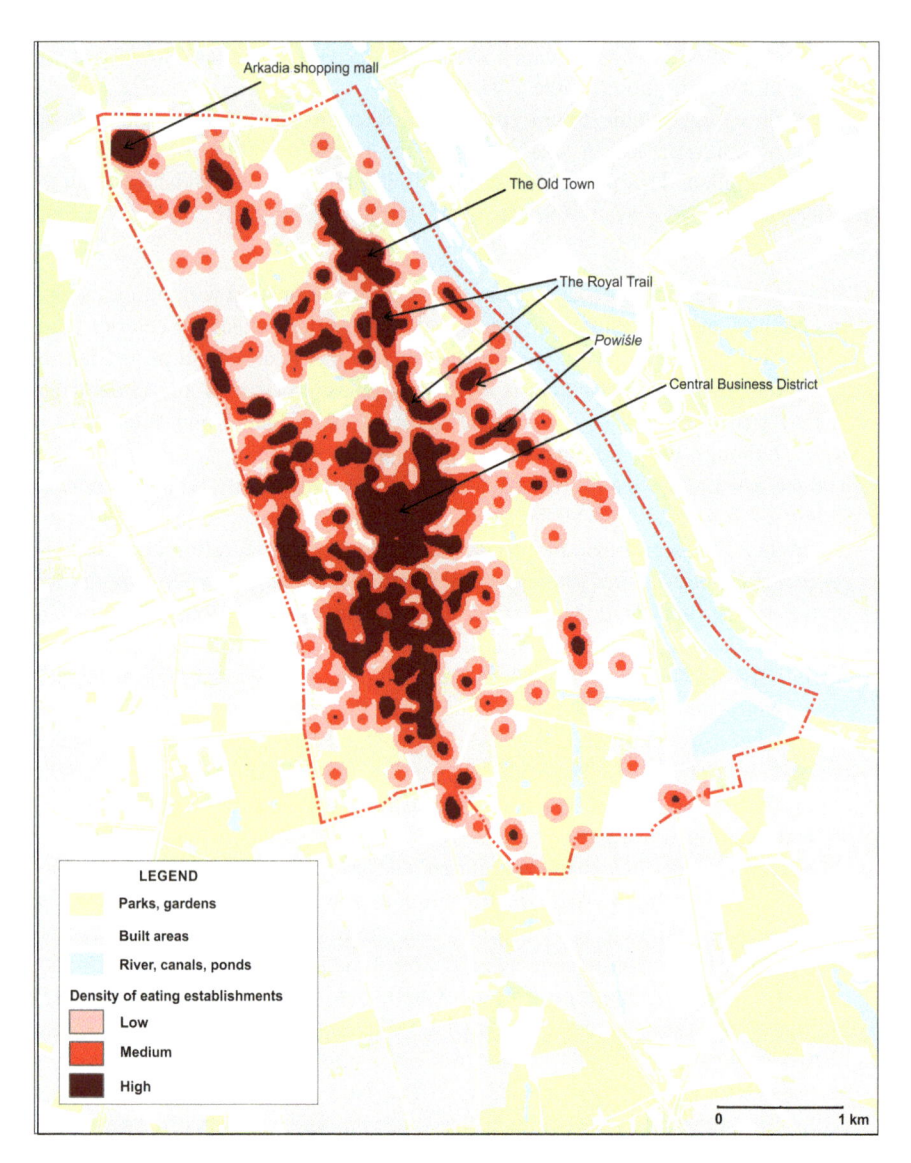

Fig. 7.4 Concentrations of eating establishments in Warsaw's central borough of Śródmieście in 2017. Source: Marta Derek's research

7.4 Gastronomy in the Centre of a Historical Tourist City. The Example of Kraków (Poland)

Although eating establishments play an important role in the tourism experience (along with other tourism services), they rarely aid in the definition and delimitation of areas of tourist activity (Ashworth and Tunbridge 2000). In the centre of a historical tourist city, however, the distribution of catering services may highlight concentrations of tourism activity. Kraków, a former capital of Poland, is a good example. Its historic centre was listed as a monument on the UNESCO World Heritage List in 1978 as one of the twelve first monuments in the world. It includes a thirteenth-century merchants' town with Europe's largest market square; numerous historical houses, palaces and churches; the remnants of fourteenth-century fortifications; the medieval Jewish city of Kazimierz; the Jagiellonian University (the oldest in Poland, one of the oldest in Europe); and the Gothic cathedral where the kings of Poland were buried. It is regarded as one of the most attractive of all Central European cities and it has used these strengths to its advantage (Ashworth and Tunbridge 2000).

The historical centre of Kraków (Dzielnica I Stare Miasto) covers 5.6 km^2 (1.7% of the total area) with 33,359 inhabitants (4.7% of Kraków's population). As many as 953 catering establishments were identified during fieldwork carried out in 2017. Among them, bars were the largest group (317) followed by restaurants (281), which in total accounted for almost 63% of the gastronomic base. Cafés, pastry shops and so-called takeaways were less numerous (140 and 115 premises, respectively). There were also 24 pizzerias, 67 entertainment venues and seven independent gastronomic gardens that only operated in the summer season in places frequented by tourists and residents (Table 7.5).

Quantitative growth in the catering base during the periods 2008–2012 and 2012–2017 was about 30%. All types of catering facilities were affected, with the exception of entertainment venues, whose number increased between 2008 and 2012 but, in 2017, decreased by eight compared to 2012 (Table 7.5). Between 2008 and 2012 the greatest growth was seen in takeaways, cafés and pastry shops and pizzerias, while between 2012 and 2017 the number of takeaways (whose number almost doubled) and bars increased most.

Regarding the Old Town, there were 89 eating establishments in 1990. In 1999 this number had reached 218, an annual growth rate of 16%. This rate has never been matched since. At this time, gastronomy was the sector that developed fastest (Górka 2004). By 2008, outlets located in this part of the city represented up to 23% of all eating establishments in the whole of Kraków (Faracik et al. 2008). Since then, expansion has continued at a rate of about 5% per year (Figs. 7.5 and 7.6), reaching 572 outlets in 2017.

Table 7.5 Change in the structure of eating establishments in the historical centre of Kraków 2008–2017

Type of catering facility	2008 Number	2008 Share (%)	2012 Number	2012 Share (%)	2012/2008 growth (2008 = 100)	2017 Number	2017 Share (%)	2017/2012 growth (2012 = 100)
Restaurants	186	32.7	235	32.1	126	281	29.5	120
Bars	189	33.3	224	30.6	119	317	33.3	142
Cafés and pastry shops	75	13.2	110	15.0	147	140	14.7	127
Entertainment venues	56	9.9	75	10.2	134	67	7.0	89
Takeaways	39	6.9	60	8.2	154	117	12.3	195
Pizzerias	19	3.3	27	3.7	142	24	2.5	89
Open-air food terraces	4	0.7	1	0.1	25	7	0.7	700
TOTAL	568	100	732	100	129	953	100	130

Source: Adapted from Piziak (2011), Piziak and Zając (2012) and Magdalena Kubal-Czerwińska's own research

7 Changes in the Distribution of Gastronomic Services ...

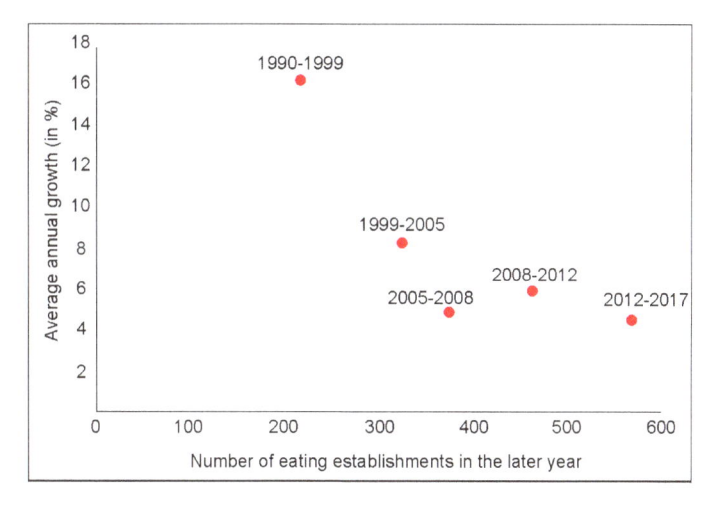

Fig. 7.5 Growth in eating establishments in the Old Town of Kraków between 1990 and 2017. Source: Magdalena Kubal-Czerwińska's own research; Brzozowska and Maciuszek (1996), Wójcik (2007), Piziak (2011), Piziak and Zając (2012)

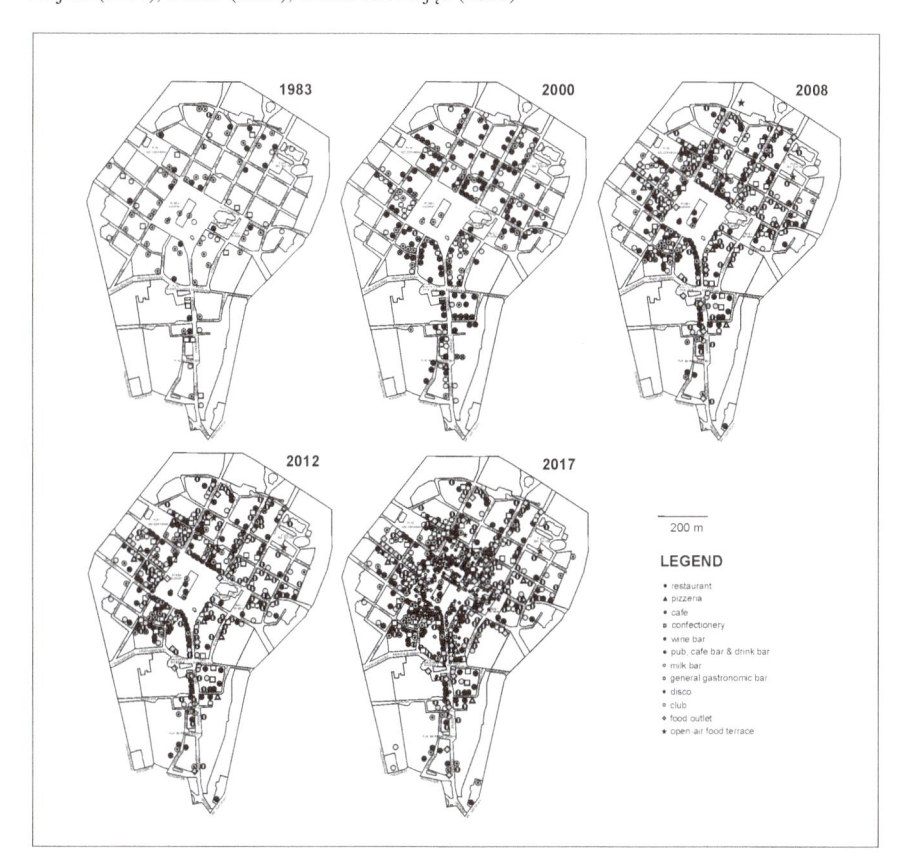

Fig. 7.6 Distribution of gastronomic facilities in the Old Town of Kraków in 1983, 2000, 2008, 2012 and 2017. Source: Piziak (2011), Magdalena Kubal-Czerwińska's own research

7.5 Gastronomy Near the Centre. The Example of Żoliborz, Warsaw (Poland)

The Żoliborz[1] district is located to the north of central Warsaw. A short distance from the downtown, its population density and land use make it part of the inner city. At the turn of the eighteenth and nineteenth centuries these areas were a place of outdoor recreation for Warsaw residents and the number of inns, taverns, etc. began to grow. Some were restaurants and cafés frequented by the intellectual elite. For example, the *Café de Bon Gout* opened in 1794 and in 1799 was renamed the *Café de Bon Esperance* (Pawłowski and Zieliński 2008, p. 77, 111). This period of prosperity ended in the 1830s. After an unsuccessful uprising against the Russian Empire (1830–1831), in 1832 the Russian authorities decided to build a fortress in the area. This was accompanied by the displacement of the population and the destruction of existing buildings. In 1918, Poland gained independence and Warsaw again became its capital. A few years later, the authorities decided to make Żoliborz a residential zone once again. In the 1920s, on the site of the former military compound and warehouses, building began. Most of the houses built in the 1920s–1930s were owned by cooperatives whose members were employees of government institutions, banks, insurance companies, journalists and military personnel. This was reflected in the names of individual settlements—Żoliborz Oficerski was inhabited by high-ranking military officers, Żoliborz Dziennikarski by journalists, and Żoliborz Urzędniczy by officials of governmental and municipal institutions.

At the end of the 1920s, west of Plac Wilsona (Wilson Square, named after American President Woodrow Wilson), a housing estate was formed that was owned by the Warsaw Housing Cooperative (Warszawska Spółdzielnia Mieszkaniowa, WSM). Initially, these homes were inhabited by blue-collar workers (47% in 1929), but high rents led to an invasion of the middle class: by 1936, blue-collar workers only made up 25% of the population (Pawłowski and Zieliński 2008, p. 184). At that time, most houses were buildings with 3–4 floors, terraced houses with 1–2 floors or single-family villas, especially in zones inhabited by journalists and military servicemen. The layout followed the Howard garden city concept and the designers were leading Polish architects. Until 1939, only areas between the Vistula River and Stołeczna Street were built up. It should be noted here that in the 1920 and 1930s the gastronomic function in Żoliborz was less visible than in other parts of Warsaw. The ground floor of modernist buildings in the area was usually designated as flats. Restaurants, cafés and pastry shops were located mainly in houses along Mickiewicza and Słowackiego streets.

The development of Żoliborz stopped during World War II, while the area was almost unaffected by the bombing that destroyed other parts of Warsaw, especially

---

[1]The name comes from the French term *Joli Bord*, in English, Beautiful Embankment. The name was chosen in the eighteenth century referring to the Piarist convent, which was situated in this area.

7 Changes in the Distribution of Gastronomic Services ...

the city centre. Post-1945, it did not have to be rebuilt, and it remained as it was before 1939. A construction boom only began in the early 1960s when new, centrally planned, housing estates began to appear in its western part. These were built to the north of the industrial district (Żoliborz Przemysłowy) adjacent to suburbs dating from before World War II (Powązki). In line with new urban concepts, each estate was to have its own services, including gastronomy. The map in Fig. 7.7 shows the situation in 1975. In this year, there were 14 eating establishments. Apart from milk bars (at Słowackiego Street, and at the intersection of Broniewskiego and Krasińskiego streets) there were a few cafés. One was on the corner of Stołeczna and Krasińskiego streets, the other opened in 1964, in the highest building in the district, on Słowackiego Street: the *Havana* café was a symbol of Polish–Cuban friendship after the Cuban Revolution in 1959. *Spotkanie* opened in 1953 on Krasińskiego Street and was often visited by families visiting green areas along the Vistula river. There were also several restaurants and bars. The *Kosmos* restaurant was the best and was located close to Plac Komuny Paryskiej (Paris Commune Square), formerly Plac Wilsona—the district's centre. The other was the *Balaton* restaurant near Mickiewicza Street in the southern part of the district. Both restaurants held dances in the evenings. Within the new housing estates (which housed around 15,000–20,000 people), there were only two dining outlets—the *Sady* milk bar and the *Rudawka* café.

The situation described above lasted until the end of the 1970s. It was only in the 1980s that new eating establishments began to appear. These new bars and cafés were increasingly owned by private entrepreneurs. An example is the *Marylka* beer bar in the green zone known as Kępa Potocka. In 1985–1988 its manager was an

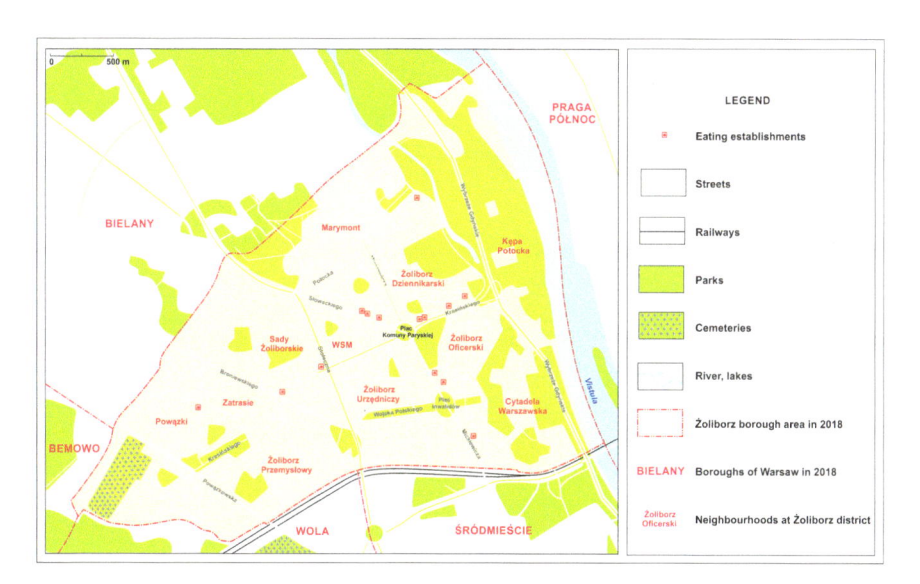

Fig. 7.7 The distribution of eating establishments in the Żoliborz district of Warsaw (Poland) in 1975. Source: Adapted by Andrzej Kowalczyk from Bystrzanowski and Dutkowski (1975)

immigrant from Afghanistan, whose wife was Polish (Marylka is a nickname for Marianna, his wife) (Kłoś 2002). This trend grew following the political and socio-economic changes that began in 1989. Initially, new eating outlets appeared spontaneously in the form of small kiosks offering fried chicken, hot dogs and pizza. These were located mainly in the vicinity of what is now the Marymont metro station. Later they started to appear in disused grocery stores, shoemakers and tailors, etc. At the same time, some of the restaurants and cafés that had operated in the 1960–1970s closed and were replaced by banks (e.g. the *Różanka* café and the *Kosmos* restaurant). This trend emerged in the mid-1990s and lasted for nearly two decades.

Research carried out in 2017 showed that there were 88 eating establishments in Żoliborz. Compared to the mid-1970s the index of change is 629 (index for 1975 = 100).

The example of Żoliborz highlights several trends regarding the localisation of gastronomy in a residential district located near the central city.

First, like many other services, catering establishments congregate near transport nodes. In many cases, they are located near central squares or crossings. A good example is Plac Wilsona. With five tram stops, 10 bus stops and a metro station, this square is the most important transport node and orientation point in the district. Shops and services are clustered here, including six catering establishments on the square itself and seven nearby.

Second, catering services are located linearly along the main axes. Słowackiego Street is a very good example and some eateries have been located here since 1975. The street is part of an axis that leads from central Warsaw to residential districts in the north west. A tram goes along it and a metro goes under it. Another good example is Mickiewicza Street, whose southern part is attractive for restaurant owners. Here, however, another factor influences the location of this cluster—the built environment.

The built environment is a very important factor in determining the existence of eateries in urban space. Some houses were designed to have commercial premises on the ground floor. This was the case for Mickiewicza Street in the 1930s, where housing owned by cooperatives included space for a shop or other services on the ground floor. When a new shopping mall opened nearby in 2004, retailers and services were pushed out and their premises were gradually taken over by gastronomy establishments (e.g. the WSM; Fig. 7.8). The situation was different in other places. Few eating establishments were located in neighbourhoods with residential buildings (single-family houses) because of a lack of suitable premises and low population density (e.g. Żoliborz Dziennikarski, Żoliborz Oficerski). The large housing estates from the Communist period (blocks of flats) did not include space for services, and eating establishments (if any) were located in separate buildings usually near transportation nodes (e.g. the Marymont Potok neighbourhood). The only type of built environment where gastronomy was planned were contemporary housing estates. This is especially apparent in Żoliborz Południowy (formerly Żoliborz Przemysłowy, see Fig. 7.8), which is a new neighbourhood built on a post-industrial site. The area is not integrated with the rest of the district and

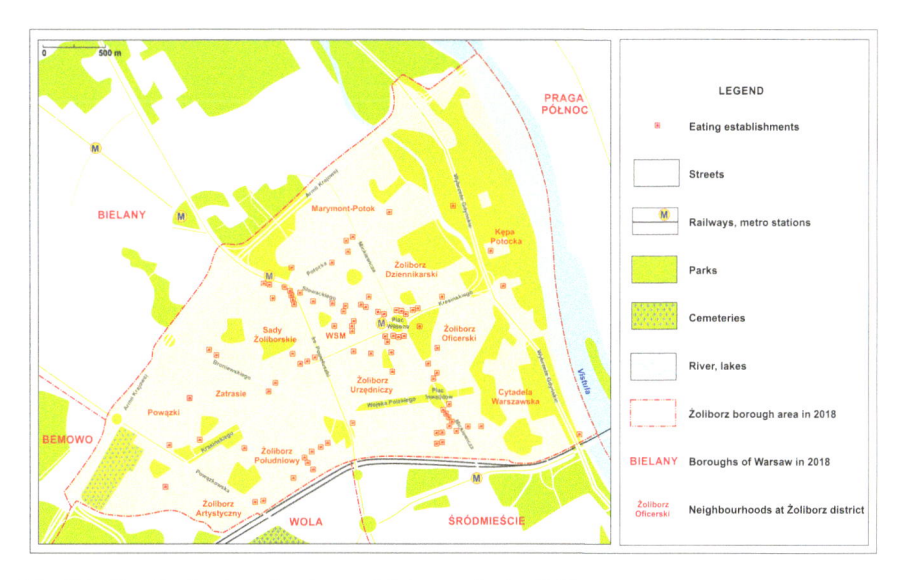

Fig. 7.8 The distribution of eating establishments in the Żoliborz district of Warsaw (Poland) in 2017. Source: Makarewicz (2018) and fieldwork by Dawid Celiński-Jakubowicz and Konrad Frączek

constitutes a separate zone with new investments built by private developers. Commercial premises, including restaurants and bars, were a planned part of the investment. Unlike the districts of the so-called 'old Żoliborz', where eating establishments attract visitors from other parts of the city, the clientele of these places mainly consists of residents living nearby.

7.6 Gastronomy Outside the Historic Centre. The Example of Podskalí, Prague (The Czech Republic)

Podskalí (a part of the Praha 2 district) in Prague is a specific area located in the New Town on the banks of the Vltava river. It is close to the second-highest point of the Vyšehrad bastion, and the view of the historical part of the city is dominated by Prague Castle which overlooks the river. Podskalí was an industrial district, inhabited by predominantly working class, with longstanding connections to the operation of timber barges on the Vltava (Newman 2012). Now it is primarily a residential, then administrative area. For tourists, it is a transit route to the Vyšehrad bastion. However, it provides a large number of gastronomic services that are used by local residents and, occasionally, tourists.

A typology of gastronomic facilities was carried out in the summer of 2009, during the economic crisis and at a time when Podskalí had not yet been developed (Fig. 7.9). A second analysis was carried out in the summer of 2017 (Fig. 7.10) in a

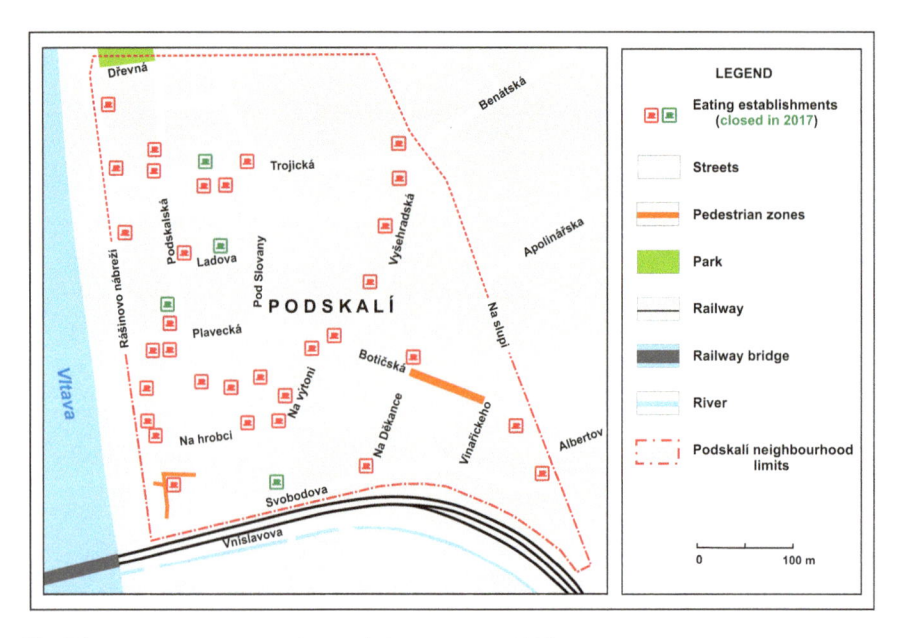

Fig. 7.9 Restaurants and bars in Podskali (Prague) in 2009

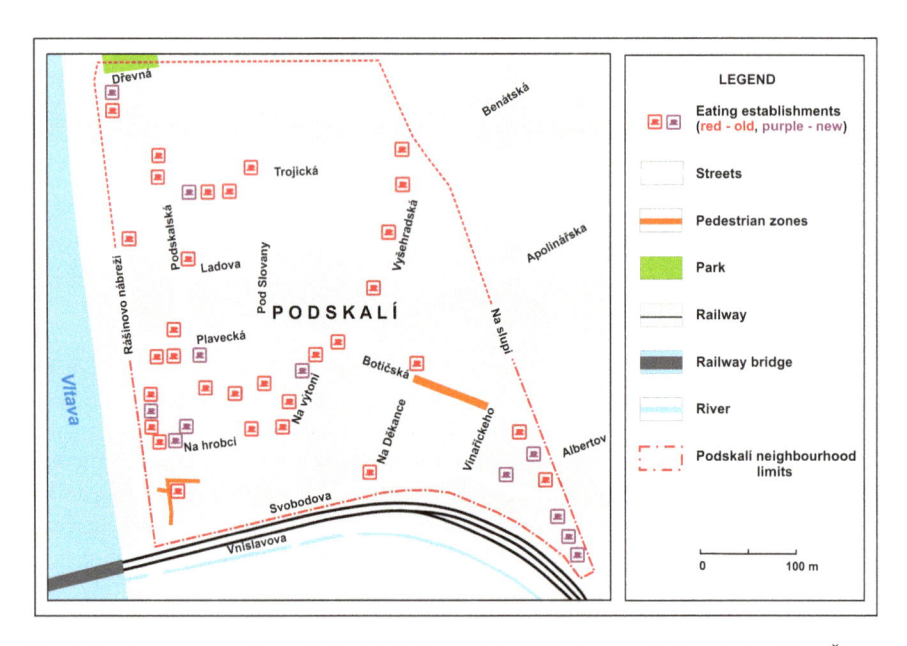

Fig. 7.10 Restaurants and bars in Podskali (Prague) in 2017. Source: Fieldwork by Jakub Šmída

completely different economic situation—a time of economic growth and increased domestic and foreign tourist traffic. Another change concerned the adjacent Vltava shore where a cycling trail, cultural programme and gastronomic facilities on moored boats had appeared. Another important change was the adoption of a law banning smoking in gastronomic facilities from June 2017. This second study found that not only was there a 10% increase in the number of facilities (from 36 to 40) but, above all, there was a qualitative shift. Only 12 of the 40 facilities were still operated by the same entity in 2017 and 2009. The number of gambling outlets and non-stop bars fell from five to one. Another change was the growth in businesses focused on high-quality or healthy food. The spectrum of services is very wide, both in terms of price, type of facility and specialisation. Of the 15 restaurants, seven served Czech cuisine, three Italian, and one each for French and Vietnamese. At lunchtime, foreigners dominate, while native Czechs are mostly employees from nearby administrative buildings or colleges. In the evening, the ratio of foreigners and locals is balanced.

These significant changes can be attributed to the attractiveness of the banks of the Vltava, which has motivated the improvement of facilities for both the local population and visitors. There has also been a major effort to relieve pressure on the historical centre, which can be considered a tourist ghetto (Dumbrovská 2017).

7.7 Conclusions

The presence of restaurants in the urban space can redefine the local context; they are part of the 'local market baskets of amenities that vary from place to place' (Clark 2003, p. 104). Their presence in the central parts of cities plays an important role in creating the urban product and attracting tourists, workers, residents and other city users. Drawing on this, gastronomy can be understood as a proxy for the centre of a city and can illustrate the changes that shift urban spaces.

Beginning in 1989, all of the cities that were analysed in this chapter (Saint Petersburg, Warsaw, Kraków and Prague), have experienced a tremendous change in the political, economic, social and cultural environment. In Poland, under communism, eating out was unpopular, mainly due to poor quality food, a lack of eating establishments, and the lack of household income. The culinary symbol of the period was the so-called 'milk bar' (*bar mleczny*), a cheap, state-subsidised, self-service canteen that offered a few dairy-based dishes. In Saint Petersburg, the catering sector was totally state-controlled, and, if people ate out at all, it was usually their workplace canteen. In the Czech Republic, the tradition of eating out was probably the most developed out of all of the three countries, but it tended to involve drinking beer rather than eating out.

The situation has changed dramatically since 1989, seen in the proliferation of eating establishments in the 1990s. In Warsaw's central district growth reached 68% between 1994 and 2003. In Kraków's Old Town the number of eateries had increased by 250% in 1999 compared to 1990. Although growth is ongoing, it is at a much slower rate than in the post-1989 period. In Kraków's historic core it has hovered at around 5% per year since 2005. An important difference between Warsaw and Kraków is the share of restaurants in the total number of eateries. In the former case, restaurants constitute almost 30% of all eating establishments, while in the latter it is almost 40%. On the other hand, takeaways make up 12.5% of establishment in Kraków, compared to only 6% in Warsaw. This can be explained by, among others, the tourist appeal of central Kraków and more business-like central district of Warsaw.

In Saint Petersburg growth in the number of eateries between 2005 and 2017 reached almost 46%, which translates into almost 4% per year. The density of eating establishments is not, however, high: six establishments per 10,000 people is far from the 150 found in Prague. Interestingly, this can be compared to 1905, when the density was eight establishments per 10,000 people, which suggests that we can expect further growth in the future.

References

Ashworth GJ, Tunbridge JE (2000) The tourist-historic city. Elsevier, Oxford
Brzozowska I, Maciuszek R (1996) Lokale gastronomiczne i rozrywkowe na Starym Mieście w Krakowie, Instytut geografii i gospodarki przestrzennej planowania, Uniwersytet Jagielloński, Kraków, MA thesis
Bystrzanowski J, Dutkowski K (1975) Wszystko o Warszawie. Wydawnictwo Sport i Turystyka, Informator, Warszawa
Censo de Locales y Actividades, 3. Actividades en locales Abiertos con Tipo de acceso Puerta de calle y Agrupados, clasificados por Actividad y Distrito, Ayuntamiento de Madrid, Madrid. http://www.madrid.es/UnidadesDescentralizadas/UDCEstadistica/Nuevaweb/Econom%C3% ADa/Empresas%20y%20Locales/censo/D2110317.xls. Accessed 20 July 2017
Clark TN (2003) Urban amenities: Lakes, Opera, and Juice Bar: do they drive development? In: Clark TN (ed) The city as an entertainment machine, Research in urban policy, Elsevier, Kidlington, 9, pp 101–140
Derek M (2013) Kierunki rozwoju usług gastronomicznych w warszawskiej dzielnicy Śródmieście, Prace i Studia Geograficzne, 52, pp 85–100
Derek M., 2018, Spatial structure of tourism in a city after transition: the case of Warsaw, Poland, [in:] Müller D. K., Więckowski M. (eds.), *Tourism in transitions. Recovering decline, managing change,* Springer, Cham, pp. 157–171
Dumbrovská V (2017) Urban tourism development in Prague: from tourist mecca to tourist ghetto. In: Bellini N, Pasquinelli C (eds) Tourism in the city. Towards an integrative agenda on urban tourism, Springer, Cham, pp 275–283
Faracik R, Kurek W, Mika M, Pawlusiński R, Pitrus E, Piziak B, Ptaszycka-Jackowska D, Rotter-Jarzębińska K, Wilkońska A, Zawilińska B (2008) Waloryzacja przestrzeni miejskiej Krakowa dla potrzeb turystyki. Raport końcowy, under the supervision of Kurek W, Mika M, Instytut Geografii i Gospodarki Przestrzennej Planowania, Uniwersytet Jagielloński, Kraków
Górka Z (2004) Krakowska dzielnica staromiejska w dobie społeczno-ekonomicznych przemian Polski na przełomie XX i XXI wieku. Użytkowanie ziemi i funkcje, Instytut Geografii i Gospodarki Przestrzennej Planowania, Uniwersytet Jagielloński, Kraków
Kaczorek A (2002) Gastronomia etniczna w Warszawie, Ethnic Gastronomy in Warsaw. MSc thesis. University of Warsaw, Faculty of Geography and Regional Studies, Warsaw

Kholodilin KA, Limonov LE (2018) CBD of St. Petersburg between 1869 and 2017: from the market to the centrally planned economy and back again, manuscript

Kłoś A (2003) Rozmowa z właścicielem baru U Araba, Gazeta Wyborcza, Available at: https://warszawa.wyborcza.pl/warszawa/1,34884,764836.html (Available online 28.06.2018)

Makarewicz A (2018) Baza gastronomiczna na Starym Żoliborzu w warszawie, manuscript of bachelor thesis. University of Warsaw, Faculty of Geography and Regional Studies

Městské části hlavního města Prahy. Souhrnné informace o 57 městských částech 2004–2016, ČSÚ v hl. m. Praze, Praha, https://www.czso.cz/documents/11236/37543548/casova_rada_MC_2016.xlsx/3c601a66-7ab8-46b0-8178-6cbdce108249?version=1.1. Accessed 20 July 2017

Naszkowska K (2013) Prezes od kawiarni, Gazeta Wyborcza, 02 Sept 2013. http://www.wysokieobcasy.pl/wysokie-obcasy/1,96856,14555716,Prezes_od_kawiarni.html?disableRedirects=true (retrived 07 June 2018)

Newman G (2012) The fractured embankment: modernity and identity at the edge of the vltava. In: Cusack T (ed) Art and Identity at the water's edge, Routledge, London and New York, pp 123–140

Pawłowski T, Zieliński J (2008) Żoliborz. Przewodnik historyczny. Wyd. Rosner & Wspólnicy, Warszawa

Piziak B (2011), Baza gastronomiczna - wielkość, struktura i rozkład przestrzenny. In: Mika M (ed) Kraków jako ośrodek turystyczny, Instytut geografii i gospodarki przestrzennej planowania, Uniwersytet Jagielloński, Kraków, pp 181–199

Piziak B, Zając T (2012) Rola bazy gastronomicznej w rozwoju turystyki przyjazdowej do Krakowa. In: Grochowicz J (ed) Szanse i bariery rozwoju turystyki krajowej i międzynarodowej, Zeszyty Naukowe, 1, Europejska Szkoła Wyższa, Sopot, pp 249–276

Prayag G, Landré M, Ryan C (2012) Restaurant location in Hamilton, New Zealand: clustering patterns from 1996 to 2008. Int J Contemp Hosp Manag 24(3), 430–450

Rokicka-Donica K (1995) Przestrzenne zróżnicowanie bazy gastronomicznej w dzielnicy Warszawa-Śródmieście, Faculty of geography and regional studies, University of Warsaw, MA thesis

Wójcik A (2007) Stan bazy gastronomicznej na Starym Mieście w Krakowie w obrębie Plant, Instytut Geografii i Gospodarki Przestrzennej Planowania, Uniwersytet Jagielloński, Kraków, MA thesis

Chapter 8
Restaurants and Bars in the Outer City

Andrzej Kowalczyk and Aleksandra Korpysz

Abstract The location of eating establishments within a city is related to the distribution of the population, and affects all areas. But this does not mean, however, that restaurants and bars in the outskirts of the city have the same characteristics as those in the downtown. Drawing on the example of Warsaw, especially its three residential neighbourhoods, this chapter reveals that the character and prices of eating establishments in the outer city largely reflect the needs of local residents. In general, there are fewer gourmet restaurants, as well as fewer establishments serving business lunches.

Keywords Outer city · Housing estates · Warsaw

8.1 Introduction

Gastronomic services in cities cater to the needs of their residents and visitors. Therefore, the location of eating establishments is related to the distribution of the population and affects all areas. This does not mean, however, that restaurants and bars in the outskirts of the city have the same characteristics as those in the downtown. Their character and prices largely reflect the needs of the local residents, although they are of a different type (e.g. restaurants or bars that are visited by people from other areas at weekends). In general, however, there are fewer gourmet restaurants, fewer establishments serving business lunches and fewer elegant restaurants that are visited with friends in the evening after going to the theatre.

A. Kowalczyk (✉)
Department of Tourism Geography and Recreation, Faculty of Geography and Regional Studies, University of Warsaw, ul. Krakowskie Przedmieście 30, 00-927 Warsaw, Poland
e-mail: akowalczyk@uw.edu.pl

A. Korpysz
Department of Tourism Geography and Recreation,
Faculty of Geography and Regional Studies, University of Warsaw,
ul. Krakowskie Przedmieście 30, 00-927 Warsaw, Poland
e-mail: aleksandra.korpysz@uw.edu.pl

© Springer Nature Switzerland AG 2020
A. Kowalczyk and M. Derek (eds.), *Gastronomy and Urban Space*,
The Urban Book Series, https://doi.org/10.1007/978-3-030-34492-4_8

8.2 Changes in the Location of Restaurants in Bródno, Warsaw (Poland)

The political and socio-economic changes that took place in Poland after 1989 caused a rapid expansion in entrepreneurship, which mainly affected large cities. The changes concerned the appearance of new eating establishments, which began to be set up by private individuals not only in the central districts of the cities, but also in their outer zones. One example of these changes is the large housing estate of Bródno, located in north-east Warsaw. In 1989 Bródno had only nine restaurants, bars (milk bars and beer bars) and cafés (Bystrzanowski and Dutkowski 1975).

At the end of the 1980s, but predominantly at the beginning of the 1990s, there were significant changes in gastronomy in Bródno. Existing restaurants, bars and cafes changed ownership, as they were privatised, and new establishments began to emerge (Fig. 8.1). The process of creating new catering establishments was largely spontaneous, and some were illegal. The economic crisis meant that the meals they served were simple, with the same dishes being offered everywhere. A significant

Fig. 8.1 Eating establishments on the Bródno housing estate in Warsaw (Poland) typical of the pre-1989 period. The Kacperek restaurant dates from the 1970s (**a**) and is different from the cafés (**b**) from the same decade, and milk bar with the same name (Kacperek). The only remaining milk bar from pre-1989 period on the Bródno housing estate (**c**)

proportion of the new catering establishments were bars and kiosks offering take-away food. The basic dishes were fried chicken, hot dogs, chips, and zapiekanka (open-faced, toasted cheese sandwiches) (Fig. 8.2). For at least a few years this type of gastronomy dominated Warsaw's city centre, as well as its poorer parts. At about the same time, pizzerias and spaghetteries began opening. Prices for pizza and pasta were higher than those in the kiosks offering zapiekanka or fried chicken. Pubs became a popular symbol of new trends. At the same time, milk bars, typical of Polish gastronomy in the 1960s–1980s, began to disappear.

It was only with the arrival of people from East Asia that bars with Vietnamese and then Chinese cuisine began to appear (Fig. 8.3). Some started up in facilities that had previously sold fried chicken and zapiekanka. This new type of gastronomy dominated until about 2005. Since then, Bródno has seen more and more bars offering Turkish cuisine, followed by Arabic. Some belong to chains (e.g. Amrit Kebab), and some are independent (e.g. Doy Doy Kebab). At the same time, outlets from the 1990s have virtually disappeared. These shifts can be attributed to both cultural and economic changes. The influx of immigrants, the increase in travel abroad by Poles for tourism and work-related purposes, television, articles in newspapers (and more recently on the web) have led to the appearance of other cuisines. At the same time, the economic development of Poland and the increase in affluence of the population of Warsaw have resulted in an increasing number of people eating outside the home.

At the June of 2018, after 28 years of political and socio-economic transformation, the number of eating establishments in the region encompassing the Bródno housing estate and neighbouring areas had grown to 95 (a nine-fold increase). At the same time, the population of this part of Warsaw remained more or less stable. This means that the appearance of new restaurants or bars was caused not by a demographic shift, but by socio-economic and socio-cultural changes. The new

Fig. 8.2 New food outlets on the Bródno housing estate in Warsaw (Poland) from the late 1980s and early 1990s. The outlet at Wyszogrodzka Street offers fried chicken (**a**). The banner at the outlet on Skrajna Street displays information about fried chicken, zapiekanka and chips (**b**)

(a)　　　　　　　　　　　　　　(b)

Fig. 8.3 The influence of foreign cuisine in gastronomy on the Bródno housing estate in Warsaw (Poland). Pizza Dominium on Kondratowicza Street (**a**). The Vietnamese bar Kim Nam at the crossroads of Rembielińska Street and Kondratowicza Street (**b**). The food outlet was probably built in the late 1990s, but the apartment house in the background was only completed in 2012. This new apartment block was built on a plot of land where there was a one-storey pavilion from the early 1970s with a grocery store and a pharmacy

housing estates that make up Bródno are diverse in terms of demographics and socio-economic status. The western, southern and northern parts (estates that were built in the 1970s and 1980s) are inhabited by elderly people (many over 65), who are often retired workers from now-defunct industrial plants and lower middle class. However, the eastern part, built after 2000, is inhabited by a middle-class population, mostly aged 30–50.

Most structures are multi-apartment buildings with either four or 10 floors. In the southern part of the area, only low-rise buildings (some predating World War II) have survived. Low-rise buildings, including partly new villas, can also still be found in Zacisze (in the east) and Ugory (in the north) neighbourhoods.

The distribution of eating outlets shows that restaurants, bars, cafes and pastry shops are concentrated in several distinct areas. The largest is in the north-east, the Atrium Targówek mall, which is covered in more detail below. In 2017, this mall hosted 21 restaurants and another, belonging to the Pizza Hut chain was nearby. The second cluster of eating establishments is in the Factory Annopol Outlet shopping centre, which has seven outlets. Although it is located on the other side of the expressway that passes to the north of Bródno, a significant proportion of the centre's shoppers are residents of the local housing estates. The third concentration is in the Galeria Renova, which has four outlets. This is situated to the south-west of a large park in the centre of the Bródno area.

Figure 8.4 shows clusters of restaurants and bars at street intersections. One example is the Kondratowicza Street and Rembielińska Street intersection. The former is the main street and the latter contains tram lines that connect this part of

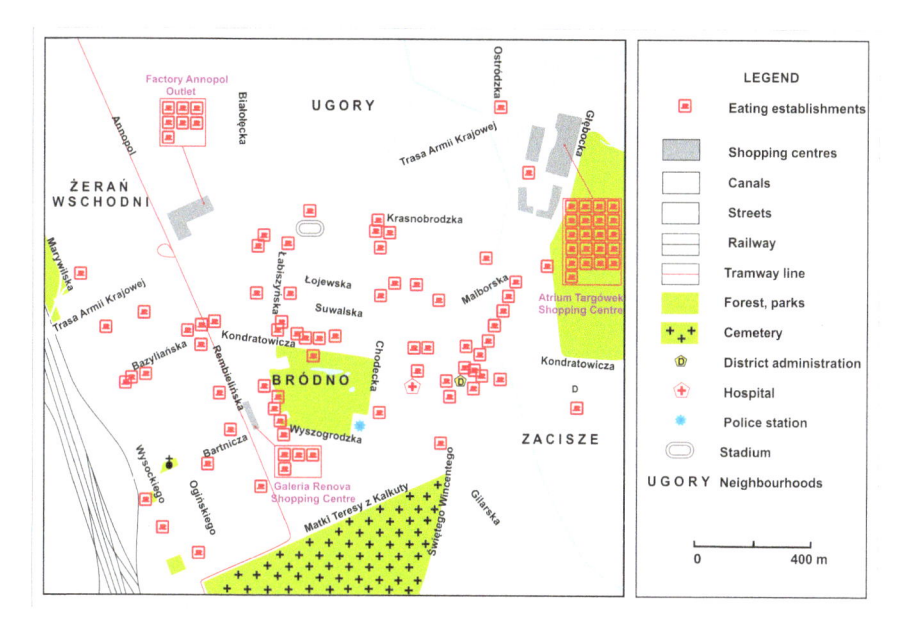

Fig. 8.4 The distribution of eating establishments in the Bródno housing estate in Warsaw (Poland) in 2018

Warsaw with the inner city (an important transport hub). Many restaurants, bars and cafés are located along the Bazyliańska Street–Kondratowicza Street axis, which crosses Bródno from west to east. Some are located near key public buildings, including one of the largest public hospitals (marked with the symbol H). This part of Bródno can be considered as its civic centre and its bus stops cover 12 routes. Another location with a high density of restaurants and bars is the part of Świętego Wincentego Street between the transport node (mentioned above) and the Atrium Targówek mall. The eating outlets located here form a line and five are located in one, 200-m section. They are all in a modern building belonging to the Tivoli estate (built in 2007–2008), opposite the Park Leśny (Forest Park) housing estate, which was built between 2005 and 2016. Both estates are inhabited by the middle classes, most are single (or young) professionals or young married couples with children. Another reason for the presence of so many eating establishments on such a short section of the street is the nearby presence of a private oncological hospital (Fig. 8.5).

The gastronomy in Bródno is very diverse. Most outlets are cheap restaurants and fast food or takeaway bars. They include pizzerias, bars with Chinese and Vietnamese cuisine, and places that offer Arabic and Turkish cuisine. The latter mainly serve döner kebabs. These types of dining establishments (and several cafés and pubs) dominate in western, northern and southern areas. However, facilities in the area to the east are mainly casual restaurants, and some are distinguished not only by their prices, interior and service quality, but also by their cuisine.

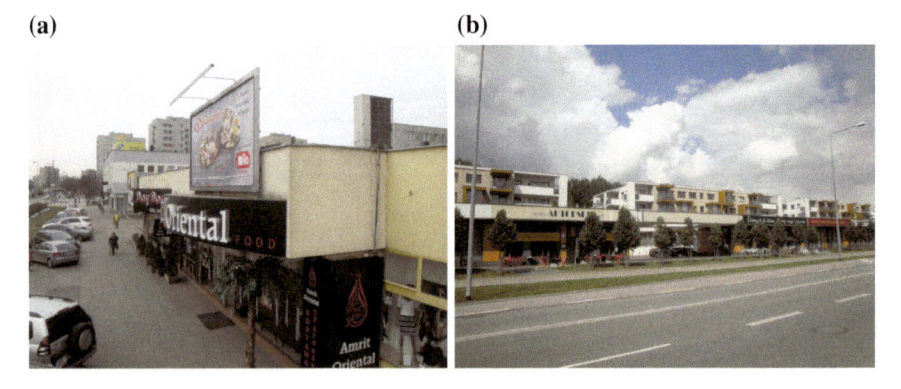

Fig. 8.5 Eating facilities along the main streets of the Bródno housing estate in Warsaw (Poland). Oriental bars border Kondratowicza Street. In the background are apartment houses for the lower middle class dating from the 1970s (**a**). Casual restaurants and bars along Świętego Wincentego Street. In the background is the new Tivoli middle-class housing estate (**b**)

At Świętego Wincentego Street, restaurants and bars offer Japanese, French, Italian, traditional Polish and fusion cuisine. The pizzeria is particularly noteworthy. The Bella Napoli restaurant was opened in 2005 and its owner is an Italian immigrant. It is the oldest pizzeria in Warsaw run by Italians. Initially, it occupied only 60 m^2, but has now expanded to over 100. The owner was born near Naples and he gained his experience preparing pizza in Italy. Currently, he has five restaurants in Warsaw and its suburbs (Jankowska 2018). Some ingredients are imported from Italy (e.g. flour) which makes the pizza particularly popular among gourmets.

The Bałkańska Dusza (Balcan Soul in English) restaurant enjoys a level of notoriety similar to that of Bella Napoli. This restaurant, which is situated in the Zacisze estate, offers dishes from the Balkans (especially Serbian and Montenegrin cuisine). Through the week (and at weekends) both restaurants attract guests from all over Warsaw, including politicians, actors and television celebrities (Fig. 8.6).

As noted above, over a quarter of all eating establishments in Bródno are situated in the Atrium Targówek shopping centre. The decision to build a large shopping centre in the north-eastern part of Warsaw was made in the mid-1990s on a 27-ha plot of former agricultural land. Originally, it was called CH Steen & Strom and its shareholders were Norway's Steen & Strom Holding A/S (37.5%), Denmark's, Thorkild Kristensen Byggeholding A/S (37.5%), and an investment fund controlled by the Danish government (25%). Construction of the mall was completed in September 1997. At that time CH Steen & Strom covered 36,000 m^2 and provided parking for 1,250 cars. The anchor store was the Carrefour hypermarket, which occupied 20,500 m^2. After a short time, CH Steen & Strom changed its name, first to CH Targówek and later to Atrium Targówek.

Figure 8.7 shows the distribution of food services in the shopping centre immediately after its opening. At that time there were six eating establishments, which were located in four zones. The McDonald's restaurant was accessible from

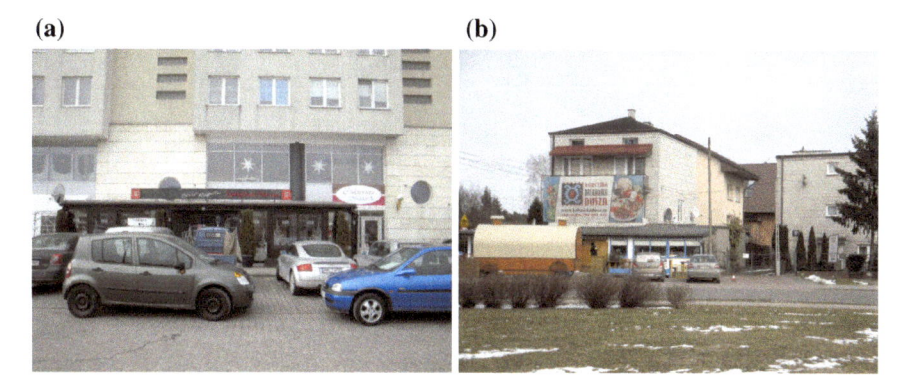

Fig. 8.6 The most famous restaurants on the Bródno housing estate in Warsaw (Poland). The Italian-owned restaurant, Bella Napoli (**a**), is situated close to the town hall. The Serbian–Montenegrin restaurant, Bałkańska Dusza, is located in a detached family house on the edge of the Zacisze neighbourhood (**b**)

both inside and outside the centre (Fig. 8.7b). Other establishments could only be entered from the shopping gallery. Other establishments included a bar, a pizzeria and a salad bar. For several years, CH Targówek did not have any casual dining (or higher category) restaurants. In 2001, it was extended to cover 65,000 m^2, with 50,000 m^2 of retail space. A new section was added to its southern side, in which, apart from the (mostly clothing) stores, there was a multiplex cinema with 12 screens. The total number of stores increased to 155 (Fig. 8.8), and parking was expanded up to a capacity of 2,900 cars. During this period, the mall was taken over by the real estate company, Foras Targówek Property sp. z o. o. Later, its owners changed again, to Atrium Poland Real Estate Management sp. z o.o., which is backed by capital from Israel.

Among the factors that influenced the decision to expand the centre was the development of new housing estates (on greenfield sites), 1–2 km north. These new estates were built for the middle classes. The other factor was the development of new suburbs outside Warsaw. The expansion resulted in major changes in the number, nature and location of dining establishments. This included a new food court, containing six restaurant and bars. The original food court was also retained. The new food court was located in the south-eastern part, near the entrance from the bus stop. Two cafes were opened in the western part, and two more adjacent to the cinema. New gastronomic facilities include bars offering oriental (Thai Wok), Mexican (Bar Gomes), international (Adi Steak House) and traditional Polish (Rejtan and Karoca) cuisine, and a fast-food fish bar. In the older part of the centre, some outlets changed ownership and cuisine. The French-style café, Fleury Michon, was replaced by the Italian restaurant Trattoria, and the fast-food bar, BarBuła's place was taken over by a restaurant offering Arabic–Turkish cuisine Hammurabi. The appearance of this casual dining restaurant is due to the arrival of new shoppers, many of them young professionals living in the new housing estates.

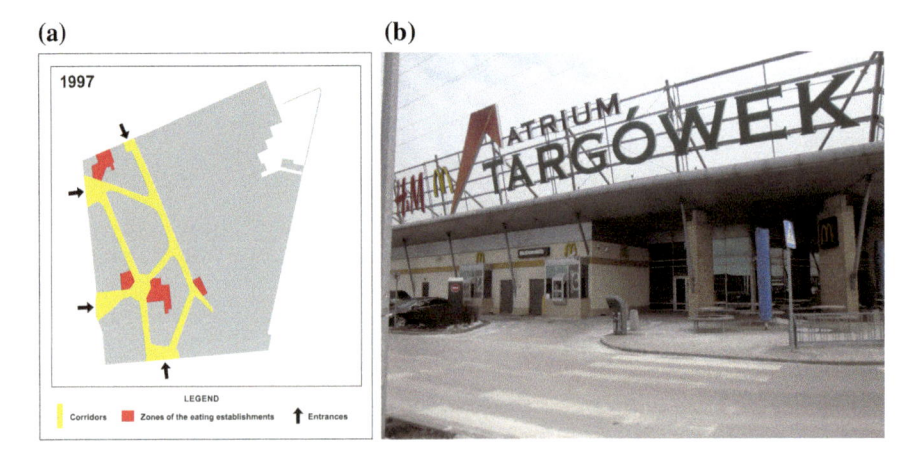

Fig. 8.7 Gastronomy at the CH Targówek (later CH Atrium Targówek) shopping centre in the Bródno housing estate in Warsaw (Poland). Distribution of eating establishments in 1997 (**a**). McDonald's restaurant to the north in 2018 (**b**). The McDrive service is on the left, the main restaurant is in the centre, with the terrace (front)

Fig. 8.8 Changes in the distribution of eating establishment in the CH Atrium Targówek shopping centre in Warsaw (Poland) 2002–2017

In the following years, the distribution and character of gastronomic establishments in CH Targówek changed very little. In 2004, a new restaurant, Grota, opened in the south-western part of the building. Its entrance was on the outside of the gallery. In addition, an ice cream shop opened in the corridor connecting the main food court with the cinema.

Further changes began in 2006 and consisted primarily in reducing the size of the original food court, closing the bar at the entrance to the Carrefour hypermarket and expanding the main food court. Changes to the old food court led to the closure of the Hammurabi restaurant and the Mangia Pizza pizzeria, which were replaced by fashion stores. The Trattoria restaurant has remained. In turn, changes in new food court consisted in the opening of new casual dining restaurants offering Japanese and international cuisine with a strong Arab–Turkish influence. A further addition was an ice cream shop in the middle of the centre. Things then remained largely unchanged for a few years.

After a few years the Sowa chain opened a pastry shop adjacent to McDonald's. A little later, two cafés were built (next to the ice cream shop), including a Starbucks outlet. During this period, more bars were opened in the main food court. These included Pizza Hut Express, a second McDonald's, a Subway outlet and the Turkish bar, Erbil Kebab. At the same time, in the summer of 2017, the Vietnamese bar, which had been built a few years earlier on the site of the Grota restaurant, shut down. These changes can be summarised as follows:

- an increase in the number of eating establishments from six (autumn 1997) to 22 (autumn 2017);
- a change in the location of the main food court, which was initially associated with the centre's expansion but then resulted from changes in the organisation of the centre's space;
- ongoing growth in the gastronomic offerings from restaurant chains, bars, cafes and pastry shops. This started with McDonald's, Fleury Michon and Dunkin' Donuts. By 2017 the centre also housed outlets from Pizza Hut, Subway, Starbucks, Tchibo, Express Kitchen, North Fish and Polish chains such as Thai Wok, Grycan, Sowa, SOKoŁYK Pijalnia Soków, Sfinks Polska S. A. and Erbil Kebab).[1]

A characteristic feature of gastronomy in the outer city is the significant share of outlets that belong to fast food chains. Some are part of international corporations, such as McDonald's, Pizza Hut and KFC, Burger King. and others to chains with a national or local range. The difference between these two categories is seen in their location. International chains are usually found in large shopping centres, near public buildings, transport hubs, etc. Most national and local chains are in less desirable locations, largely because their clients are residents of nearby houses and housing estates. The location of these restaurants and bars is also influenced by the fact that their customers often order food via telephone or the web to be delivered by car, scooter or motorcycle. Sometimes they also offer takeaway food.

An example is the Da Grasso sp. z o. o. pizzeria chain. The first Da Grasso outlet opened in Łódź in 1996 and the chain soon began to open restaurants in Warsaw (*O nas* 2017). They were built in the outer city, sometimes on the periphery (Fig. 8.9). Although customers can eat in, most of the company's turnover comes from home delivery. Interestingly, some facilities are located in large housing estates, while others are located in neighbourhoods with detached and semi-detached houses. This is the case in the peripheral boroughs of Rembertów, Wawer and Włochy and in Targówek. The decision to locate a facility in Targówek resulted from the fact that,

---

[1] In 2017, Atrium Targówek began to expand further. The new part (in the south-west) will house a number of new restaurants and bars.

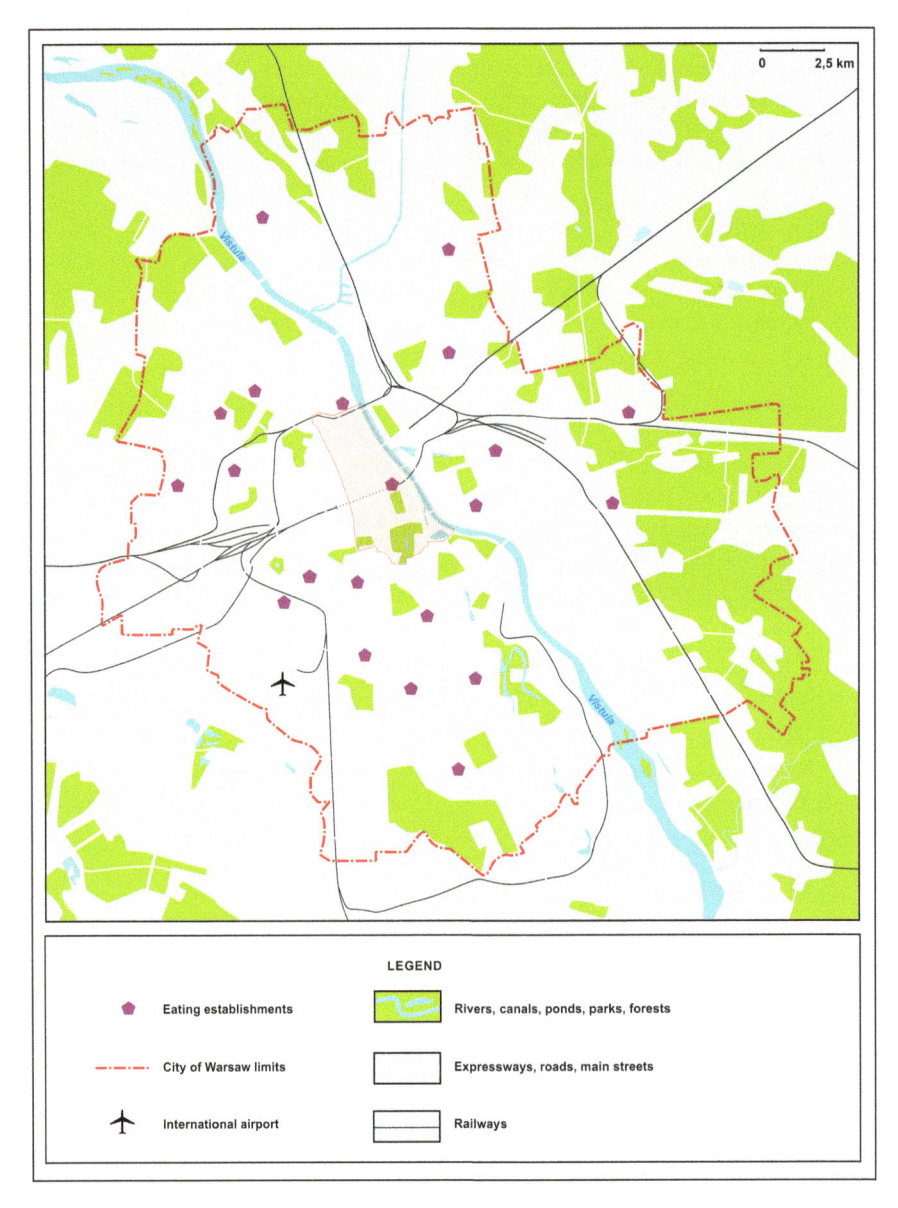

Fig. 8.9 Distribution of the Da Grasso chain pizza parlours in Warsaw (Poland) in July 2017. The areas bound by red line are the downtown area (Śródmieście borough)

at that time, there were no restaurants or bars within 500 metres. Similarly, several years after the opening of the pizza parlour at Gilarska Street, there are still no restaurants in the area, because residences are single-family houses with gardens (Fig. 8.10 and Table 8.1).

Fig. 8.10 Pizza parlour at Gilarska Street in the Zacisze neighbourhood in Warsaw-Targówek borough

Table 8.1 Distribution of Da Grasso sp. z o. o. pizza parlours in the urban zones of Warsaw in July 2017

Zone	Establishments	
	Number	Per cent
Downtown	1	5
Inner city	8	38
Outer city	12	67
Total	21	100

Source: Adapted from *Menu* (2018)

8.3 Eating Facilities in Peripheral Residential Areas of Warsaw (Poland): Tarchomin and Nowodwory

The area discussed in this section is part of the Białołęka borough, the second largest (by area) in the city of Warsaw.[2] It is bounded to the south by Maria Skłodowska-Curie Bridge, to the west by the Vistula river, to the north by the border with the city and to the east by Modlińska street (one of the main roads). This area consists of two neighbourhoods: Tarchomin and Nowodwory.[3] It is a typical residential area, which only significantly developed during Poland's political and economic transformation. Multifamily buildings are very common here (mainly gated communities). This area is rather incoherent in terms of its architecture and urban planning, which can be traced to its rapid, chaotic development by several different developers. A significant majority of the population are migrants from other parts of the country, many of whom are quite young. They are attracted by property prices, and many people buy their first flat here. As most residents commute to work in other parts of the city, this borough is also referred to as the 'bedroom of Warsaw'.

In the past it was a rural area, only integrated into the city in 1951. For a long time it was underinvested and somewhat isolated from the other parts, largely because of the lack of transport links. In the 1950s, there was only one bus line, which expanded to two in the 1970s (Sołtan 2002). The first bus line, on Światowida street, currently the main axis of Tarchomin and Nowodwory, appeared in the early 1990s, and the bus terminus Nowodwory was built in 2001. Nowadays, residents can choose between several daily bus lines, several night buses and two tram lines. A significant improvement in the connection with the rest of the city was influenced by two recent investments. In 2012, the Maria Skłodowska-Curie bridge was opened. This improved travel to the other side of the Vistula river, including the first station of the Warsaw metro. Another important transport investment was the tramline. The first stage of this investment was completed in 2014, followed by the second in 2017.[4]

In the 1970s the first housing estates started to appear. These have substantially changed the landscape. Nowodwory started to expand at the turn of the twenty-first century. The intensive development of multifamily housing continues to this day and new housing investments appear every year. Construction is focused on undeveloped areas, where investments are being made in new infrastructure, but the density of existing areas is also increasing. The opening of the first shopping mall in

---

[2] The whole borough covers an area of 73 km^2 (2017) and is inhabited by approximately 116,000 people (2016) (*Miasto Stołeczne Warszawa*, http://www.um.warszawa.pl, accessed 30 April 2018). Population density is nearly 1,600 inhabitants per square kilometre.

[3] Municipal Information System—http://zdm.waw.pl/miejski-system-informacji (Accessed 30 April 2018).

[4] The next stage will be the extension of the tram line from the Nowodwory terminus to Modlińska street.

8 Restaurants and Bars in the Outer City

this area, Galeria Północna, in 2017 was another significant change. Surveys have shown that there are 36 eating establishments—42.9% of all facilities in the two neighbourhoods.

Turning to gastronomy, there were already several inns in the area in the nineteenth century, when Tarchomin and Kępa Tarchomińska were both villages with several hundred inhabitants (Trelewicz 2003). There are currently over 80 facilities[5] (Fig. 8.11). Concentrations are found on the main transport routes. Both Światowida street and Odkryta street have a considerable number of housing estates, and there are public transport lines along them. On Światowida Street, the greatest wealth and diversity of dining options occurs in the Galeria Północna shopping mall. There are also several places near the Nowodwory metro terminus (kebabs, pizzerias, patisseries, etc.). In total, over 45% of facilities in the area are either directly on, or in

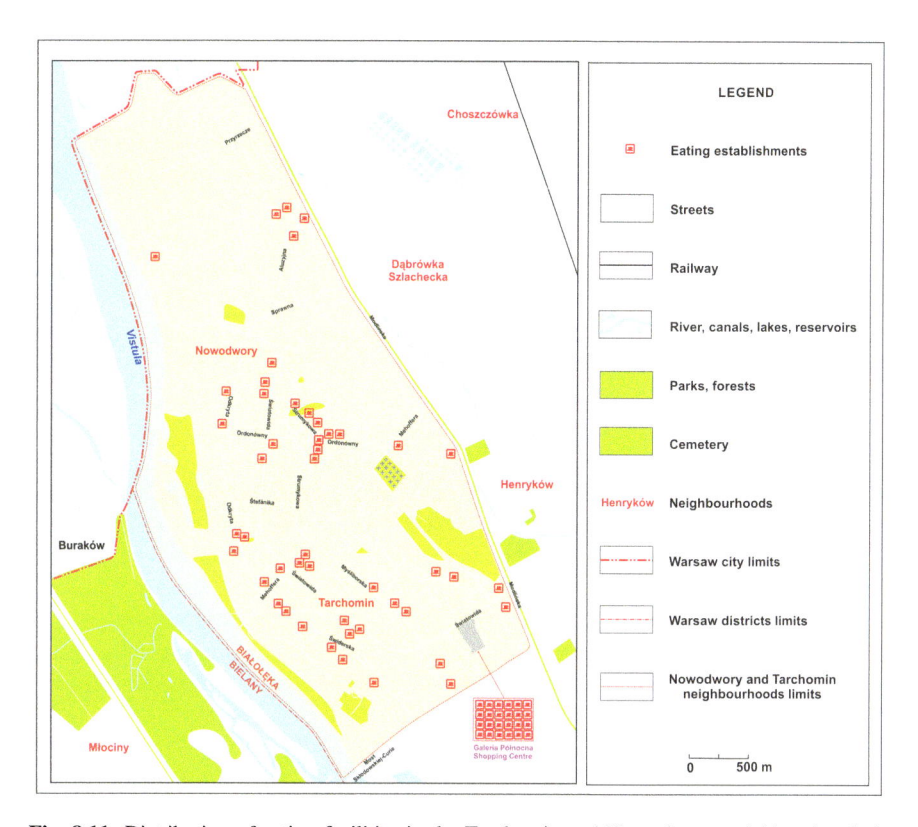

Fig. 8.11 Distribution of eating facilities in the Tarchomin and Nowodwory neighbourhoods in Warsaw (Poland) in 2018

---

[5]Based on an inventory carried out in April 2018.

close proximity to its main axis. On Odkryta street, close to the Vistula River, there are a few outlets in and around a small shopping centre (patisserie, sushi and Italian cuisine). There is another cluster near the Tesco hypermaket on Światowida Street and in the vicinity of the intersection of Strumykowa, Ordonówny and Pasłęcka streets. The Galeria pod Dębami, a local shopping centre also houses several eating establishments in addition to separate facilities. The available options are dominated by pizzerias, kebabs and places serving homemade dinners. However, the number of coffee bars and pubs seems to be disproportionate to the size and population of the area. All facilities are public (they are not located in gated communities). Sometimes they operate on the ground floors of blocks of flats, sometimes they are free-standing buildings. Many are fairly new, and there is a high turnover: when one fails it is usually quickly replaced by another.

8.4 Conclusions

The above-described examples show, that the location of eating services in outer parts of the city (in this case Warsaw) refers to central-place theory (at transport nodes) and the model of the city proposed by Hoyt (along main streets) (see Sect. 5.1). This is due to the layout of the streets and the character of the buildings. This arrangement is reflected in publications on urban space and, as Harvey writes in one of his earlier works, '…the way in which the space is fashioned can have a profound impact on social processes'. (Harvey 1970: 49). The correctness of this statement can be clearly seen on the example of the part of Kondratowicza Street and the Świętego Wincentego Street in the Bródno housing estate, where restaurants and bars are located in two-storey pavilions built especially for commercial functions (shops and gastronomy).

Another observation related to the described above Bródno housing estate is the issue of eating establishments located in the CH Atrium Targówek shopping centre. The survey shows that a large part of them belongs to global restaurant chains, such as Pizza Hut, McDonald's or Starbucks. Although such restaurants and bars resemble those described by Augé (1995) as 'non-places', field observations show that they are treated by guests (especially teenagers and children) similarly to classic 'places'. This directly refers to the view of Massey (1999), who sees the 'place' through external connections to the rest of the world.

References

Augé M (1995) Non-places. Introduction to an anthropology of supermodernity (English trans by Howe J). Verso, London–New York

Bystrzanowski J, Dutkowski K (1975) Wszystko o Warszawie. Informator. Sport i Turystyka, Warszawa (Maps drew by Ostrowski J, Ostrowski W)

Harvey D (1970) Social processes and spatial form: an analysis of the conceptual problems of urban planning. Pap Reg Sci25, 1(April):47–69
Jankowska E (2018) Enzo Rossi: Ludzie mówili - nic tu nie ma, poza ślepą drogą, a ten wariat otwiera pizzerię. Metro Warszawa. http://metrowarszawa.gazeta.pl/metrowarszawa/7,141634,23075516,enzo-rossi-o-pizzy.html#Z_BoxLokWawImg. Accessed 27 Feb 2018
Massey D (1999) Power-geometries and the politics of space-time. Hettner-Lecture 1998, Hettner-Lectures 2, Department of Geography, University of Heidelberg, Heidelberg
Menu (2018). https://www.dagrasso.pl/menu. Accessed 20 Feb 2018
Miasto Stołeczne Warszawa. https://www.um.warszawa.pl. Accessed 30 April 2018
Miejski system informacji. https://zdm.waw.pl/miejski-system-informacji. Accessed 30 April 2018
O nas. https://www.dagrasso.pl. Accessed 01 July 2017
Sołtan A (eds) (2002) *Historia Białołęki i jej dzień dzisiejszy*. Gminny Ośrodek Kultury Warszawa-Białołęka. Warszawa
Trelewicz M (2003) *Różnorodność warszawskiej Białołęki*. Promotor, Warszawa

Chapter 9
Suburbanisation and Gastronomic Services on the Outskirts of Warsaw (Poland): Piaseczno

Andrzej Kowalczyk and Sylwia Seremak

Abstract Urbanisation causes functional and spatial changes in the areas around cities, also called metropolitan areas. There are several reasons why eating establishments appear in suburban areas: (1) they are situated in settlements that have been independent towns or larger villages, (2) bars or restaurants appear in new settlements whose emergence is closely related to the process of suburbanisation, (3) dining outlets appear along the most important roads and at intersections, (4) shopping centres, logistic centres, office parks, etc. in the suburban area are often accompanied by bars and cafeterias and (5) suburbs with a developed leisure function, restaurants, bars and cafés appear in places intended for outdoor recreation. This chapter draws upon the example of the district of Piaseczno (south of Warsaw), which consists of six municipalities. Three of them directly adjoin Warsaw and are parts of its suburbs, while the three others are largely rural, although they now include many new housing estates. It is reflected in the distribution of eating establishments.

Keywords Suburbs · Metropolitan area · Gastronomic services · Warsaw · Piaseczno district

9.1 The Increasing Importance of Gastronomy in Suburban Areas

Urbanisation, and above all the territorial development of cities, causes major functional and spatial changes in the areas around cities. These changes also apply to gastronomy. In historical times (Antiquity, the Middle Ages), there were inns

---

A. Kowalczyk (✉) · S. Seremak
Department of Tourism Geography and Recreation, Faculty of Geography and Regional Studies, University of Warsaw, ul. Krakowskie Przedmieście 30, 00-927 Warsaw, Poland
e-mail: akowalczyk@uw.edu.pl

S. Seremak
e-mail: sylwia.seremak11@gmail.com

© Springer Nature Switzerland AG 2020
A. Kowalczyk and M. Derek (eds.), *Gastronomy and Urban Space*,
The Urban Book Series, https://doi.org/10.1007/978-3-030-34492-4_9

and other establishments associated with gastronomy on the outskirts of cities. Their number and importance clearly increased with the spatial development of cities, which began in the nineteenth century. This phenomenon grew stronger in the twentieth century with the emergence of suburbanisation. As Parham suggests (2015b, p. 120), 'The city and its surrounding productive countryside have historically enjoyed a symbiotic relationship, which has been critical in shaping urban growth and development'. This problem was noticed at the beginning of the nineteenth century by Johann Heinrich von Thünen, author of 'The Isolated State' (1826). His views expressed in the concentric model of land use have had a great influence on the development of location theory and 'over the years, the model has attracted attention from some of the most well recognized names in geographical research' (O'Kelly and Bryan 1996, p. 473).

Elsewhere, Parham claims that '...urban peripheries, and the wider regions influenced by, or becoming urban settlements, are the loci for a series of food-related, spatialized issues. Among others these include problems of urban sprawl, the presumption of primacy for urban development in the context of the changing nature of farming on urban edges with the advent of a post-productivist agricultural model, the argued need to protect and localise food-sheds, and the transforming practices of peripheral, conurbation and rural food consumption and gastronomic tourism' (Parham 2015b, p. 118).

Dining options in the suburban area are mainly located in shopping centres and along major roads (Newman 2014). Newman's research at Stevenson, a suburb of Vancouver, British Colombia (Canada), showed that gastronomic establishments were randomly distributed around the city centre, while the only noticeable concentration was seafood restaurants along the Fraser River (Newman 2014, p. 10). A similar survey was conducted in Washington DC (Leslie et al. 2012). The aim of these studies, however, was not just to describe the phenomenon but, above all, to explore the relationship between the distribution of food outlets and the quality of food and lifestyle.

There are several reasons why eating establishments appear in suburban areas. First, they are situated in settlements that have been independent towns or larger villages. As in the past, they are mainly used by residents. Second, bars or restaurants appear in new settlements whose emergence is closely related to the process of suburbanisation. Third, dining outlets appear along the most important roads and at intersections. Fourth, shopping centres, logistic centres, office parks, etc. in the suburban area are often accompanied by bars and cafeterias (Parham 2015a, pp. 226–229). Finally, in suburbs with a developed recreational function, restaurants, bars and cafés appear in places intended for outdoor recreation.

One example of a large urban area, where the suburban zone is well-developed in the context of gastronomy is the Victoria–Kowloon metropolitan area in the Hong Kong Special Administrative Region of the People's Republic of China. Many places are visited by Hong Kong residents and visitors not only for tourism, but also for lunch or dinner. The more famous include Stanley (Fig. 9.1) on Hong Kong island, Sai Kung in the continental part of Hong Kong (the New Territories) and Tai O on the island of Lantau.

Fig. 9.1 Fashionable restaurants in the suburbs of Victoria (Hong Kong Special Administrative Region of the People's Republic of China). **a** Restaurants on the waterfront at Stanley. **b** Jumbo Kingdom, or the Jumbo Floating Restaurant, in Aberdeen was established in 1976 by Dr. Stanley Ho, one of the richest people in East Asia. The restaurant can accommodate up to 2,000 diners. Source: Anna Kowalczyk 2011

9.2 Gastronomy in Suburban Areas—An Example from the Peripheries of Warsaw

9.2.1 The Geographical Distribution of Eating Establishments in Piaseczno

In big cities, restaurants and bars are mostly located in areas that do not play a tourist or recreational role, but are more important for housing. Examples include the Piaseczno district in the southern part of the metropolitan area of Warsaw (Seremak 2017). Piaseczno covers over 621 square kilometres and in June 2016 had a population of 179,496. It consists of six municipalities: Góra Kalwaria, Lesznowola, Konstancin-Jeziorna, Piaseczno, Prażmów and Tarczyn. Around 47.81% of the population lives in the towns of Góra Kalwaria, Konstancin-Jeziorna, Piaseczno and Tarczyn. In 2016 there were 228 catering establishments, nearly half of them in the municipality of Piaseczno.

The first thing that can be seen from Table 9.1 is the big differences between the municipalities of Piaseczno in terms of the number of inhabitants, area and population density. In Piaseczno municipality population density is 632.9 people per square kilometre. A relatively high population density is also found in Konstancin-Jeziorna (in the east) and Lesznowola (in the west). This is because these three communes directly adjoin Warsaw and are part of its suburbs. Residential buildings, logistics centres, industrial facilities and roads dominate in many places despite the presence of arable fields, forests and meadows. They can thus be considered as urbanised areas, while the three remaining municipalities (Góra Kalwaria, Prażmów and Tarczyn) are largely rural, although they now include many new housing estates.

Table 9.1 Geographical characteristic of the Piaseczno district in 2016

Municipality	Population	Area (km^2)	Density of population (per km^2)
Góra Kalwaria	26,393	144.1	1832
Konstancin-Jeziorna	24,773	78.3	316.4
Lesznowola	25,129	69.3	362.6
Piaseczno	81,207	128.3	632.9
Prażmów	10,616	86.5	122.2
Tarczyn	11,378	114.3	99.5
Piaseczno district	179,496	620.8	289.1

The comparison of the data in Tables 9.1 and 9.2 shows that municipalities with the lowest ratio of people to eating establishments are Lesznowola (584.4 people per establishment), Konstancin-Jeziorna (651.9), Piaseczno (804.0) and Tarczyn (812.7), while the smallest saturation—highest ratio—is in Prażmów (2123.2). If we express the population of each municipality as a percentage of the total population of Piaseczno district, we can compare that to the percentage share of eating establishments, an indication of the 'surplus' of establishments. The highest surplus values are in Lesznowola (14.0% vs. 18.9%) and Konstancin-Jeziorna (13.8% vs. 16.7%), and the lowest are in Piaseczno (45.2% vs. 44.3%) and Tarczyn (6.3% vs. 6.1%). Values for the two municipalities in the southern part lie in the middle: Góra Kalwaria (14.7% vs. 11.8%), and Prażmów, the most rural of the municipalities (5.9% vs. 2.2%).

Overall, the largest concentration of eating outlets occurs in the centre of Piaseczno town, notably the town square. To the west of the centre, there are multifamily housing estates and local retail outlets, in the vicinity of which the gastronomic restaurants (mainly bars) are concentrated. Eating facilities are also located along the main streets in the town centre.

Table 9.2 Distribution of eating establishments in Piaseczno in 2016 by municipalities and type of outlet (survey conducted in 2016 by Sylwia Seremak)

Municipality	Types of eating establishments					Total	
	Restaurants	Bars	Cafés	Pubs	Ice cream parlours	Number	%
Góra Kalwaria	15	9	2	–	1	27	11.8
Konstancin-Jeziorna	23	9	6	–	–	38	16.7
Lesznowola	14	24	5	–	–	43	18.9
Piaseczno	46	33	15	4	3	101	44.3
Prażmów	2	3	–	–	–	5	2.2
Tarczyn	8	4	1	–	1	14	6.1
Total	108	82	29	4	5	228	100.0

Source: Seremak (2017)

The next biggest concentrations of gastronomic services are two shopping centres in the northern part of the city. This confirms the views of Parham that in the suburban areas the main tasks related to satisfying food needs are fulfilled by shopping malls, hypermarkets, etc. (Parham 2015a, pp. 144–156). In the Fashion House Outlet shopping centre, food and beverage facilities are located in the central part of the building, in a specially designated zone, where tables are shared by customers of all the bars and ice cream parlours. Further north, close to the boundary with Warsaw, is the CH Auchan shopping centre. Here, dining establishments are definitely more widely dispersed throughout the centre. A high concentration of eating places also occurs in Józefosław village, near the southern border of the Kabacki Forest bordering directly onto Warsaw. It has the largest population after Piaseczno town. According to the 2011 census, 7,130 people live there. Józefosław is mainly characterised by a new, single- and multifamily housing development. There are eight gastronomic points in the village: six restaurants and two cafes, situated on the main streets. At their intersection there is a small shopping centre with a café belonging to the Carte d'Or Cafe chain.

Clusters of gastronomic establishments also occur in the villages of Zalesie Górne and Zalesie Dolne, close to recreational facilities. Zalesie Górne (3,340 inhabitants in 2011) is the second-largest village in the municipality. The proximity of the large forest and the Wisła Recreational Centre influence the recreational character of the village, where it is possible to rent equipment, use the swimming pool and play paintball. In Zalesie Górne there are seven establishments: three restaurants, two bars, a pub and a café. They are mainly located in the centre of the village.[1] Since May 2017, the Centre has had new investors who plan to add food trucks and a bar offering cocktails and craft beers.

Some eating establishments in the Piaseczno municipality are located along roads that pass through it. There are three such roads: route 79 (Góra Kalwaria–Piaseczno–Warsaw) on which there are two shopping centres (discussed above) and bars at petrol stations on route 721 (Lesznowola–Piaseczno–Konstancin-Jeziorna) and route 722 (Prazmów–Piaseczno), both of which mainly have bars. In the north west, the Lesznowola municipality lies adjacent to Piaseczno. Lesznowola has 43 eating establishments. One cluster occurs at Mysiadło village (3,610 inhabitants). Bordering directly with Warsaw it is characterised by new housing estates. At Mysiadło there are nine dining facilities: five restaurants, three bars and a café, located on the main streets of the village.

A cluster of eating outlets also occurs in Łazy village (population 2,335). It is one of the largest settlements in the Lesznowola municipality. Expressway 7 (Gdańsk–Warsaw–Kraków) passes through it. Food and beverage outlets are located in the centre of the village and along the main transit highway. There are seven eating outlets: three restaurants, three bars and a café.

---

[1]In summer there are small bars and food stands at the Wisła Recreational Centre, but they were not operating during the 2016 inventory and were not included in the analysis.

There is an interesting cluster of establishments in Wólka Kosowska. Although only small (population 769), it is a very important commercial centre. Since the 1990s, it has hosted clothing wholesalers, suppliers of electronic equipment, furniture, etc. making it the largest logistics centre in the Warsaw area for goods imported from Asia. The Chinese Shopping Centre lies on the north side of the road that runs through the village, and on the south side there are warehouses. The village has 13 bars offering Chinese, Vietnamese and Turkish cuisine. Most are situated in and around warehouses.

It is important to briefly mention the history of Konstancin–Jeziorna. The town was created in 1969 as a result of the merger of two settlements, Konstancin and Jeziorna. Konstancin (or Skolimów–Konstancin), founded in 1897, was created in accordance with the concept of Howard's garden city, as a place for the rich inhabitants of Warsaw to relax. At that time Konstancin was connected with Warsaw by a narrow gauge railway. Initially, the city was dominated by second homes, but in 1920s–1930s many owners moved to Konstancin permanently. This was made possible by a good rail connection (45 min travel time) and the increasing ownership of cars. A large part of the town's population during that period were military officers, senior officials in ministries, famous actors and writers, owners of factories and bankers, etc.

Research from 2016 showed that there were 38 dining establishments in the municipality; 36 were in Konstancin–Jeziorna town (17,184 inhabitants). The presence of such a large number of eating outlets is partly due to its specific function, it was formerly a spa town. In addition, a significant number of restaurants, bars and cafés are associated with the fact that the town is home to people with high incomes: millionaires (owners of the largest Polish industrial and commercial companies, directors of the largest banks, TV and broadcasting stations etc.), television celebrities, actors, famous writers, journalists and scientists. In addition, the town's residents include several ambassadors, together with representatives of foreign banks and insurance companies. The largest concentration of dining outlets is at the Park Zdrojowy (Spa Park in English) and along the main streets. The Park is perceived by visitors and residents not only as a place for rest and recreation, but as a symbol of the town and most eating establishments are causal restaurants.

The second concentration is the shopping centre, Stara Papiernia, (the Old Paper Mill—it is located in a paper mill that dates from the eighteenth century) in the northern part of the city. Here, there are four eating establishments, which are scattered among the other service and commercial stores. There is also a cluster of dining venues on the main transit street, which runs alongside the Stara Papiernia. The gastronomic character of Konstancin–Jeziorna is distinguished by the location of several (usually luxury) restaurants in historic villas and buildings dating from the 1920s that are connected with the history of the town. This underlines its prestigious nature, both past and present. The example of Konstancin-Jeziorna confirms the views of the authors who believe that such places are visited by '… visitors who are primarily motivated by the opportunities to experience peri-urban landscapes, enjoy locally focused restaurants…' (Parham 2015b, p. 122).

To the south lies Góra Kalwaria municipality (population 11,868) which, in 2016, had 27 eating places. The greatest concentration is in the town centre and along route 79 (Góra Kalwaria–Piaseczno). Somewhat surprisingly there was only one restaurant in the village of Czersk, which has a fourteenth-century castle that is situated above the picturesque Vistula River valley and is frequently visited by tourists. On the west side, the municipality of Prażmów is adjacent to the municipality of Góra Kalwaria. The inventory showed only five eating establishments, and only one restaurant was located in Prażmów village, which was the seat of local authorities. To the west of Prażmów itself (and to the south of Lesznowola municipality) lies the municipality of Tarczyn. The number of eating places here is slightly higher (14 in total), which is because the town (population 4,105) is the seat of local authorities and situated on expressway 7. Before the expressway bypassed the town, there were several restaurants and bars in the town square, and many have been preserved. Other outlets are by the expressway, mainly gas station.

In summary, then, we can see that the eating establishments in Piaseczno district are primarily concentrated in the following areas:

- in Piaseczno itself. This is the largest town and the seat of district authorities. Here, there are two clusters:
  - in the commercial and administrative centre and the densely built-up residential and service zone near the downtown;
  - on new housing estates (a zone west of the centre of Piaseczno town);
- at Józefosław village, in new, detached and semi-detached housing estates in the north-eastern part of Piaseczno municipality;
- in the largest shopping centres, CH Auchan, Fashion House Outlet;
- in smaller settlements that are the seat of local authorities: Góra Kalwaria, Lesznowola and Tarczyn;
- in settlements with special functions:
  - at Konstancin-Jeziorna, a town with spa and housing functions;
  - in areas with outdoor recreation: Zalesie Górne and Złotokłos in Piaseczno municipality,
  - at Wólka Kosowska (Lesznowola municipality), a settlement with a well-developed commercial function;
- along the major transit routes: expressway 7, routes 79, 721 and 722.

A detailed analysis of the distribution of eating outlets in Piaseczno district shows that 60% of all facilities are located within 5 kilometres of Warsaw's borders. This figure is influenced by the towns of Piaseczno and Konstancin–Jeziorna. Józefosław and Mysiadło also lie in this area, and they are distinguished by their typical suburban features and good transport links with Warsaw. In other zones the number of restaurants, bars etc. is similar: 15% of all dining facilities lie within 5–10 kilometres; 12% within 10–15 km; and 13% within 15–20 km. The map in Fig. 9.2 shows that at a distance of over 20 kilometres the only clusters are at the towns of Góra Kalwaria and Tarczyn.

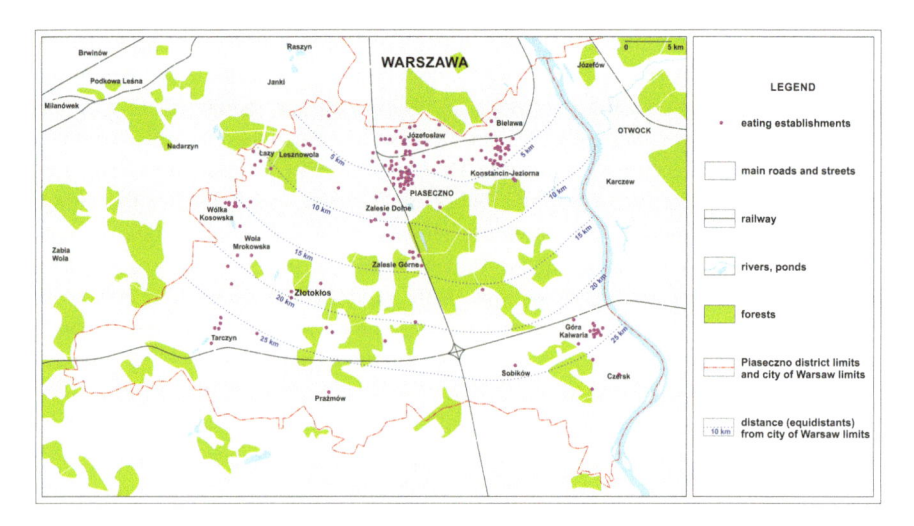

Fig. 9.2 Distribution of eating establishments in the Piaseczno district (Poland) in 2016

9.2.2 Types of Eating Establishments in Piaseczno

Table 9.2 shows that out of 228 eating establishments in the district, restaurants accounted for 48% and bars for 35%. However, in each municipality, the number of individual types of dining facilities is slightly different. Although they are usually dominated by restaurants, there are more bars in Lesznowola, partly due to the concentration of Asian bars in Wólka Kosowska.

Some dining outlets offer ethnic cuisine. Although 39 restaurants and bars specialise in Polish cuisine (21% of the total number), while 33 (36% of all ethnic restaurants) serve Italian cuisine. Other types of cuisine are much less frequent. In 2016, there were 15 Turkish restaurants and bars, 10 Vietnamese, nine Japanese, six Chinese and eight others with dishes from different East Asian countries. Others include American (three fast-food restaurants: McDonald's, KFC and a burger bar), Indian (two restaurants and bars) and Georgian, Serbian, Syrian and Cambodian.

The biggest number of dining establishments offering ethnic cuisine is in the city of Piaseczno. The other cluster is in Lesznowola municipality. Bars serving Vietnamese and Chinese cuisine in Wólka Kosowska are very popular not only with residents, but also people who visit from Warsaw. Their atmosphere and interior design are unattractive, due to the nearby market halls. They mainly cater to the Asian community that works and lives here, and consequently offer an opportunity to sample authentic Chinese and Vietnamese cuisine. Cafés are also important. Clusters of cafes occur at key leisure locations (e.g. the town square in Piaseczno, Park Zdrojowy in Konstancin–Jeziorna). Other clusters can be found in shopping centres, which usually belong to chains that include Green Coffee Nero, So! Coffee and Vincent.

Some eating establishments are situated in interesting historical buildings. For example, the Odjazd and Zawrotnica restaurants in the centre of Piaseczno town are inside the old narrow gauge railway station (the names of these restaurants mean 'departure' and 'train's turnpike'). In Konstancin–Jeziora, the Park Café & Rest and the A.D. 1899 restaurants are situated in historic buildings. The first is in a neo-classical villa from the 1930s, and the second is in a house built in 1899. The Oycowizna inn, in Lesznowola village is also interesting. This is in a large building covered with straw thatch. It is partly built of wood and partly of brick. The wooden part was completely moved, preserving the original elements of a rural cottage from 1905. The gastronomic infrastructure of the Piaseczno district is not unique: other districts around Warsaw have similar characteristics. On the one hand, there are eating establishments that have existed there for many years and, on the other hand, there are many restaurants, bars and cafés that have opened in recent decades as the result of the suburbanisation process.

9.3 Conclusions

The situation in the Piaseczno district confirms the thesis about the rapid development of the gastronomic function in the areas located in the suburbs of large agglomerations. The research also confirms Parham's observations regarding contemporary suburbanisation, according to which food establishments include '…big box food stores, hypermarkets, fast food outlets and chain restaurants, petrol station forecourt "road pantries" and the food courts of outlet and megamalls' (Parham 2015b, p. 124). Although the Piaseczno district is part of the metropolitan zone of Warsaw, which was the most suburbanised, the growth of gastronomic services can also be noticed in other districts neighbouring Warsaw. However, what differs Piaseczno from them is a more diverse socio-economic structure of the residents. In the vicinity of Piaseczno, next to the settlements populated by the middle class, there are enclaves of wealth inhabited by millionaires, representatives of foreign corporations, politicians, etc. Undoubtedly, this is the same as the presence of the only resort in the Warsaw region (Konstancin-Jeziorna spa town) that determines a significant number of restaurants and bars with a high standard of services.

But the Piaseczno district, mainly due to the geographical location (not only the proximity of Warsaw, but also the short distance to international airport and expressways) is also a cluster of logistic centres, retail centres, etc. These factors influence the existence of many fast-food restaurants and bars in the area of Piaseczno, as well as outlets with ethnic cuisine. In the latter case, it is related to the large wholesale centre in Wólka Kosowska, which mainly belongs to the Chinese, Vietnamese, and immigrants from Turkey.

From the theoretical point of view, the situation the Piaseczno district clearly refers to the concentric zone theory by Burgess (1925) and the sector theory by

Hoyt (1939), which were created to describe the city's internal structure as such, but may also be useful for the study of the suburban area. At the same time, the increased concentration of eating establishments in larger urban centres is confirmed by the central place theory (see Chap. 2 for more details).

References

Burgess EW (1925) The growth of the city: an introduction to a research project. In: Park RE, Burgess EW, Mc Kenzie RD (eds) The city. The University of Chicago Press, Chicago and London, pp 47–62
Hoyt H (1939) The structure and growth of residential neighborhoods in American cities. Federal Housing Administration, Washington
Leslie T, Frankenfeld C, Makara M (2012) The spatial food environment of the DC metropolitan area: clustering, co-location, and categorical differentiation. Appl Geogr 35(1–2):300–307
Newman L (2014) Commensality, sustainability, and restaurant clustering in a suburban community. Suburb Sustain 2(2), Article 2. http://scholarcommons.usf.edu/cgi/viewcontent.cgi?article=1016&context=subsust. Accessed 20 March 2018
O'Kelly M, Bryan D (1996) Agricultural location theory: von Thünen's contribution to economic geography. Prog Hum Geogr 20(4):457–475
Parham S (2015a) Food and urbanism: the convivial city and sustainable future. Bloomsbury Publishing, London–New Delhi–New York–Sydney
Parham S (2015b) The productive periphery: food space and urbanism on the edge. In: Cinà G, Dansero E (eds) Localizing urban food strategies. Farming cities and performing rurality, 7th international Aesop sustainable food planning conference proceedings, Torino, 7–9 October 2015, Politecnico di Torino, Torino, pp 118–130
Seremak SA (2017) Rozmieszczenie i charakterystyka usług gastronomicznych w powiecie piaseczyńskim. MA thesis, Wydział Geografii i Studiów Regionalnych, Uniwersytet Warszawski

Chapter 10
Eating Establishments in Smaller Cities and Towns in Poland (on Selected Examples)

Małgorzata Durydiwka

Abstract This chapter highlights that the number and structure of eating establishments in smaller cities and towns is consistent with the hierarchical relationship predicted by central place theory. It also highlights other factors that affect the development of gastronomy in smaller cities and towns. One is the tourist function, which is shaped by different types of tourism; others include the city's location, its history and contemporary development. This chapter seeks to explore, among other things, the popularity of ethnic cuisines, vegan and vegetarian food, and eating establishments in redeveloped buildings. It looks at the culinary landscape in some smaller Polish cities and towns and discusses the creation of culinary trails as a way to promote sightseeing in the city.

Keywords Tourism · Small cities and towns · Poland · Revitalisation · Promotion

10.1 Introduction

Eating establishments are one of the characteristics of urban space. But the character and the dynamics of changes in the development of catering services in cities depend on many factors, including: the size of the city or town itself (in terms of its population), its location, its economic and administrative position, and the level of the development of its tourist function. At the same time, the development of catering services can be essential for the economy of the city (Au and Law 2002), especially in the context of the development of tourism. In some cities (e.g. Burgos) strategic plans for catering services are being worked out, and they are regarded as a tool for the development of tourism and for the creation of new jobs (Gastronomic cities. City strategy on gastronomy as tool for tourism and employment development. Feasibility study—Fermo 2014). The presence and diversity of eating

---

M. Durydiwka (✉)
Department of Tourism Geography and Recreation, University of Warsaw, ul. Krakowskie Przedmieście 30, 00-927 Warsaw, Poland
e-mail: mdurydiw@uw.edu.pl

© Springer Nature Switzerland AG 2020
A. Kowalczyk and M. Derek (eds.), *Gastronomy and Urban Space*,
The Urban Book Series, https://doi.org/10.1007/978-3-030-34492-4_10

establishments contributes to a higher evaluation of the level of attraction of the city for tourists. But the locals also use the eating establishments. In countries such as France, Spain, Greece and the Czech Republic, eating out is very popular. It is becoming more popular in Poland too. According to the report "Polska na talerzu" ("Poland on the plate" in English)[1] (2015), nowadays 88% of Poles eat out, with 14% doing so at least once a week, as compared to 2005 when only 23% stated that they ate out. Eating out has a social context for Poles: it is an opportunity to meet with friends and family. Many people organise parties for various occasions (birthdays, name days, communions, examinations) in eating establishments, which is an indication of both convenience and prestige. It should be stressed, however, that compared to other nations, Poles still spend little on eating out. According to Eurostat data, Poles spend 3.2% of their household budgets for accommodation and catering services, while Czechs, for instance, spend 8.4%, Slovaks, 5.4%, and Germans, 5.3% (Fronczak 2016).

In the era of the centrally planned economy, catering services were largely state-controlled and planned. Since they did not create new material values, their development was not regarded as a priority. Their shortage was particularly visible in smaller cities and towns (Szymańska 2003). Nowadays this situation is clearly improving, which can be noticed not only in towns with a distinct tourism function but also in towns with other social and economic functions. To a large extent, this is due to changes in the property structure of eating establishments, and to the entrance into the market of facilities owned by various chains, including the use of franchising by new chains (e.g. Chłopskie Jadło).

For the purposes of this study, the location and character of eating establishments—restaurants, bars and cafés—in selected Polish cities and towns (with a population below 500,000) have been analysed. The establishments have been inventoried using web sources, such as Google Maps, Zomato, TripAdvisor, and city and town hall web sites. Available studies that deal with the development of catering services in cities and towns have also been used, as well as results of the author's inventories and own observations.

10.2 Structure of Eating Establishments in Smaller Cities and Towns

Data from the National Statistical Office shows that in the past decade there has been a gradual decrease in the number of bars and food stands in favour of restaurants. This is particularly visible in towns and cities, irrespective of their size (Gheribi 2013). An analysis of revenue from catering services (up from 17.7 million

---

[1]The report *Polska na talerzu* deals with the culinary customs of Poles. It has been prepared on the basis of a survey conducted in March 2015 by IQS on behalf of MAKRO Cash & Carry using the CAWI methodology, using a sample of 1000 Polish web and internet users above 15 years of age.

Table 10.1 Total number of eating establishments in Poland in 2005 and 2016

Category of eating establishments	2005	2016
	Number	
Restaurants	9716	20,018
Bars	40,834	20,844
Food stands	34,572	24,083

Source: National Statistical Office

PLN in 2005 to 30 million PLN in 2016) shows that the value of the catering market is increasing (Table 10.1).

However, the development of catering services in smaller towns and cities measured by both the number of facilities and their diversification with respect to the type of facility and the kind of dishes (cuisine) served, is not uniform. It is more intensive in cities than regional centres and in areas that have a better-developed tourism function. This is also reflected in the frequency of use of services. Fronczak (2016) points out that the larger the town, the higher the percentage of people who use such services; the same is true for education levels.

An analysis of the number and diversity of eating establishments shows that catering services can be discussed in terms of democratisation (Barrère et al. 2014). For instance, in Gdynia, a town with a population of 250,000, which is an important economic, academic and tourist centre in northern Poland, there are over 220 eating establishments, including 86 restaurants, 19 pizzerias, 53 bars (35 fast-food type) and 39 cafés (http://gdynia.pl/turystyczna-pl). In Białystok, a town with a population of 300,000, which is an important regional centre in north-eastern Poland, and at the same time a city with a fairly well-established tourism function (educational tourism), there are 176 eating establishments, including 105 restaurants and 21 bars. As the town only had two restaurants and a few bars in the 1970s, this can be regarded as a significant development (Mniszewski and Werpachowska 2009). Some establishments are franchises, owned by global chains such as McDonald's (five), KFC (five) and Pizza Hut (one), or by domestic chains such as Da Grasso (three), Biesiadowo (one) and Savona (five) (Table 10.2).

In smaller, sub-regional centres there are other new developments. One example is in the town of Siedlce (population 77,000) in the eastern part of the Mazovian municipality. Studies by Ornowski (2003) and Próchnicka (2017) have shown an increase in the number of eating establishments in the town, particularly restaurants (Table 10.3).

In smaller towns with a well-developed tourism or health resort function, such as Kazimierz Dolny,[2] Krynica-Zdrój and Władysławowo, catering facilities are one of the more important factors that shape this function. Of course, the number and type of catering facilities vary, depending on the tourist character of the town. For

---

[2]The tourist function in Kazimierz Dolny began to develop in the late nineteenth century, when the town became a favourite summer resort visited by people from Warsaw, Lublin and other Polish cities. Nowadays, Kazimierz, with its reconstructed historic heritage, is regarded as an architectural gem, a Mecca for artists and an attractive tourist centre (Historia 2017).

Table 10.2 Eating establishments in Białystok (Poland) by type in 2017

Type of eating establishments	Number	%
Restaurant	105	59.7
Bar	21	11.9
Fast foods	19	10.8
Pub	11	6.3
Café	10	5.7
Bistro	6	3.4
Wine bar	2	11
Others	2	1.1
Total	176	100.0

Source: https://www.zomato.com/pl/bialystok/restauracje

Table 10.3 Eating establishments in Siedlce by type in 2003 and 2016

Type of eating establishments	2003 Number	2016
Restaurant	7	24
Pizzeria	9	12
Bar	21	12
Kebab bar		8
Fast food		3
Pub	10	6
Café and café-bar	2	8
Others	8	4
Total	57	77

Source: Adapted from Ornowski (2003) and Próchnicka (2017)

instance, in Kazimierz Dolny, an important centre of landscape-oriented tourism that is dominated by weekend tourism, there are 77 eating establishments, mainly restaurants (32), bars (21) and cafés (11) (Ługowska 2014). In Krynica-Zdrój, a town located in Beskid Sądecki with a spa tradition going back over a century, which has recently become an important centre for recreation (skiing) and business tourism, there are 56 establishments, including 32 restaurants, five bars and 16 cafés (Malitka 2014). In Władysławowo, a typical seaside resort, there are 43 eating establishments, including nine restaurants, five pizzerias, 17 bars, five fish grills and seven cafés (Gastronomia 2017). Tourist towns that have a fairly high turnover of tourists—many of whom are families with children or groups of young people—are characterised by a high proportion of bars. There are two main reasons for this: first, the relatively low prices, and second, the short waiting time. While in larger urban centres and transit-oriented towns, fast food facilities are common, they are much harder to find in many smaller tourist towns.

Catering services can be characterised by their diversification, for instance, the creation of ethnic establishments. In most European cities modern developments in catering services coincided with the inflow of immigrants, mostly from Asia (Turks, Kurds, Hindus, Pakistani, Chinese, Vietnamese, Thai) (Mniszewski and

Werpachowska 2009). In smaller Polish cities and towns it was rare for ethnic restaurants and bars to be managed by immigrants but, in recent years, this has changed, which undoubtedly contributes to improvements in the quality of food. The most popular establishments serve Polish and Italian food, which corresponds to the tastes of Poles, most of whom prefer facilities with Polish (56%) and Italian food (52%) (Polska na talerzu 2015). In Białystok, for instance, there are 67 establishments serving Polish cuisine (38.1% of all eating establishments in the city). Kamler (2017) identified 37 eating establishments that served regional food, which is one-fifth of the total. Here, we are talking about Podlasie region cuisine, which combines influences from Lithuanian, Belorussian, Tartar, Russian, Ukrainian, and Jewish cuisines, and includes characteristic dishes like potato babka and, stuffed dumplings (*kołduny*) and baked stuffed pastries (*pierekaczewnik* and *czeburaki*). In a way, this is a response to growing interest in traditional regional cuisine among Poles. Similar tendencies can be observed in smaller tourist towns, where many restaurants serve traditional local food. One example is *Lemko* cuisine with its traditional Lemko *pierogi*, stuffed cabbage rolls (*kiszenice*) and cabbage soup (*warianka*) in Krynica-Zdrój. Another is Podhale region cuisine with dishes such as sour cabbage soup (*kwaśnica*), whey and cottage cheese soup (*zuwka*), and grilled smoked cheese (*oscypek*).

Larger urban centres are characterised by the presence of national and ethnic restaurants. For instance, in Białystok, establishments serving Italian food constitute about 20% of all eating establishments—mostly pizzerias. On the other hand, few restaurants serve Far Eastern food (around 20); most often, these are sushi bars and restaurants (eight in total). Sushi bars are becoming more popular, however, and their number is clearly growing, especially in multi-function centres, where the population is mostly young and educated. In Siedlce, kebab bars are fairly popular; they constitute 10% of all eating establishments in the town.

A characteristic feature of smaller urban centres is a shortage of establishments serving vegetarian and vegan food. While in Gdynia, for instance, there are over 60 eating establishments regarded by the website TripAdvisor as vegetarian-friendly[3] and around 30 that serve vegan food they are rare in smaller towns (Table 10.4).

10.3 Characteristic Features of the Location of Eating Establishments in Smaller Cities and Towns

The location of eating establishments in urban space depends on many factors. According to Wilk (1984, after Kowalczyk and Derek 2010), these include: the size of the town; its spatio-functional structure; transportation (and communication)

---

[3]It is worth adding that vegetarian food as a fashion is a key episode in the culinary history of Gdynia. The main advocate and promoter of this phenomenon was the Green Way chain of bars, founded in 1997 in Gdynia. One of the first establishments was opened in Abraham street and it is still in business today (http://www.kulinarnagdynia.pl/historia-szlaku/lata-90, accessed 30 Nov 2017).

Table 10.4 Eating establishments serving vegetarian and vegan food in selected smaller cities and towns in Poland in 2017

Selected smaller cities and towns	Number of eating establishments serving	
	Vegetarian food	Vegan food
Gdynia	61	29
Białystok	54	4
Siedlce	3	–
Słupsk	13	1
Krynica-Zdrój	7	–
Kazimierz Dolny	11	4
Władysławowo	5	–

Source: Adapted from TripAdvisor (Accessed 26 Nov 2017)

routes; the layout and function of service facilities; and the location of larger population groupings. An initial analysis of the distribution of eating establishments in smaller towns and cities in Poland shows that they are concentrated mostly in central parts, where trade, services, entertainment facilities, and the main tourist attractions are found, especially historical areas. In Gdynia and Białystok, for instance, about 40% of eating establishments are located in the very centre of the town, i.e. in the most tourist-oriented urban space. In Siedlce, the share is higher, about 60% (Fig. 10.1). A similar situation exists in Słupsk and Krynica-Zdrój (almost half of all eating establishments). In the remaining parts of town, catering facilities are more dispersed. Therefore, the density of establishments decreases with the distance from the town centre. At the same time, a growing number of eating establishments are being opened in residential neighbourhoods on the outskirts, notably in Gdynia and Białystok.

The localisation of eating establishments in small and medium towns and cities is affected by both their distance from the city centre and main communication routes. The topography of the area and its spatial layout can mean that the distribution of establishments is radial (e.g. Krynica-Zdrój) or concentric-radial (e.g. Słupsk).

In towns and cities that are regional or sub-regional centres, a fairly large proportion of establishments are in shopping centres, where entire floors (or parts thereof) are often designed as food courts. The location of shopping centres in urban space varies. Some are located almost in the town centre (e.g. Galeria Siedlce in Siedlce, CH Batory in Gdynia and the Alfa, Atrium and Galeria Jurowiecka centres in Białystok). Others are situated on the outskirts (e.g. CH Riviera in Gdynia and CH Zielone Wzgórze in Białystok).[4] In Galeria Siedlce, for instance, there are 15 establishments, which is one-quarter of all such establishments in the

---

[4]Independent of their location with respect to the city centre, they are easily accessible by public and private transport.

Fig. 10.1 Eating establishments in Siedlce (Poland) in 2017. Source: Adapted from Próchnicka (2017, p. 51), underlay map by Andrzej Kowalczyk

town. Shopping centres situated in the central district of Białystok (CH Alfa, CH Atrium and Galeria Jurowiecka) also contain about one-quarter of all facilities. In contrast, centres situated in the central district of Gdynia (Batory and Gdynia Waterfront) account for only 6%.

In terms of precise location, there is a distinct tendency to open facilities in historical buildings, often modified for the purposes of catering. This can be seen in towns and cities of all sizes and across a range of functions (Table 10.5). The range of historical buildings used is very diverse, running from residential (e.g. the restaurant in Abraham's Little House[5] in Gdynia, Fig. 10.2), to administrative (e.g. the Bagażownia in Sopot, in the historical building of the former luggage depot at the railway station).

Eating establishments are also an important element of revitalised urban spaces.[6] The economic character of the city, both past and present, determines the way the revitalised facilities and entire town districts are developed. A good example of this

---

[5]Antoni Abraham (1869–1923) was a well-known Kashubian social activist, nicknamed the 'King of the Kashubian people', and an advocate of the Polish character of Pomerania. The house at 30 Starowiejska street in Gdynia was never owned by him, but he and his family lived there during the last three years of his life. In 1936, a commemorative plaque was installed on the building (Fryc 2016).

[6]The present study deals only with the spatio-functional aspects of revitalisation.

Table 10.5 Eating establishments in historic buildings—selected examples from small cities and towns in Poland

Selected small cities and towns	Examples
Gdynia	Restaurant Sempre in Abraham's Small House (1904)
	Restaurant Główna Osobowa in a tenement house, from the period of Gdynia modernism
Sopot	Restaurant Bagażownia in the historical building of the former luggage depot at the railway station
Białystok	Restaurant Peper's in a building from the first half of the nineteenth century
	Esperanto Cafe in the historical Town Hall
Siedlce	Restaurant brewery Brofaktura in the historical Market Hall, early twentieth century
Słupsk	Restaurant Anna de Croy in the sixteenth-century Pomeranian Dukes' Castle
	Restaurant Biały Dom in a manor house from the interwar period
Wisła	Restaurant Zdrojowa in Dom Zdrojowy (mid-1930s)
Cieszyn	Restaurant Kamienica Konczarowskich in a 16th-century tenement
	Wine bar U Czecha in a building from the turn of the seventeenth century
Krynica-Zdrój	Restaurant Dwóch Świętych in the guest house Małopolanka (1920s)
	Restaurant Pod Zieloną Górką in a historical villa from 1850

Fig. 10.2 Sempre restaurant in the co-called Abraham's Small House in Gdynia. Entrance to the restaurant (**a**). Interior of the restaurant (**b**). Source: Mateusz Kaczmarek 2018

is the town of Ustka, where health resort and recreational tourism—the dominant economic sector—have imposed spatio-functional changes on districts that were linked to the declining maritime industry (cargo port, wharf, fish processing plant). At the same time, the status of Ustka as a health resort, along with decisions about

Fig. 10.3 Eating establishments in revitalised buildings at Ustka (Poland) in 2017. Restaurants in new tourist facilities at Ustka (**a**). The restaurant Pod Dębem in the wattle-and-daub house in the old part of Ustka (**b**). Source: Małgorzata Durydiwka 2018

spatial management and guidelines from the historic preservation office have restricted the construction of new facilities and the renovation and adaptation of existing ones. Another example is the post-industrial facilities and land lots at the Słupia river, which have been adapted to tourism and recreation (Poczobut 2010). This has led to the construction of guest houses and high-end apartments for tourists. Their lower floors are often occupied by eating establishments, mostly restaurants and cafés (Fig. 10.3a). Revitalisation work has also begun in the oldest, central district of Ustka, which used to be a predominantly residential neighbourhood. Wattle-and-daub houses have been renovated and adapted to new functions, often with catering facilities (Fig. 10.3b).

Another good example is Zabrze, in the Upper Silesia region where post-industrial facilities now fulfil new functions. Although the history of Zabrze (coal mines and steel mills) may appear to preclude the introduction of catering functions, part of the revitalisation of the entire Guido mine complex included a pub in the former nineteenth-century Pump Room, 320 m below ground, which is the deepest-located establishment of this type in Europe. It serves, among other drinks, the locally brewed beer called Guido and hot snacks (Fig. 10.3). It also features a long counter lit from below, a projector for television broadcasts and wireless internet access. A restaurant was also opened in one of the buildings of the boarding house at the mine's Centre for the Mechanisation of Mining. Another restaurant has also been opened in the Maciej pit (http://www.szybmaciej.pl, accessed 27 Nov 2017), part of the old Concordia mine and in the former bathhouse of the Zabrze steel mill. The restaurant was included in the culinary guide published by Gault & Millau. Elsewhere, a modern restaurant with a sky bar (35 m above ground) is planned in the tower of a former meat factory (Fig. 10.4).

Fig. 10.4 Eating establishments in the buildings after revalorisation and restructurisation. Interior of eating establishments in former coal mine Guido in Zabrze (Poland). Source: Małgorzata Durydiwka 2016

10.4 Selected Factors for the Promotion of Eating Establishments in Smaller Cities and Towns

Using catering services is gradually becoming more popular in Poland. More eating establishments are being created and a range of actions is being taken to promote them. One example is the creation of culinary trails. The first such trail was created in 2008 in Gdynia: the Centre of Gdynia Culinary Trail (*Szlak Kulinarny Centrum Gdyni*, SKCG). Its purpose was to promote sightseeing in the city, as well as its rich culinary history. The project includes promoting Gdynia restaurants and their offerings and encourages restaurant owners to work together. The trail also introduces Gdynia's as Poland's 'window to the world' in terms of its culinary history and identity. All of this is designed to enhance the attractiveness of Gdynia restaurants in the eyes of the residents and tourists. The trail currently includes over 40 eating establishments located in the centre of the city. It was one of the first urban trails to appear in an Android smartphone app and is now available in several free apps that use geolocation (http://arg.gdynia.pl/projekty.html, accessed 01 Dec 2017).

In the early 1920s, fashionable restaurants such as Casino, Polska Riwiera and Bodega were opened in Gdynia, catering to the growing number of summer visitors. The old inn, Pod Dębem (Under the Oak Tree), mentioned in sixteenth-century records, was also very popular. When, in the 1930s, Gdynia port was being developed and the city began to grow quickly, several diners and restaurants opened catering to builders and new residents, not only from the nearby Kashuby region, but also from most other parts of Poland. The social cross section of the population was wide: from elites (architects, lawyers, engineers) to poor people from the villages and small towns of the Podkarpacie region. The food on offer had to reflect a wide range of tastes, preferences and financial means. Undoubtedly, Gdynia was a major cultural and culinary melting pot in independent Poland in the period 1918–1939. One of the first inns in the newly built town was opened in the early 1920s by August Skwiercz, in Abraham's Little House at 1 Starowiejska Street. It has been modernised many times since and currently houses the Sempre restaurant.

In the interwar period, the culinary history of Gdynia was closely bound to its famous Polish transatlantic liners. The twin ships, MS Batory and MS Piłsudski operated mostly on North Atlantic routes (New York and Montreal). The ships' chefs gradually introduced culinary novelties and world trends into their food range. Nowadays, the ambience of pre-war Gdynia can be found in the Dobry Adres restaurant and the Cyganeria café, with their period accessories and interiors, and a range of photos and press reprints.

After the war, some old restaurants re-opened (e.g. Café Bałtyk and Europa), and new establishments appeared (e.g. Lido, Cyganeria, Adria or Patria). In 1945, Gdynia had 82 eating establishments, but when authorities decided to close private businesses for political reasons their number decreased (to 33 by 1950). Only after the October Thaw, in 1956, did their number start to grow again. In 1969–1971, one of the most popular restaurants was located in MS Batory, which was withdrawn from service and moored near Kościuszko Square.

In 1973, there were 128 eating establishments, 101 of them were nationalised (excluding cafeterias and canteens within institutions). Among the very few privately owned services were the cafés Cyganeria and Mariola. Some of the most famous Gdynia restaurants of the 1970s and 1980s were Polonia, Róża Wiatrów (both still in business) and Myśliwska, which was famous for its game dishes.

During the socialist period, its inhabitants had easier access to colonial wares, which were beyond the reach of most Poles. This influenced the development of their culinary tastes and their interest, in later years, in international food. Although the 1980s saw the development of fast-food stands with casseroles and hotdogs, *gyros* (nowadays replaced by Turkish and Arab kebab variants) were more popular. After 1989, many establishments went bankrupt, unable to compete in the free market, but were soon replaced by new ones.

Nowadays, a renaissance of well-known Gdynia bistros can be observed (e.g. Kwadrans, AMD, Anker) as well as the intensive development of new ones (Anker, Przystanek). These facilities sit somewhere between bars and restaurants, which makes them particularly popular among the city centre's white-collar workers, who are usually in a hurry. The late 1990s and the beginning of the next millennium witnessed the development of ethnic restaurants. At that time, Italian (La Vita, Da Vinci, La Gondola), Indian (Taj Mahal), Mexican (Pueblo), Chinese (Moon), Greek (Santorini, El Greco) and, later, French (Petit Paris) and Thai (Thai Hut) restaurants were opened. In the last two or three years Japanese restaurants have been opened, as sushi becomes increasingly popular. Fusion cuisine can also be found at establishments such as Trafik and Charlie.

Source: Based on http://www.kulinarnagdynia.pl/historia-szlaku (Accessed 03 Dec 2017).

One of the main image-related activities of the SKCG is the organisation of the annual Culinary Weekend in Gdynia in September. During the event, restaurants offer specially prepared mini-meals for 5 or 10 PLN, as part of their happy hours. The event also includes live cooking demonstrations, competitions and culinary education for children and seniors (http://arg.gdynia.pl/projekty.html, accessed 03 Dec 2017). It is interesting to note that some of the SKCG's eating establishments are part of the Gdynia Modernism Trail, which promotes the architectural style from the period when Gdynia was built.

In 2010, Białystok set up its own trail, the Białystok Culinary Trail (BSK)[7] at the initiative of the City Hall. It contains over 30 restaurants, located mostly in the city centre, serving traditional regional and Polish cuisine and other European and world cuisine. It also organises culinary workshops and competitions. The trail was immensely popular with both tourists and locals, especially in the beginning, and in 2011 it was ranked third in the national competition for the Best Tourist Product (http://bialystok.naszemiasto.pl/tag/najlepszy-produkt-turystyczny-2011.html, accessed 25 Sept 2017). Unfortunately, nowadays the initiative has become largely forgotten. The BSK's role has been taken over by the Podlasie Culinary Academy, which not only promotes local products and dishes, but also vegan, vegetarian and molecular cuisine (https://www.pb.pl/bialostockie-smaki-i-kulinarna-akademia-869796, accessed 06 Oct 2017).

Another interesting initiative is the Podlasie Breakfast of Champions. Established in Bialystok in 2015 this is an open-air culinary picnic, which takes place in the summer in the gardens of the Branicki Palace. Its ambience attracts

---

[7]This is the second culinary trail in Poland after Gdynia.

people of all ages in large numbers. The dishes on offer come from the best restaurants, and include sweet cakes and pastries, excellent coffee. Visitors can also buy the regional food that made the Podlasie Voivodship famous. Gluten-free, vegan and fitness-related (or health-related) products are also available. The beautiful surrounding gardens, the gentle background music, the recreational zone with its deck chairs, and the children's play area all combine to create a unique atmosphere. The event promotes the culinary richness and heritage of the region at its finest, covering traditional and contemporary trends.[8]

Gastronomic services are an indicator of the development of contemporary cities. An analysis of the number and structure of eating establishments in smaller cities and towns shows a certain hierarchical relationship. This can be explained using Christaller's central place theory which argues that the number of eating establishments and their diversification grows with the city's position in the administrative hierarchy. But there are other factors that affect this relationship. One is the tourist function, which is shaped by the different types of tourism, depending, to a large extent, on the city's location, but also its history and contemporary development. Other factors include the role of the city as a transit location.

Some eating establishments become showpieces; these tourist destinations are a must-see for tourists (Piziak 2011). In this case, they are a tourist attraction in their own right, and a particularly important one, because an increasing number of tourists are foodies. Culinary showpieces are not only typical for large cities (e.g. the Wierzynek restaurant in Kraków) but can increasingly be found in smaller towns and cities. Usually, these eating establishments are:

- known for serving regional, national, or ethnic cuisine, e.g. the Tejsza restaurant in Tykocin,[9] specialising in Jewish cuisine, or the Akuku restaurant in Kazimierz Dolny, specialising in Old Polish cuisine;
- characterised by their interesting interior design, e.g. the Główna Osobowa bar bistro in Gdynia;
- situated in historically and/or architecturally unique places, e.g. the Szyb Maciej (Maciej Pit) restaurant in Zabrze and Piwnica Ratuszowa (Town Hall Cellar) in Świdnica,[10] the oldest restaurant in town.[11]

---

[8]http://www.podlaskieit.pl/index.php (Accessed 12 Oct 2017).

[9]Tykocin is a small town in Podlasie with a population of almost 5000. Its rich history is forever related to the Jewish minority.

[10]Świdnica (Lat. *Svidnica,* Ger. *Schweidnitz,* Czech *Svídnice*) is a town in Lower Silesia with a population of 60,000, which was known in the Middle Ages for brewing beer. In cities such as Prague, Wroclaw, Heidelberg and Kraków Swidnica caves are open, serving beer brewed in Świdnica. In the Middle Ages, the cellars of the local town hall housed an inn serving local beer.

[11]This is located in the cellars of the local town hall. To emphasise the historic ties, Świdnica beer and game dishes are still served (http://www.piwnicaratuszowa.pl, accessed 20 Dec 2017).

10.5 Conclusions

In the context of the localisation of eating establishments in smaller cities and towns in Poland, two marked tendencies can be observed. First, their distribution is closely related to the distribution of what Christaller, calls central services (transport, trade, tourism, administration, etc.). Therefore, they are concentrated:

- in central districts of towns, which are usually historic centres, with historic buildings that attract tourists;
- along main communication routes;
- near railway and coach stations;
- in shopping centres and hypermarkets.

These locations have good spatial accessibility and, in the first three cases, are often in the central area. At the same time, the development of large shopping centres and malls (including shopping arcades) has led to eating establishments being opened in the outskirts of cities. Another clear trend is the relationship to population distribution. In smaller cities, such as Gdynia, this means that eating establishments are often located in housing estates. This is largely the result of changes in the lifestyle of Poles, manifested by the increasingly frequent use of catering services, among other things. An analysis of eating establishments in smaller cities and towns cannot ignore their economic aspect in the strict sense of the word. On the one hand, eating establishments influence gross national product; on the other hand, they impact the job market. In tourist-oriented cities and towns the development of catering services attracts tourists thanks to a growing interest in regional cuisine. This, in turn, can drive the development of other businesses, creating a multiplier effect.

References

Au N, Law R (2002) Categorical classification of tourism dining. Ann Tour Res 29(3):819–833
Barrére C, Bonnard Q, Chossat V (2014) The profiles of Western World gastronomy. In: IX Annual Gastronomerica conference joined with XXI econometric conference, Lyon, France
Fronczak J (2016) 500+ wpłynie pozytywnie na branżę HoReCa. In: Raport 2016. Rynek gastronomiczny w Polsce, BROG Marketing Sp. z o.o. S.K., Warszawa, pp 12–14
Fryc K (2016) W zabytkowym Domku Abrahama w Gdyni otwarto restaurację. http://trojmiasto.wyborcza.pl/trojmiasto/1,35612,20246220,w-zabytkowym-domku-abrahama-w-gdyni-otwarto-restauracje.html. Accessed 23 Nov 2017
Gastronomia (2017) http://www.wladyslawowo.info.pl/gastronomia. Accessed 20 Oct 2017
Gastronomic cities. City strategy on gastronomy as tool for tourism and employment development. Feasibility study—Fermo (2014) European Regional Development Found. http://www.urbact.eu/sites/default/files/media/gastronomic_cities_feasibility_study_fermo.pdf. Accessed 20 Nov 2017
Gheribi E (2013) Konsument i przedsiębiorstwo na rynku usług gastronomicznych. Black Unicorn, Jastrzębie Zdrój
Historia (2017) http://www.kazimierz-dolny.pl/turystyka/historia. Accessed 15 Dec 2017
http://arg.gdynia.pl/projekty.html. Accessed 01 Dec 2017

http://bialystok.naszemiasto.pl/tag/najlepszy-produkt-turystyczny-2011.html. Accessed 25 Sept 2017
http://gdynia.pl/turystyczna-pl. Accessed 20 Oct 2017
https://pl.tripadvisor.com. Accessed 26 Nov 2017
http://trojmiasto.wyborcza.pl/trojmiasto/1,35612,20246220,w-zabytkowym-domku-abrahama-w-gdyni-otwarto-restauracje.html. Accessed 25 Nov 2017
http://www.kulinarnagdynia.pl/historia-szlaku/lata-90. Accessed 30 Nov 2017
https://www.pb.pl/bialostockie-smaki-i-kulinarna-akademia-869796. Accessed 06 Oct 2017
http://www.piwnicaratuszowa.pl. Accessed 20 Dec 2017
http://www.podlaskieit.pl/index.php?mact=News,cntnt01,detail,0&cntnt01articleid=1198&cntnt01returnid=58. Accessed 12 Oct 2017
http://www.szybmaciej.pl. Accessed 27 Nov 2017
https://www.zomato.com/pl. Accessed 14 Aug 2017
Kamler K (2017) Wielokulturowość w przestrzeni Białegostoku. Wydział Geografii i Studiów Regionalnych, Uniwersytet Warszawski, Warszawa, MA thesis
Kowalczyk A, Derek M (2010) Zagospodarowanie turystyczne, Wydawnictwo Naukowe PWN, Warszawa
Ługowska P (2014) Funkcja turystyczna Kazimierza Dolnego i jej przestrzenne zróżnicowanie. Wydział Geografii i Studiów Regionalnych, Uniwersytet Warszawski, Warszawa, MA thesis
Malitka D (2014) Rozwój funkcji turystycznej i uzdrowiskowej Krynicy-Zdrój. Wydział Geografii i Studiów Regionalnych, Uniwersytet Warszawski, Warszawa, MA thesis
Mniszewski M, Werpachowska J (2009) Kulinarny Białystok, czyli jak to jest u nas z gastronomią. http://www.poranny.pl/obserwator/art/5186948,kulinarny-bialystok-czyli-jak-to-jest-u-nas-z-gastronomia,id,t.html. Accessed 20 Sept 2017
Ornowski Ł (2003) Siedlce jako miejsce spędzania czasu wolnego dla mieszkańców. Wydział Geografii i Studiów Regionalnych, Warszawa, Uniwersytet Warszawski, MA thesis
Piziak B (2011) Baza gastronomiczna – wielkość, struktura i rozkład przestrzenny. In: Mika M (ed) Kraków jako ośrodek turystyczny. Instytut Geografii i Gospodarki Przestrzennej, Uniwersytet Jagielloński, Kraków, pp 181–198
Poczobut J (2010) Rewitalizacja najstarszej części Ustki. In: Muzioł-Węcławowicz A (ed) Przykłady rewitalizacji miast. Instytut Rozwoju Miast, Kraków, pp 133–150
Polska na talerzu (2015) MAKRO Cash & Carry, Warszawa
Próchnicka A (2017) Infrastruktura czasu wolnego Siedlec i jej wykorzystanie przez mieszkańców. Wydział Geografii i Studiów Regionalnych, Uniwersytet Warszawski, Warszawa, MA thesis
Szymańska W (2003) Dostępność przestrzenna usług gastronomicznych w mieście na przykładzie Słupska i Koszalina. Słupskie Prace Geograficzne 1:31–38
Wilk J (1984) Gastronomia. In: Nowakowski M (ed) Kształtowanie sieci usług, Instytut Kształtowania Środowiska, Państwowe Wydawnictwo Naukowe, Warszawa, pp 248–262

Chapter 11
Ethnic Cuisine in Urban Space

Marta Derek

Abstract This chapter draws attention to restaurants and bars offering ethnic cuisines in the urban space. It offers a general overview of factors that are thought to be responsible for this widespread phenomenon from a historical perspective, ranging from migration to the unification of global trends. As well as analysing changes in dominant cuisines, it also explores the distribution of different kinds of cuisines across the urban space. It focuses mainly on two cities: Amsterdam (The Netherlands), as an example of a Western European city with a long tradition of immigration and a long history of serving ethnic cuisines; and Warsaw (Poland), as an example of a Central European city, where after years of communism ethnic gastronomy only began to expand at the beginning of the 1990s. Although the history of the development of ethnic cuisines in the two cities is very different, their current situation is similar: Italian cuisine dominates, while Japanese cuisine is one of the most popular. Migration, once considered to be one of the most important influences on eating habits, is being increasingly replaced by globalisation.

Keywords Ethnic gastronomy · Migration · Standard deviational ellipses · Amsterdam · Warsaw · Madrid

11.1 Introduction

Although ethnic eateries have been popular in cities through the ages, it was not until the nineteenth century that the phenomenon gained in strength. In the United States, this was mainly due to immigration from Europe and, later, from Asia, the Caribbean and Latin America. As Gabaccia (1998, p. 7) notes, 'during the long nineteenth century, successive waves of Irish, British, German, Scandinavians, Slavs, Italians, Jews, Chinese, Japanese, and Mexicans changed the face, and the

---

M. Derek (✉)
Department of Tourism Geography and Recreation, Faculty of Geography and Regional Studies, University of Warsaw, Ul. Krakowskie Przedmieście 30, 00-927 Warsaw, Poland
e-mail: m.derek@uw.edu.pl

eating patterns, of American farmlands and cities'. In Europe, ethnic-style cuisines were introduced by early migrants from overseas colonial territories. The earliest establishments were around sea ports (e.g. London, Liverpool, Hamburg, Amsterdam) where newcomers settled and worked (Leung 2010). In the aftermath of crisis of 1929 many of these workers lost their jobs, and some entered the food business as restaurant owners or street food vendors (van Otterloo 2009). In the United Kingdom the presence of ethnic cuisines began as early as the end of the nineteenth century when the first wave of Chinese and Indian immigrants arrived (Leung 2010). In Germany there were a few Italian and Chinese restaurants but, as emphasised by Möhring (2008), these places were exceptions rather than the rule. Nevertheless, in the 1900s consumers bought much of their food from entrepreneurs 'of their own kind' (Gabaccia 1998).

The real development of ethnic eateries in cities of Western Europe and the United States started in the post-war period, although immediately after the Second World War Chinese restaurants had suffered a reversal because of the Korean War and the position of the 'two Chinas' (Roberts 2002). There were over 6,000 Chinese restaurants in the United States by 1960, with 600 in New York City, Newark and the surrounding states and, in 1957, sushi bars began to appear (Roberts 2004; Lee et al. 2014). In the 1950s, Indian restaurants were already fashionable in London. Research by Möhring (2008) showed that ethnic restaurants began to become widespread in Cologne in the late 1960s. While in 1950 there were only three (one Italian and two Chinese), by 1955 there were five, in 1960 eleven and in 1968 fourteen.

Ethnic eateries became more apparent in the 1970s and the sector continued to grow in the 1980s. In many western European countries that was a result of the oil crisis in the beginning of the 1970s (Möhring 2008; Ram et al. 2002). Job loss from deindustrialisation and the rise of unemployment hit immigrant workers hardest (Rekers and van Kepmen 2000). This situation occurred in Britain, where the transformation of the labour market created the general conditions for ethnic entrepreneurialism, mainly from South Asia. A supply of potential business entrants emerged, many of whom invested in catering and retail where less capital and expertise is needed than in many other sectors (Ram et al. 2002). Growth continued until 1991, when the sector was hit by the economic recession (Ram et al. 2002). In Germany, the rise of unemployment among foreigners, mainly Turks, resulted in döner production, which increased rapidly in the 1970s and gained real momentum in the 1980s (Caglar 1995). Between 1975 and 1985, the number of ethnic restaurants in Germany doubled from around 20,000 to around 40,000; in 1985, every fourth restaurant was run by a non-German (Möhring 2008). In the Netherlands, the number of Chinese–Indonesian restaurants rose from 225 in 1960 (11.7% of the total number of restaurants in the country) to 1,916 in 1982 (32%, a much faster growth rate than for the sector in general) (Rijkschroeff 1998). The United States saw the appearance of restaurants offering ethnic cuisine accelerate during the 1970 and 1980s (Turgeon and Pastinelli 2002). By 1980, according to Zelinsky, the number of Chinese restaurants in the United States and Canada had risen to 7,796, representing 29% of all ethnic and regional restaurants listed in telephone directories (cited in Roberts 2002, p. 165).

The situation was different for Japanese cuisine, which today is one of the most popular worldwide. In Europe, the first establishments began to emerge in the 1970s, and the spread of the fashion for Japanese food in cities across Europe was encouraged by the United States. As pointed out by Ćwiertka (2005, p. 242), 'the pioneering Japanese expatriates recall, with a high dose of nostalgia and pride, the culinary deprivation that they were forced to experience in Europe only a few decades ago'.

But it was not only cuisines deriving from outside the European culinary tradition that boomed in the 1970s. This was also the moment when Italian cuisine began to enjoy a worldwide reputation. Its success was connected to two developments: the large-scale emigration of Italians to North America, and its compatibility with modern eating habits and cooking styles (Hjalager and Antonioli Corigliano 2000). It is currently one of the most popular cuisines in Europe and the United States.

In urban areas, catering businesses run by foreigners were initially situated in decaying inner cities or red-light districts that were unattractive for native entrepreneurs. In Germany, for example, areas that were unsuitable for a *gutbürgerliches* (German) restaurant provided ethnic restaurants with a new clientele. In addition to other migrants, people (especially young clubbers) were attracted to these eateries with full meals served at moderate prices, late at night when 'traditional' restaurants were closed (Möhring 2008).

Nowadays restaurants serving ethnic cuisines are spread across all urban areas. In a globalised world, their distribution is no longer connected to areas where migrant concentrate, especially in the case of the most popular cuisines. This trend can be observed in Madrid (Spain), where the spatial concentration of ethnic restaurants and bars is very high in the case of the Japanese cuisine, slightly lower for Italian cuisine (which is the most popular), and almost non-existent in the case of Chinese cuisine (Fig. 11.1, Tables 11.1 and 11.2). The phenomenon may be explained by the fact that in Madrid, as in many other cities, Japanese restaurants located in the city centre are among the more expensive, and are visited by upper and middle classes (corporate employees, officials) and foreign tourists. Japanese cuisine tends to be expensive due to the high quality of the products, the techniques used to prepare food, the way it is served, etc. As Ćwiertka (2001) suggests, sushi suits the taste of the busy, professional elite and a sushi bar provides a perfect venue for a business lunch. In the case of Madrid's Chinese gastronomy, the situation is different. Restaurants and bars are more diversified in terms of price, menu, interior and atmosphere, service, etc. On the one hand there are luxury and expensive restaurants in the centre and, on the other hand, cheap bars on the outskirts of the city. Interestingly, there are few Chinese restaurants or bars in districts with a Chinese community (Usera, Tetuán and Lavapiés neighbourhoods in the southern part of Centro). This can be explained by the fact that unlike many other large agglomerations, Madrid lacks a so-called Chinatown.

(a)　　　　　　　　　(b)　　　　　　　　　(c)

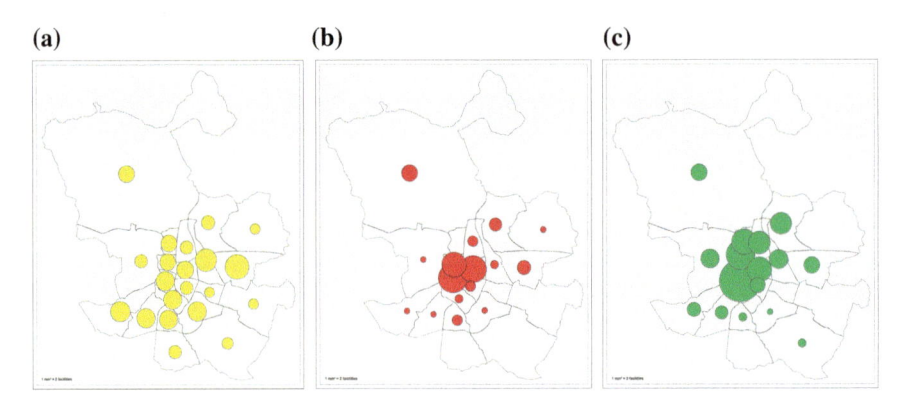

Fig. 11.1 Differences in the distribution of restaurants and bars in Madrid by ethnic cuisine in 2013. **a** Chinese cuisine. **b** Japanese cuisine. **c** Italian cuisine. Source: data gathered and mapped by Andrzej Kowalczyk, basing on information published in *Páginas Amarillas/Páginas Blancas 2014. Madrid Capital/Zona Centro, Consulta la guía de restaurantes más completa*, 1//11/2013, Connecti S.A.U. Impeso por Ediciones Informatizadas, S.A., Madrid, pp. 427–433

Table 11.1 Distribution of restaurants and bars recommended for ethnic cuisine in Madrid (Spain) in 2013 (by districts)

Districts	Ethnic cuisines					
	Chinese		Japanese		Italian	
	Number	%	Number	%	Number	%
Arganuzela	10	5.9	2	2.0	8	3.8
Barajas	3	1.8	1	1.0	–	–
Carabanchel	11	6.5	1	1.0	5	2.4
Centro	11	6.5	26	26.0	54	25.7
Chamartín	5	3.0	3	3.0	15	7.1
Chamberí	8	4.7	17	17.0	25	11.9
Ciudad Lineal	15	8.9	2	2.0	10	4.7
Fuencarral-El Pardo	8	4.7	8	8.0	8	3.8
Hortaleza	6	3.5	4	4.0	14	6.7
Latina	12	7.1	1	1.0	6	2.8
Moncloa-Aravaca	5	3.0	1	1.0	10	4.7
Moratalaz	3	1.8	–	–	–	–
Puente de Vallecas	11	6.5	1	1.0	1	0.5
Retiro	5	3.0	3	3.0	7	3.3
Salamanca	9	5.3	21	21.0	18	8.5
San Blas	17	10.0	6	6.0	8	3.8
Tetuán	8	4.7	–	–	18	8.5
Usera	10	5.9	3	3.0	2	0.9
Vicálvaro	3	1.8	–	–	–	–

(continued)

Table 11.1 (continued)

Districts	Ethnic cuisines					
	Chinese		Japanese		Italian	
	Number	%	Number	%	Number	%
Villa de Vallecas	4	2.4	–	–	2	0.9
Villaverde	5	3.0	–	–	–	–
MADRID	169	100.0	100	100.0	211	100.0

Source: Data gathered by Andrzej Kowalczyk, basing on information published in *Páginas Amarillas/Páginas Blancas 2014. Madrid Capital/Zona Centro, Consulta la guía de restaurantes más completa*, 1//11/2013, Connecti S.A.U. Impeso por Ediciones Informatizadas, S.A., Madrid, pp. 427–433

Table 11.2 Statistical measures showing spatial variation of location of restaurants and bars recommended for ethnic cuisine in Madrid (Spain) in 2013

Statistical measures	Ethnic cuisine		
	Chinese	Japanese	Italian
Standard deviation (δ)	3.860	7.191	11.954
Arithmetic mean (μ)	8.048	4.762	10.048
Coefficient of variation (V)	48.0	151.0	119.0

Source: Data gathered and calculated by Andrzej Kowalczyk, basing on information published in *Páginas Amarillas/Páginas Blancas 2014. Madrid Capital/Zona Centro, Consulta la guía de restaurantes más completa*, 1//11/2013, Connecti S.A.U. Impeso por Ediciones Informatizadas, S.A., Madrid, pp. 427–433

11.2 Ethnic Gastronomy in a Western European City: Amsterdam

In 2013, up to 11% of all restaurants in the Netherlands were in Amsterdam (Terhorst & Erkuş-Öztürk 2015). The city is an international port and capital of a country that established colonies worldwide. It is therefore a good example of a city where ethnic cuisines have had opportunities to develop. After 1911, when many Chinese people started working for Dutch shipping companies, Chinese immigrants were already running restaurants in the harbour districts (Van Otterloo 2009). Apart from a Chinese clientele, early patrons of these first ethnic restaurants included artists, bohemians, Dutch repatriates and military personnel from the colonies who appreciated Chinese cuisine (Rijkschroeff 1998). After the Second World War, the Netherlands had two colonial territories: the East Indies (Indonesia) and Dutch Guiana (Surinam). Between 1945 and 1960 (i.e. the period when ethnic eateries started to develop in the cities of Western Europe), the number of Indonesian restaurants grew from none to nine, while that of Chinese–Indonesian restaurants rose from two to 44. In 1965 there were, however, only eight Italian restaurants (Van Otterloo 2009).

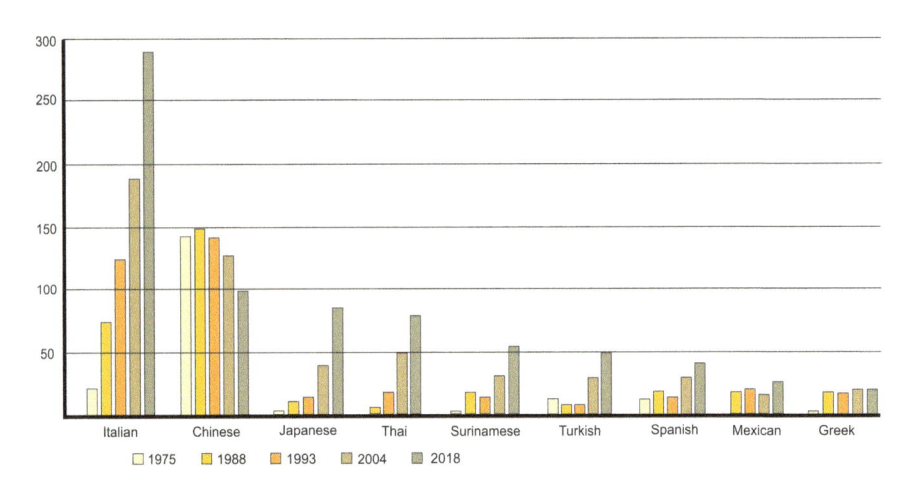

Fig. 11.2 Eating establishments serving ethnic cuisines in Amsterdam between 1975 and 2018
Source: *Horeca in Amsterdam: minder cafés, meer restaurants*, 2005; personal communication with Onderzoek, Informatie en Statistiek of the City Hall in Amsterdam; https://www.iens.nl/restaurant+amsterdam (Accessed 10 May 2018)

In the 1980s, a variety of cuisines sprang up rapidly. In 1989 as many as 33% of immigrant entrepreneurs in Amsterdam worked in the catering trade (De Feijter, Sterckx & Gier, cited in Van Otterloo 2009). For the first time, cuisines such as Mexican, Thai, Lebanese, Peruvian, etc. appeared in Amsterdam. Surprisingly, this was not the case for Japanese cuisine, which was becoming increasingly popular in many other European capitals (Ćwiertka 2001). The Dutch did not, however, resist the invasion of Italian cuisine: between 1975 and 1993 the number of Italian restaurants increased six-fold, from 20 to 125 (Fig. 11.2).

As a result of these new developments, the number of restaurants offering traditionally popular Chinese cuisine, which had existed in this area since the beginning of the 20th century, began to decrease. Although it remained popular, the number of restaurants serving it dropped by one-third between 1988 and 2018 (Fig. 11.2). The percentage of Chinese and Indonesian restaurants among all ethnic restaurants in Amsterdam dropped from 20% in 1996 to 9.7% in 2013. At the same time, the number of restaurants serving cuisines other than Dutch—French and Chinese–Indonesian—increased from 51% in 1996 to 70.2% (Terhorst & Erkuş-Öztürkm 2015, p. 93). Italian and Japanese developed particularly quickly, while Moroccan, Surinamese and Turkish were also popular. Although the Moroccan restaurants have only really come into the picture in recent years, the two others have already decreased and then increased again in recent years (Fig. 11.2). Turkish restaurants enjoyed a new lease of life at the end of the 1990s and the beginning of the 2000s. When Turkish restaurants, and their belly dancers, became less popular after the 1970s, many Turks switched to Italian pizzas. However, they have returned (Fig. 11.2) and Turkish pizza is increasingly found on pizzeria menus (*Horeca in Amsterdam...*, 2005), which is only one of many examples of blurring ethnic culinary barriers. The percentage of outlets serving Surinamese cuisine grew between 1993 and 2018. I n his case, the first wave of growth was at the end of the 1970s, a few years after Surinam independence (1975) when a large number of Surinamese settled in the Netherlands (Rijkschroeff 1998). In 1983 as many as 96 out of 214 Surinamese entrepreneurs in Amsterdam (45%) were concentrated in the 'cafés and restaurants' sector. Most (54) were originally Creole (Boissevain and Grotenbreg 1986, p. 8).

Japanese cuisine has seen spectacular growth in Amsterdam after 2000. Until that time, very few outlets served this kind of food, and it belonged to a category of exclusive ethnic cuisines, only known and available to a very small sector of Dutch society. The majority of Japanese restaurants in Holland were set up and run by Japanese and targeted overwhelmingly at a Japanese clientele. At the beginning of the twenty-first century these proportions have changed, and now Westerners constitute the majority of customers in Japanese restaurants. This has been reinforced by ongoing popularity of teppanyaki-style places and modern sushi bars (Ćwiertka 2001). As suggested by Ćwiertka (2001, p. 16), Japanese cuisine in Europe 'is on its way to becoming an eclectic, modern culinary experience rather than the exotic curiosity. Sushi bars, in particular, project a global, futuristic image through the websites that they use to communicate with or attract clients'.

In 2018, Amsterdam restaurants served about 40 different national cuisines, among which the most popular were Italian, French and Middle Eastern (https://www.iens.nl/restaurant+amsterdam; 10.05.2018). They differed significantly in terms of their location within the urban space of the city. In case of the restaurants that opened between 2002 and 2012 (511), more than half of Italian and Japanese restaurants (53% in each case) are concentrated in the historic core. However, this falls to 4% for Turkish restaurants and 33% for Chinese. On the other hand, 89% of Turkish restaurants and 40% of Chinese restaurants established between 2002 and 2012 are located in neighbourhoods other than the historic core and the Amsterdam South borough. These data are partly the result of a de-concentration process: restaurants move from the historic core to other neighbourhoods, especially to Amsterdam South which is within the ring road (Terhorst & Erkuş-Öztürkm 2015).

11.3 Ethnic Gastronomy in a Central European City: Warsaw

Unlike Amsterdam, Warsaw does not have a long history of serving ethnic cuisines. Although before the Second World War there were restaurants serving Greek, Italian, etc., under communism the ethnic catering sector was almost non-existent. Although there was a 'recommendation' to open restaurants featuring cuisines from other socialist countries in big cities (one restaurant per country) the number of these establishments was limited. Between the 1950 and 1970s, restaurants serving Russian (Trojka), Chinese (Szanghaj), Hungarian (Budapest, Balaton), Bulgarian

(Sofia) and Cuban (Hawana) cuisines were opened in Warsaw. It should be noted that this trend was common in many central European capitals, where restaurants called Warschau, Bukareszt, Budapest, or Habana were opened (Brzostek 2010; Derek 2017).

Ethnic gastronomy began to expand at the beginning of the 1990s. Cetnarska and Kowalczyk (1995) showed that in 1994 there were 45 establishments serving ethnic cuisines in Warsaw. Most (40; 89%) were concentrated in the city centre. Seven years later, research by Kaczorek (2002) indicated that their number had increased to 422; a more than ten-fold increase. These ethnic eateries offered 29 different cuisines and constituted 20% of all catering establishments in the city. A large number (130; 31%) congregated in the central district.

A study conducted by Derek in 2013 focused on the central borough of Warsaw, Śródmieście (Derek 2013; 2017). A total of 210 establishments serving ethnic cuisine were identified (including 131 restaurants, 60 bars, four cafés, seven takeaways and eight pizzerias that could not be qualified as Italian cuisine). This translates into 62% growth in 12 years, around 5% on average. The same research was repeated in 2017 (Table 11.3) and showed a total of 247 outlets serving ethnic food, representing almost 18% growth compared to 2013 (around 4.5% per year on average). This data highlights that the number of ethnic eateries has continued to grow. Between 2003 and 2017 growth has remained at about 4.5–5% annually, which is significantly more than the growth of eating establishments in Śródmieście in total. The percentage of ethnic outlets among all eating establishments grew from 19.4% to 21.2%. Surprisingly, this supports the work of Kaczorek (2002), who obtained similar results for the whole of Warsaw.

An important change has occurred regarding the dominant types of cuisine. Although the number of outlets serving Italian cuisine fell, it still dominates in Warsaw. The second most important is Japanese, which is a recent phenomenon. In 2017 it overtook Vietnamese and Asian cuisine, whose share significantly fell. Finally, the number of outlets serving French food also fell. Of the four dominant cuisines that constitute more than 61% of all the ethnic outlets in the city, Japanese was the only one that grew between 2013 and 2017.

Other cuisines that developed quickly between 2013 and 2017 were Georgian (a threefold growth, but starting with only two restaurants), Thai and Indian. The number of outlets serving Thai and Indian food has more than doubled in these four years. Growth was also seen in establishments serving American and Spanish cuisine, reflecting a global burger fashion (Table 11.3). This shows that although growth in the number of ethnic restaurants is steady, their profile has changed significantly.

The geographic distribution of ethnic outlets was measured using a directional distribution (standard deviational) ellipse. This measure showed that the distribution of outlets serving ethnic cuisine was more concentrated in space than non-ethnic establishments (Fig. 11.3), although the pattern does not apply to Polish cuisine, which is clearly skewed towards the Old Town and the Royal Route (see Chap. 6). Indian and Japanese establishments are most concentrated in space (Fig. 11.4). Their dispersion along the north–south axis is the smallest among all ethnic

Table 11.3 Eating establishments serving ethnic cuisines in the central borough of Śródmieście (Warsaw) in 2013 and 2017. In grey: sectors where growth exceeded the average for all establishments serving ethnic cuisines

Type of cuisine	2013	Share in 2013 (in %)	2017	Share in 2017 (in %)	Growth 2017/2013 (2013 = 100)
Italian	48	22.9	45	18.2	94
Japanese	30	14.3	40	16.2	133
Vietnamese and Asian	41	19.5	39	15.8	95
Middle Eastern and Turkish	28	13.3	28	11.3	100
Thai	6	2.9	15	6.1	250
Indian	6	2.9	12	4.9	200
American	6	2.9	10	4.0	167
Mexican	7	3.3	8	3.3	114
French	10	4.8	8	3.2	80
Spanish	4	1.9	7	2.8	175
Georgian	2	1.0	6	2.4	300
Other ethnic cuisines*	22	10.5	29	11.7	132
Total ethnic cuisines	210	100	247	100	118
Total ethnic cuisines	68		60		88
Other eating establishments	805		865		107
TOTAL	1083		1172		108

cuisines. In the first case, the mean for the 12 restaurants serving Indian cuisine is located furthest south among all ellipses indicating that they congregate in the southern part of Śródmieście. The district is a typical business zone, with offices and public authorities. In the second case, the limited dispersion of the 40 establishments serving Japanese food is surprising, as it is the second-most frequently served ethnic cuisine in Śródmieście. Most of these outlets congregate in a compact area of the central district, with the mean localised almost at its central point (the metro station Centrum).

Establishments serving Middle Eastern, Turkish and Italian cuisine are the most widely dispersed. Their popularity is huge—Italian cuisine dominates, while Middle Eastern and Turkish is the fourth most popular in terms of number of establishments (Table 11.3). Moreover, they are popular with mainstream consumers, a fact that may influence their wide distribution.

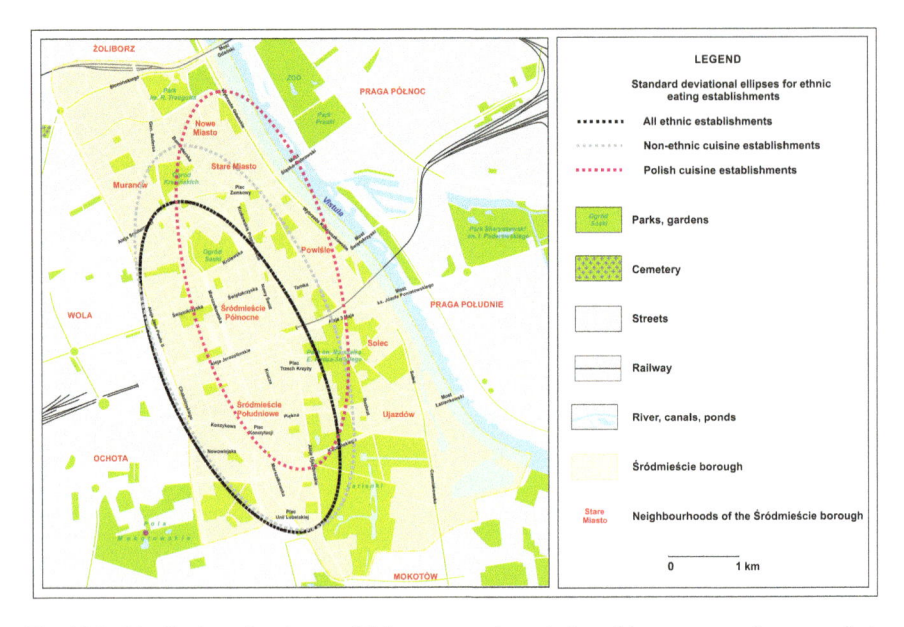

Fig. 11.3 Distribution of eating establishments serving ethnic cuisines compared to non-ethnic establishments and those serving Polish cuisine in Śródmieście borough (Warsaw) in 2017 Source: Marta Derek, underlay map by Andrzej Kowalczyk

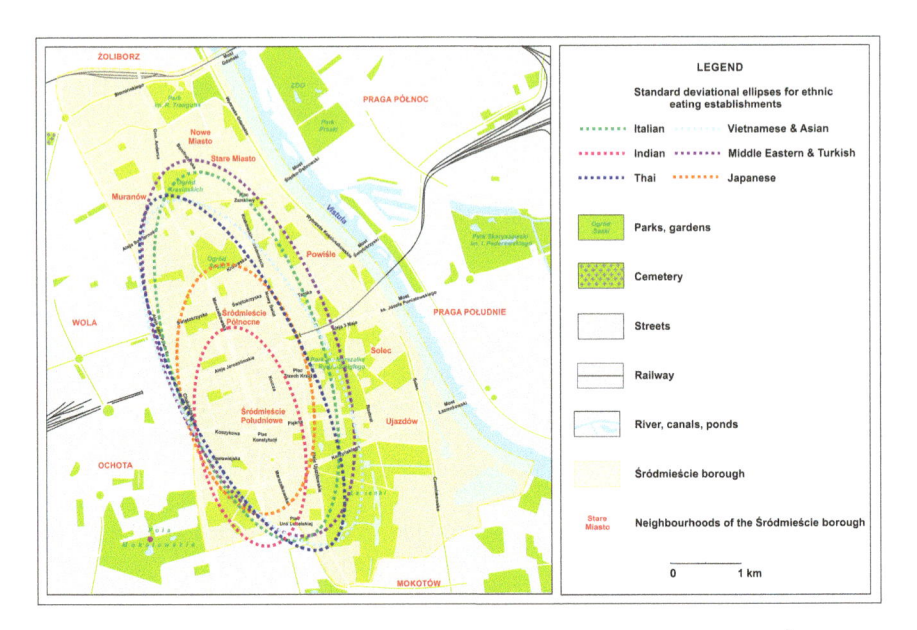

Fig. 11.4 Distribution of eating establishments serving different ethnic cuisines in Śródmieście borough (Warsaw) in 2017 Source: Marta Derek, underlay map by Andrzej Kowalczyk

The range of distribution of establishments serving Vietnamese (and Asian) and Thai cuisine is most similar to the standard deviational ellipse for all ethnic restaurants. Means are very close to the central point of the borough. This means that the location of these two cuisines is the most typical for all ethnic establishments, regardless of their number: 39 outlets serve Vietnamese and Asian cuisine and 15 serve Thai food (Fig. 11.4).

11.4 Conclusions

The development of ethnic eateries in Amsterdam and Warsaw has been very different. In Amsterdam, Chinese migrants opened up the first catering outlets at the beginning of the twentieth century. After the Second World War, Chinese–Indonesian restaurants appeared and, after a period of growth in the 1970 and 1980s, by the 1990s the number of Chinese restaurants 'equalled the number of post offices' (Ćwiertka 2001, p. 17). At the same time ethnic gastronomy in Warsaw sprang up after almost 50 years of communism. It has continued to grow since 1990, with the greatest growth between 1994 and 2001. In most cases (with the exception of Vietnamese; Derek 2017) it has been unconnected to the immigration of ethnic groups to Poland.

Although their histories are very different, their current situation is similar. In both cases, Italian food dominates. This, however, is a global trend, and some researchers no longer consider Italian cuisine as ethnic. Japanese food is very popular in both cities. In the Netherlands, the Japanese food boom hit in 1999, at a time when Europe had already peaked (Ćwiertka 2001). In Poland, it was even later: the first Japanese restaurant (called Tokyo), was established in 1992, and by 2003 there were still only four such restaurants in the central borough (Brożek 2007). However, in 2013 there were 30 in Śródmieście alone. The owner of one Japanese restaurant, interviewed for a culinary magazine five years later, suggested that between 2003 and 2008 more than 120 'sushi establishments' had been opened in Warsaw and its surroundings (Ignacionek 2008). As Bestor (2000, pp. 56–57) points out, 'from an exotic, almost unpalatable ethnic specialty, then to haute cuisine of the most rarefied sort, sushi has become not just cool, but popular'. Similarly, Ćwiertka (2001) suggests that an era of Japanese food as an ethnic cuisine has shifted towards an era of 'global sushi'; the phenomenon termed by Scarpato and Daniele (2002, p. 301) as 'global sushization'. Ćwiertka (2001) sees a certain analogy with pizza and Italian cuisine, suggesting that the managers of modern sushi restaurants see it as a trendy snack with great potential for global expansion. The author states that 'in this context, the Japanese identity of sushi is thus comparable to the Italian identity of pizza' (Ćwiertka 2001, p. 16). In both Warsaw and Amsterdam, the popularity of Chinese-like cuisine has fallen (in Poland Chinese and Vietnamese food are very often served in one bar, called oriental or 'Asian') and it is being replaced by Thai food. An increase in the number of Middle Eastern and Turkish eateries is another common trend.

In conclusion, it appears that both in the post-colonial, western European city of Amsterdam, and the post-communist, central European city of Warsaw, ethnic gastronomy reflects global trends rather than the local situation and its history. Migration, once considered to be one of the most important influences on eating habits, is being increasingly replaced by globalisation.

References

Bestor TC (2000) How sushi went global, Foreign policy, 121, November–December, 54–63

Brożek M (2007) Zmiany w przestrzennym zróżnicowaniu bazy gastronomicznej w dzielnicy Warszawa-Śródmieście na przestrzeni lat 1994–2003, Wydział Geografii i Studiów Regionalnych, Uniwersytet Warszawski, Warszawa, MA thesis

Brzostek B (2010) PRL na widelcu. Baobab, Warszawa

Boissevain J, Grotenbreg H (1986) Culture, structure and ethnic enterprise: the Surinamese of Amsterdam. Ethn Racial Stud 9(1), 1–23

Caglar AS (1995) McDöner. Döner Kebap and the social positioning struggle of German Turks. In: Costa JA, Bamossy GJ (eds.) Marketing in a multicultural world: ethnicity, nationalism, and cultural identity, Sage, London, pp. 209–231

Cetnarska H, Kowalczyk A (1995) Infrastruktura turystyczna i paraturystyczna. In: Warszawa jako centrum turystyczne: raport o stanie turystyki, Uniwersytet Warszawski, Wydział Geografii i Studiów Regionalnych – Biuro Zarządu Miasta Stołecznego Warszawy, Wydział Kultury Fizycznej i Turystyki, Warszawa, pp 23–46

Ćwiertka KJ (2001) Japanese food in Holland: the global trend spreads. Food Cult 3, 15–19

Ćwiertka KJ (2005) From ethnic to hip: circuits of Japanese cuisine in Europe. Food Foodways 13 (4), 241–272

Derek M (2013) Kierunki rozwoju usług gastronomicznych w warszawskiej dzielnicy Śródmieście, Prace i Studia Geograficzne 52, 85–100

Derek M (2017) Multi-ethnic food in the mono-ethnic city: tourism, gastronomy and identity in central Warsaw. In: Hall DR (ed.) Tourism and geopolitics: issues and concepts from Central and Eastern Europe, CABI, Wallingford, pp 223–235

Gabaccia DR (1998) We are what we eat. Ethnic food and the making of Americans, Harvard University Press Cambridge, MA

Hjalager AM, Antonioli Corigliano M (2000) Food for tourists – determinants of an image. Int J Tour Res 2(4), 281–293

Horeca in Amsterdam: minder cafés, meer restaurants, 2005, Fact sheet, Onderzoek, Informatie en Statistiek (OIS), Gemeente Amsterdam, nummer 2, maart 2005

Ignacionek E (2008) Sushi – pomysł na biznes, Poradnik Restauratora, 10, http://www.poradnikrestauratora.com.pl/archiwum/pazdziernik-10-2008,Kultura—Sushi—pomysl-na-biznes,Rok-2008,17,72.html. Accessed 04 Jan 2013

Kaczorek A (2002) Gastronomia etniczna w Warszawie, Wydział Geografii i Studiów Regionalnych, Uniwersytet Warszawski, Warszawa, MA thesis

Lee JH, Hwang J, Mustapha A (2014) Popular ethnic foods in the United States: a historical and safety perspective. Compr Rev Food Sci Food Saf 13(1), 2–17, https://onlinelibrary.wiley.com/doi/abs/10.1111/1541-4337.12044. Accessed 06 July 2018

Leung G (2010) Ethnic foods in the UK. Nutr Bull 35(3), 226–234, https://onlinelibrary.wiley.com/doi/epdf/10.1111/j.1467-3010.2010.01840.x. Accessed 06 July 2018

Möhring M (2008) Transnational food migration and the internalization of food consumption: ethnic cuisine in West Germany. In: Nützenadel A, Trentmann F (eds.) Food and globalization: consumption, markets and politics in the modern world, Berg, Oxford and New York, pp 129–150

Ram M, Jones T, Abbas T, Sanghera B (2002) Ethnic minority enterprise in its urban context: South Asian restuarants in Birmingham. Int J Urban RegNal Res 26(1), 24–40

Rekers A, Van Kempen R (2000) Location matters: ethnic entrepreneurs and the spatial context. In: Rath J (ed.) Immigrant business: the economic, political, and social environment, Palgrave MacMillan, New York, pp 54–69

Rijkschroeff BR (1998) Etnisch ondernemerschap. De Chinese horecasector in Nederland en in de Verenigde Staten van Amerika, Labyrint Publication, Capelle a/d Ijssel

Roberts JAG (2002) China to Chinatown: Chinese food in the west. Reaktion Books, London

Scarpato R, Daniele R (2002) New global cuisine: tourism, authenticity and sense of place in postmodern gastronomy. In: Hjalager AM, Richards G (eds.) Tourism and gastronomy, Routledge, London, UK, pp 296–313

Terhorst P, Erkuş-Öztürk H (2015) Urban tourism and spatial segmentation in the field of restaurants: the case of Amsterdam. Int J Cult, Tour Hosp Res 9(2), 85–102

Turgeon L, Pastinelli M (2002) "Eat the world": Postcolonial Encounters in Quebec City's Ethnic Restaurants. J Am Folk 115(456):247–268

Van Otterloo AH (2009) Eating out 'ethnic' in Amsterdam from the 1920s to the present. In: Nell L, Rath J (eds.) Ethnic Amsterdam. Immigrants and urban change in the twentieth century, Amsterdam University Press, Amsterdam, pp 41–59

Chapter 12
Tourist Experience and Change in Culinary Tastes. An Example of Polish Students in Warsaw

Katarzyna Gwiazdowska and Andrzej Kowalczyk

Abstract Travelling and culinary experiences of people on nutrition, ethnic cuisine, etc. have an impact on culinary habits. Tourists who encounter the local population at the place of destination start to be interested in the habits and culinary traditions of other nations. This familiarity with other culinary traditions becomes not only a source of new gastronomic experiences, but also the beginning of changes in eating preferences. This phenomenon can be noticed especially in the case of young people, e.g. students. This chapter presents the results of a study that sought to discover the culinary preferences of a group of students at the University of Warsaw. The study aimed to determine whether student tourism is linked to their increased interest in, and knowledge of, ethnic cuisines. It revealed that although eating in ethnic restaurants is popular among students, the relationship between travelling and eating ethnic food is not particularly strong. The location of the 121 ethnic restaurants and bars visited by participants was also explored.

Keywords Tourist experiences · Gastronomy · Students · Warsaw

12.1 Changes in Culinary Tastes in Poland—A General Overview

One of the issues addressed by the literature on gastronomy is the impact of travel and the culinary experiences of tourists on nutrition, ethnic cuisine, etc. Such research is guided by nutrition sciences, cultural anthropology and the geography of

---

K. Gwiazdowska · A. Kowalczyk (✉)
Department of Tourism Geography and Recreation, Faculty of Geography and Regional Studies, University of Warsaw, ul. Krakowskie Przedmieście, 30, 00-927 Warsaw, Poland
e-mail: akowalczyk@uw.edu.pl

K. Gwiazdowska
e-mail: katarzyna.alicja@gmail.com

© Springer Nature Switzerland AG 2020
A. Kowalczyk and M. Derek (eds.), *Gastronomy and Urban Space*,
The Urban Book Series, https://doi.org/10.1007/978-3-030-34492-4_12

culture and tourism. Culinary traditions and nutrition are inextricably linked to culture (Fumey and Etcheverria 2004; Fumey 2007) and are part of the national (or regional) heritage. Although eating preferences are impermanent and—like culture—change over time, globalization has increased the pace of change. Eating preferences not only evolve between generations, but also in the same individual who, in a relatively short time, can fundamentally change their model of nutrition. This phenomenon can be seen in Poland. Although pasta (known as noodles) and pizza began to become popular (still known as noodles) in the 1970s, the real culinary revolution only took place in the 1990s.

However, this does not mean that Poles have abandoned their national traditions. A survey requested by Lidl Polska Sklepy Spożywcze sp. z o. o. (the Polish branch of the German company) showed that although Poles (especially older people) mostly ate traditional dishes, ethnic cuisines were becoming increasingly popular. Polish cuisine was preferred by 81% of people aged 60 and over and rural residents (81% of respondents lived in the countryside). However, in the 15–19 age group, the proportion of those who preferred Polish cuisine was only 60%, and among those with higher education it fell to slightly over 50%. Italian cuisine was popular among 12% of respondents—notably among residents of cities with over 500,000 inhabitants, people with higher education, those holding managerial positions (it was chosen by 34% of managers), and young people aged 15–19. Although the study did not address the issue of change in culinary tastes and food preferences, the results may indirectly indicate the existence of connections between culinary preferences and tourism activity, since the biggest changes are seen in residents of large cities, people with higher education and young people.

12.2 Impact of Tourism on Changes in Culinary Preferences

Many studies have shown that the culinary experience of tourists is of increasing importance; it has been linked to both with migration and the increasing number of tourist destinations. Tourists who encounter the local population start to be interested in the habits and culinary traditions of other nations, and visit other countries in order to get to know, in situ, other cuisines (Hall and Sharples 2003), a practice that has become known as 'culinary tourism' (Long 1998; Kowalczyk and Gwiazdowska 2015).

This familiarity with other culinary traditions becomes not only a source of new gastronomic experiences, but also the beginning of changes in eating habits. As noted above, globalization is an important factor affecting the shape of culinary tourism. On the one hand, culinary traditions are transferred from one place (region, country) to another; on the other hand, globalization is a kind of threat, because it can lead to the universalization of eating preferences and the disappearance of local culinary traditions (Friedberg 2003, p. 4).

When visiting other places, regions or countries, tourists are often interested in experiencing new flavours and practices, such as basic products and additions to their dishes, ways of preparing dishes, related customs, etc. Tourists who come into contact with previously unknown food products and ready-made dishes become culinary tourists (Cohen and Avieli 2004, pp. 758–760). Mass media also play an important role in promoting change in culinary preferences. This applies to both daily and weekly newspapers, and monthly magazines, as well as television (for example, the Travel Channel, the National Geographic Channel, the Discovery Channel or Planete+), and especially the Internet, where social networks are just one source of information that combines travel with the kitchen.

12.3 Students Culinary Preferences—A Case Study from Warsaw

The study referred to above shows that changes in culinary preferences affect young people most. These young adults have not yet established long-term nutritional habits. Wojciechowska (2011, p. 81) showed that among 300 students at Warsaw universities, 72% of respondents regularly ate outside the home, in 64% of cases at fast-food bars and other establishments offering fast-food dishes. Thirty-eight percent preferred Polish cuisine, but 31% liked Italian cuisine. Japanese and Indian dishes were popular, as was Arabic and Turkish cuisine (between 6 and 9%). However, only 3% regularly ate Chinese or Vietnamese cuisines. While the popularity of Japanese cuisine (generally assumed to take the form of sushi) is easy to understand, it is difficult to find an answer to the question of why Chinese and Vietnamese cuisines are less popular than Indian cuisine? It may be that unlike Indian cuisine, which is still a novelty in Poland, catering establishments specializing in Chinese and Vietnamese dishes offer a few familiar dishes, often limited to pieces of meat in a 'sweet and sour sauce', soup with 'won ton' dumplings or 'spring rolls'. However, there is another hypothesis. It is possible that the relatively high popularity of Indian cuisine (possibly confused with Pakistani cuisine) results from frequent trips made by young Poles to Great Britain, where Indian/Pakistani dishes are not only often offered in restaurants and bars (e.g. take-away outlets), but they are also cheap.

In 2014/2015, a questionnaire was administered at the University of Warsaw, which aimed to determine whether student tourism was linked with their increased interest in, and knowledge of, ethnic cuisines. The sample consisted of 330 students at the Faculty of Geography and Regional Studies, made up of 231 women (70%) and 99 men (30%). As the sample consisted of students, 319 (91.5%) respondents were in the 19–24 age group. Among these, the vast majority (70.6%) lived in Warsaw or its surroundings (20.3%).

One of the first questions was: 'Have you traveled abroad?' In 299 (90.6%) of cases, the answer was positive and only 31 (9.4%) had not. Among those who had travelled, 104 (34.8%), had travelled once per year, 39 people twice (13.0%), and 46 (15.4%) more than twice. One hundred and ten made less than one trip per year (36.8%). Another key question was, 'Are you interested in getting to know the cuisine of other countries?' Here, 90% of answers were positive (297) and only 10% were negative. A comparison of these results suggests that foreign trips are conducive to interest in the culinary traditions of other countries. This was verified using Pearson's $\chi 2$ compatibility test (Table 12.1).

This result can be interpreted as follows: the probability of a statistical relationship between travelling abroad and interest foreign cooking is moderate. This is unsurprising, as it is unlikely that in this time of globalization (notably through television, films, press and books), universal access to the Internet and the increasing number of immigrants, travel would be the sole factor underlying increased interest in foreign cooking. Respondents expressed interest in 65 different cuisines, while the vast majority named a specific country, some were more general (Asian, European, Middle Eastern, Latin American, etc.). Consistent with previous research, respondents were most interested in Italian cuisine (202, 23.5%). Significantly fewer people expressed interest in Asian, Latin American and other European cuisines (Table 12.2).

Respondents were also asked, 'How often do you consume dishes of cuisine other than Polish?' Most (146, 44.2%) ate ethnic dishes several times per month. Significantly fewer (58, 17.6%) reported that they did it once per month, and the same number (46, 13.9%) ate them several times each week or less often than once per month. Although nearly a third of respondents ate these meals outside the home (95, 32%), a similar number (91, 30.6%) cook them themselves or eat out and at home (41, 13.8%). It should be noted that a relatively large group of people (33, 11.1%) eat foreign dishes while visiting friends. This indirectly suggests that eating in ethnic restaurants is popular. This is probably due to the relatively low price of Chinese, Vietnamese, Turkish and Arabic cuisine (but not Japanese, Thai, Indian, Greek or Italian cuisine), as well as the abundant supply of such facilities. This thesis was confirmed by answers to the question, 'Do you think that Warsaw lacks ethnic cuisine facilities?' Over three-quarters of the sample thought that there were plenty of ethnic food outlets in Warsaw; however, Mexican was noted as being relatively rarely represented.

A further question was, 'If you go to an ethnic restaurant, how do you choose it?' Most respondents said it was 'the desire to learn new, unknown flavours from

Table 12.1 A cross tabulation analysis of the questions, 'Have you traveled abroad?' and, 'Are you interested in getting to know the cuisine of other countries?'

Frequency of travelling	Interest in learning about other cuisines		
	Yes	No	Sum
Travelling once a year	189	25	214
Travelling at least twice a year	81	4	85
Sum	270	29	299

Source: Survey conducted by Gwiazdowska in 2014/2015

Table 12.2 Ethnic cuisines most often chosen by students at the Faculty of Geography and Regional Studies of the University of Warsaw (named by at least 10 respondents)

Cuisine	Answers Number	%
Italian	202	23.5
Indian	76	8.8
Chinese	64	7.4
Japanese	58	6.7
Mexican	55	6.4
Spanish	51	5.9
Greek	43	5.0
French	40	4.7
Thai	30	3.5
Arabic	19	2.2
Turkish	18	2.1
Georgian	15	1.7
Hungarian	15	1.7
American	13	1.5
Vietnamese	12	1.4

Note Vaguely defined cuisines are not listed. This was true for Asian (19), Balkan (13) and Mediterranean (11) cuisines
Source: Survey conducted by Gwiazdowska in 2014/2015

different cultures', 'a recommendation from friends/relatives', and 'price', along with 'looking for flavours discovered during a trip', and 'willingness to learn new, unknown flavours of different cultures. These results highlight the significant interest in foreign cooking among the sample. It is likely that this is related to the sample characteristics, as geography tends to attract students who are interested in travelling and learning about the world, open to other cultures, and are more knowledgeable in this area than the average young person in Poland. It should also be noted that a relatively large percentage of the sample prepare foreign meals themselves. This suggests a partial adoption of culinary preferences from other cuisines and a combination of traditional Polish cuisine with ethnic cuisines.

The final question concerned the location of the 121 ethnic restaurants and bars visited by respondents. Figure 12.1 shows that students mainly visited establishments located in the city centre. This is understandable, not only because of the large number of ethnic restaurants in the downtown area, but also because of the location of the Faculty of Geography, which is on the main campus of the University of Warsaw and close to the Presidential Palace, the historical Old Town and other tourist attractions. Many different types of restaurants and bars, including those offering ethnic cuisine, are nearby. Students also visited restaurants and bars in inner city neighbourhoods adjacent to the city centre. The outlying borough of Ursynów (in southern Warsaw) is another area with a high number of restaurants visited by students. Although relatively far from the city centre, it has almost 150,000 inhabitants and is known to be upper-middle-class. Many residents of the

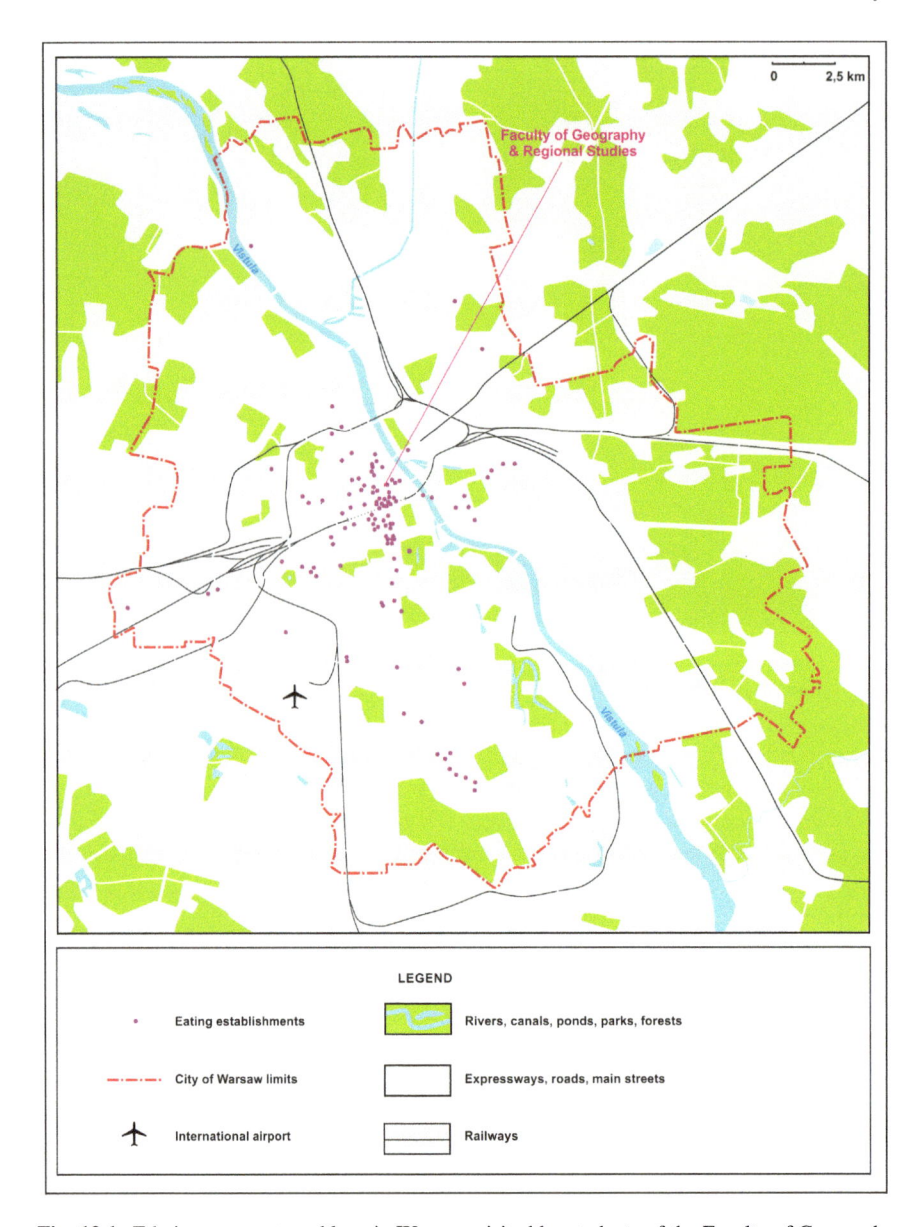

Fig. 12.1 Ethnic restaurants and bars in Warsaw visited by students of the Faculty of Geography and Regional Studies of University of Warsaw (Poland)

new housing estates in this borough are professionals and students, which has led to the expansion of eating facilities, including restaurants and bars offering East-Asian, Indian, Mexican or Italian cuisine.

12.4 Conclusions

The results of research carried out among students of the University of Warsaw showed that they eagerly spend their holidays travelling abroad. A significant part of them had left Poland at least once a year. Their tourist experience has an impact on culinary tastes, as over 60% reported that at least once a month they eat dishes associated with other cuisines. The research confirmed that Italian cuisine is the most popular culinary tradition. This phenomenon is also reflected in surveys conducted in Poland on the national sample. Also in foreign literature, Italian cuisine is the one that enjoys the greatest popularity.

The research shows that cuisines of East Asia are relatively popular among Polish students. This is due to several reasons. First of all, some East-Asian dishes have gained popularity around the world, so their existence in Poland only reflects global trends (e.g. sushi) (Ćwiertka 2005). Second, the number of restaurants and bars with Chinese or Indian cuisine in Warsaw is constantly growing. Third, the prices of dishes in Chinese, Vietnamese or Turkish bars are often much cheaper than in similar bars with Polish (or many others) cuisine. Fourth, the research was conducted among students of geography. Probably the specificity of geographical studies means that they are more open to other cultures and travel abroad more often. Finally, respondents visit not only restaurants and bars located in the centre of Warsaw, but also located in the outer parts of the city. This shows that dining outlets associated with other culinary traditions have become part of the 'landscape' (or *servicescape*; Ferrari 2019) of Warsaw and have become an integral part of the urban space.

References

Cohen E, Avieli N (2004) Food in tourism: attraction and impediment. Ann Tour Res 31(4): 755–778

Ćwiertka KJ (2005) From ethnic to hip: circuits of Japanese cuisine in Europe. Food Foodways 13 (4):241–272

Ferrari F (2019) Servicescape and gastronomic tourism. In: Dixit KS (ed) The Routledge handbook of gastronomic tourism. Routledge, Oxon and New York, pp 161–168

Freidberg S (2003) Editorial. Not all sweetness and light: new cultural geographies of food. Soc Cult Geogr 4(1):3–6

Fumey G (2007) *Penser la géographie de l'alimentation (Thinking food geography)*. "Bulletin de l'Association de géographes français", 84e année, no. 1 (mars). Cartographie géomorphologique (2)/Géographie de l'alimentation, pp 35–44

Fumey G, Etcheverria O (2004) Atlas mondial des cuisines et gastronomies. Autrement, Paris

Hall CM, Sharples L (2003) The consumption of experiences or the experience of consumption? An introduction to the tourism of taste. In: Hall CM, Sharples L, Mitchell R, Macionis N, Cambourne B (eds) Food tourism around the world. Development, management and markets, Butterworth-Heinemann, Oxford, Burlington, pp 1–24

Kowalczyk A, Gwiazdowska K (2015) *Zmiany upodobań żywieniowych i zainteresowanie kuchniami etnicznymi – przyczynek do turystyki (kulinarnej?)/Changes in culinary customs and attention to ethnic cuisines - contribution to (culinary?) tourism*. Turystyka Kulturowa 9 (wrzesień) 6–24. http://turystykakulturowa.org/ojs/index.php/tk/article/view/639. Accessed 20 Dec 2016

Long LM (1998) Culinary tourism: a folkloristic perspective on eating and otherness. Southern Folklore 55(3):181–204

Wojciechowska D (2011) *Rozwój bazy gastronomicznej, a zmiany nawyków żywieniowych Polaków na przykładzie Warszawy*, Wyższa Szkoła Turystyki i Rekreacji w Warszawie, Warszawa, BA thesis

Chapter 13
The Food Supply Chain in the Restaurant Industry: A Case Study from Warsaw, Poland

Andrzej Kowalczyk

Abstract In gastronomy, as in other businesses, products used in restaurants or bars undergo a long process with many stages. It begins with the raw materials, followed by refining, processing and assembly. At the last stage the final product reaches the restaurant guests. The process is referred to as a supply chain and there are several ways to manage it. This chapter pays attention to the geographical aspects of this phenomenon. From this point of view, an important consideration is that suppliers often have their headquarters and warehouses in the suburbs, and buyers, primarily restaurants and bars, are usually located in the city centre. The chapter gives an example of a Polish company operating in Warsaw, which provides supplies to catering establishments, analysing the geographical distribution of its suppliers and customers.

Keywords Supply chain · Food · Restaurant · Warsaw · Supplier · Customer

13.1 Introduction

Most food and non-food products using in gastronomy undergo a long process with many stages.[1] It begins with the raw materials, followed by refining, processing and assembly, usually undertaken by different companies. Eventually, the final product reaches the consumer, typically households or restaurant guests. The process is referred to as a supply chain and there are several ways to manage it. Mentzer et al.

---

[1] In this chapter we omit the ways food is produced or grown, stored, and distributed to the very first supplier. We do not develop the "sustainability" of the restaurant system and do not go into ethical problems of purchasing unsustainable food (for more details on these problems see for example: Steel 2013, Hellier 2019).

A. Kowalczyk (✉)
Department of Tourism Geography and Recreation, Faculty of Geography and Regional Studies, University of Warsaw, ul. Krakowskie Przedmieście 30, 00-927 Warsaw, Poland
e-mail: akowalczyk@uw.edu.pl

© Springer Nature Switzerland AG 2020
A. Kowalczyk and M. Derek (eds.), *Gastronomy and Urban Space*,
The Urban Book Series, https://doi.org/10.1007/978-3-030-34492-4_13

Table 13.1 Factors determining the choice of a supplier of goods to an eating establishment in Poland (%)

Factors	February 2017	January 2018
Quality of the goods	89	90
Price of the goods	77	79
Speed, punctuality of receiving the goods	25	21
A wide range of goods	17	15
Ability to deliver goods to the client	11	9
Proximity of the supplier	9	10
Convenient payment terms	5	3
Bonus from the supplier	5	2

Source: Adapted from Frontczak (2017, p. 7) and Szot (2018, p. 10)

(2001, p. 2) define supply chain management as '…systematic, strategic co-ordination of the traditional business functions within a particular company and across businesses within the supply chain, for the purposes of improving the long-term performance of the individual companies and the supply chain as a whole'. This definition is very general and does not explain the situation in the gastronomic industry. However, the authors state that a supply chain is not only the flow of products, but also the flow of services, finances and information from a producer to a consumer. According to Murphy and Smith (2009, p. 213), '…the most important benefits of supply chain management include: (a) improved co-ordination from supplier to buyer, (b) reduced lead times, (c) greater productivity and efficiency, (d) smaller inventories, (e) increased delivery reliability, and (f) lower costs. With respect to restaurants, a properly managed supply chain also supports the chef's ability to build relationships with suppliers, identify new sources, and manage supplier relationships'. A survey conducted in Poland indicates that the decisive factors in choosing a product supplier are quality and price. Delivery speed and timeliness are less important. Even less important are the delivery of goods to the customer and the proximity of the supplier (Table 13.1).

Table 13.2 presents various ways of organizing the supply chain for restaurants, bars, etc. in Poland for the period 2014–2016. This shows that an increasing number of dining establishments have products delivered directly to the premises (27% in 2014 compared to 40% in 2016). However, Table 13.3 shows that it

Table 13.2 Delivery practices to eating establishments in Poland in 2014–2016 (%)

	2014	2015	2016
All or almost all goods are delivered to us	27	38	40
Most goods are delivered to us, but we collect some ourselves	28	22	21
Half are delivered, we collect half	29	23	21
We collect all or almost all goods ourselves	7	8	10
We collect most goods ourselves, but some are delivered	10	9	8

Source: Adapted from Frontczak (2017, p. 6)

Table 13.3 Delivery practices to eating establishments in Poland in 2017 by type of establishment (%)

The way of delivering goods	Hotel restaurant	Restaurant	Pizzeria	Fast food	Café, teahouse	Bar, pub
All or almost all goods are delivered to us	48	34	41	42	34	43
Most goods are delivered to us, but we collect some ourselves	18	15	20	19	16	25
Half are delivered, we collect half	20	30	31	27	30	17
We collect all or almost all goods ourselves	5	10	4	5	6	5
We collect most goods ourselves, but some are delivered	9	10	4	7	14	9

Source: Adapted from Szot (2018, p. 10)

depends on the type of eating outlet. In 2017 most goods were delivered by suppliers to hotel restaurants, bars and pubs, along with pizzerias and fast-food outlets.

The logistics literature describes various supply chain models. According to Kock (2013, p. 18), the most popular are Supply Chain Management (SCM), Supplier Relationship Management (SRM) and Customer Relationship Management (SRM). The first is the most important and describes all of the activities within and around a food supply chain, which together create a product or service. The literature focuses on the delivery and logistics behind the delivery of locally produced foods to hotels and restaurants and examines problems related to supply, prices, reliable food differentiation, quantity and quality and on-time delivery. Other factors that may be important for the selection of a supplier by a restaurant owner or chef are the uniqueness of products, the supplier's location, personality and previous experience of the buyer. From the suppliers' point of view, they tend to favour buyers who do not argue over the proposed price, place regular orders, make regular payments, and who are credible and easy to work with. Others factors are the location of the restaurant, as well as previous experience with the buyer.

Research shows that the preferences of the owner, manager or chef of a restaurant play a significant role in the food supply chain. As mentioned above, it is customary to choose a supplier (or seller) based on the availability of a given product, its price, quality, and the ability to deliver the right quantity at a given time. Sometimes the choice of supplier results from the experience of managers or chefs at their previous workplace. Some studies have found that chefs are responsible for establishing links between a restaurant and its suppliers (Kock 2013, p. 97; Murphy and Smith 2009).

From the geographical point of view, an important consideration is that suppliers often have their headquarters and warehouses in the suburbs and buyers, primarily restaurants and bars, are usually located in the city centre. Figure 13.1 illustrates this problem. It shows the location of a restaurant in Warsaw owned by Strefa

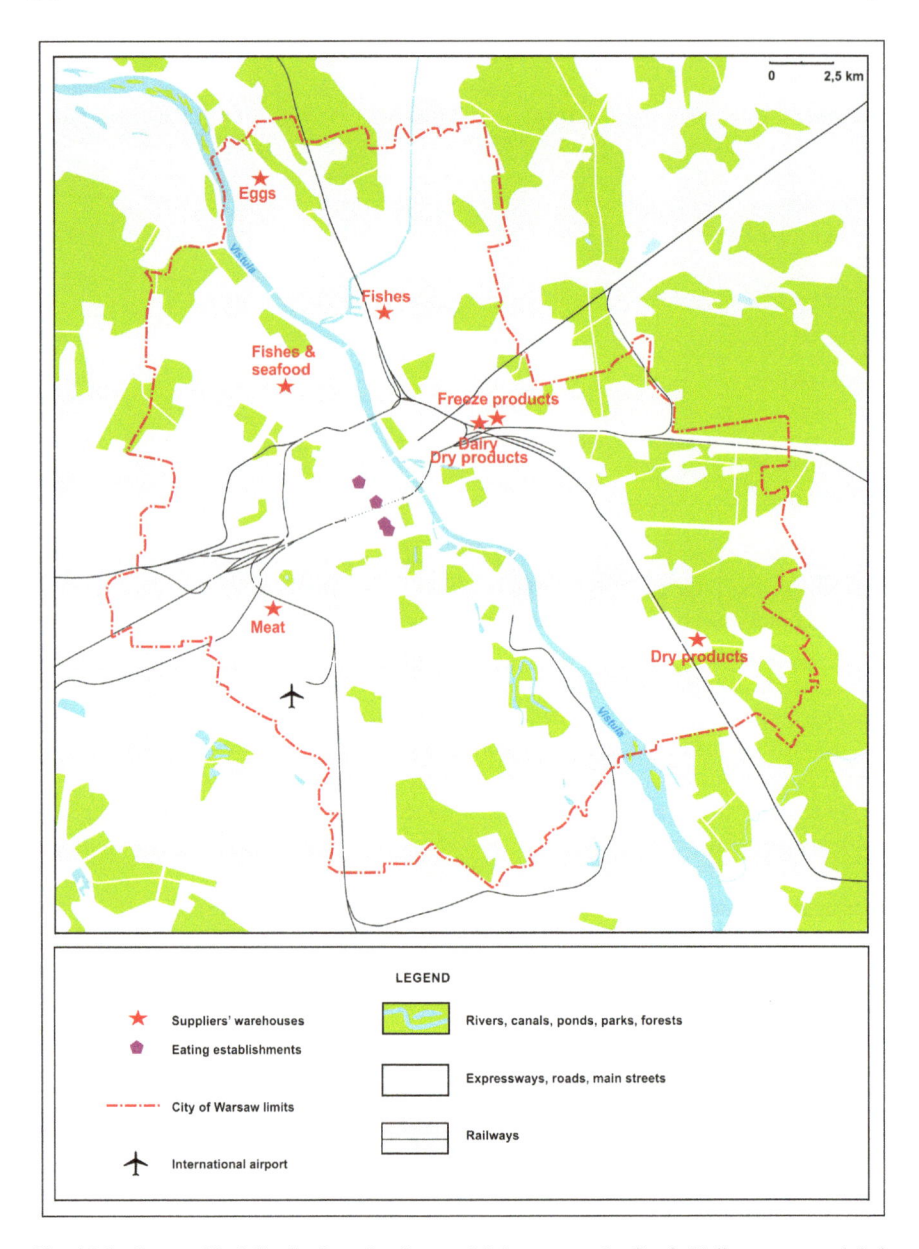

Fig. 13.1 Geographical distribution of eating establishments run by Strefa Kulinarna s.c. and their suppliers in Warsaw (Poland) in 2017

Kulinarna s.c. (established in 2012) and wholesalers (some goods are supplied from outside Warsaw, and even other regions of Poland).

Other important suppliers provide restaurants and bars with kitchen equipment, cleaning products, etc. In other words, supply chains are an essential part of the hospitality, as well as retail, industries. Companies that sell food products are particularly important, as they generally have everyday contacts with restaurants and therefore are an integral part of the system. However, the food supply chain does not have to be limited to a specific city. Most companies supply businesses over a large area and their zones of activity can cover the wider region, often the whole country, and sometimes internationally. While food manufacturing is concentrated in a limited number of global companies, many other parts of the supply chain are dominated by small- and medium-sized enterprises. Some of these companies only operate on a local scale, but others cover much larger areas.

In Indonesia, Telfer and Wall (2000) distinguish several levels of institutions participating in a hierarchically organized supply chain. At the first, bottom level, are producers and collectives. The next stage is local collectors who buy food and deliver it to local markets. The third level is distant markets and wholesale markets. The highest level is small local suppliers, trading companies and supermarkets; here the restaurant is purchased (this is the case for the Sheraton Senggigi Beach Resort on the island of Lombok). All of these components create an endogenous system that covers the area near the restaurant. However, the latter authors (Fig. 2 on p. 436) also distinguish a network of external links, both within Indonesia and beyond. In the case of internal links, the lowest level is the same as the case described above. The difference appears only at a higher level, where the local market replaces the Indonesian market, from which products are delivered to food processing companies. The situation is similar in the case of foreign connections, where foreign markets, food processing companies and food exporters as well as Indonesian importers appear on the higher level. Both systems provide food for trading and transport companies. The food supply network presented by Telfer and Wall addresses the phenomenon of providing food to restaurants in a holistic way, and their model is applicable in various situations. However, in the context of a city, it can be simplified and the following main components can be distinguished:

- unprocessed (e.g. farmers) and processed food producers,
- intermediaries between food producers and wholesale traders,
- wholesale traders and wholesale + retail traders,
- final recipients (restaurants)

Sometimes, retail traders and intermediaries between wholesalers and consumers exist as subsystems. Finally, it should be noted that the problem of the restaurant supply chain has been studied in detail. An example comes from Calgary (Canada) where the object of the analysis was to deliver food from local farmers to the River Café restaurant (Newman et al. 2013, pp. 14–20). However, the very specific scope of this research (the problem of local food and organic farming) makes it difficult to transpose the results to the situation described below.

13.2 Wholesalers in the Food Supply Chain—An Example from Warsaw (Poland)

One of the components of political and socio-economic change in Poland in the early 1990s was an increase in private sector entrepreneurship, including gastronomy. New restaurants and bars (often previously administered by the state and local authorities) opened, and new companies began to supply restaurants with food products, kitchen equipment, furniture, clothing, cleaning products, etc. The first large company operating in the sector was Elektromis (a Polish company established in 1987, it opened its first warehouses in 1990), which in 1993 changed its name to Eurocash.[2]

Political and economic transformation was followed by foreign capital flows. Two companies deserve special attention in this context: MAKRO Cash & Carry (from The Netherlands) and Selgros Cash & Carry (from Germany). The former launched operations in 1994, and the latter in 1997. At around the same time (1995) Eurocash was taken over by the company Jerónimo Martins (Portugal) and, since 2003, has been called Eurocash Cash & Carry. All three companies specialize in the cash and carry trade, and their main clients are the owners of small shops, restaurants and bars. Although the scope of their activities is similar, there are some differences in the way they operate and their warehouse locations, which can be clearly seen if the analysis covers the region of Warsaw. The clearest differences concern Eurocash Cash & Carry, which in Warsaw has only one wholesale outlet (in the south-western suburbs), and around nine others located in towns found within a radius of 40–50 km from the centre of Warsaw. These locations are due to the fact that their clients are mainly the small shops and eating establishments that are typically found in smaller cities and villages. This distribution is unlike MAKRO and Selgros, which are mostly located either in the city itself or in towns adjacent to it.

In Warsaw, the first MAKRO Cash & Carry hall began operating in the western part of the city in 1994. The second outlet opened in 1995 (in Ząbki, a town a few hundred metres from the north-eastern limits of Warsaw), and in 2006 a third warehouse opened in the north-west part of the city.[3] Selgros Cash & Carry has three warehouses, one located just beyond the city's southern borders. Although the company opened its first outlet in 1997 in Poznań, it was followed in 1998 by a

---

[2]*Elektromis*, https://pl.wikipedia.org/wiki/Elektromis (Accessed 10 Feb 2018); *Eurocash*, https://pl.wikipedia.org/wiki/Eurocash (Accessed 10 Feb 2018); *Historia*, http://eurocash.pl/grupa-eurocash/historia.html (Accessed 10 Feb 2018).

[3]*Historia MAKRO w Polsce*, https://www.makro.pl/o-makro/historia-makro-w-polsce (Accessed 10 Feb 2018).

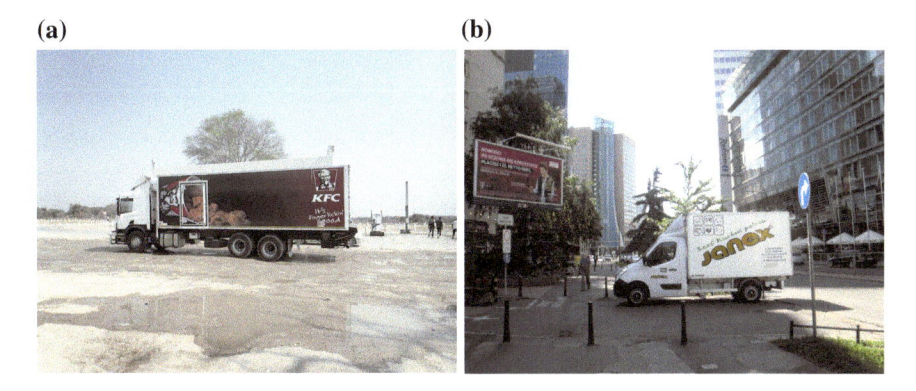

Fig. 13.2 Two types of transport used to deliver supplies to restaurants. Divundu (Namibia)—a truck used to supply long-distance supplies between cities, in this case to KFC's restaurant in Katima Mulilo (**a**). Warsaw (Poland)—a cargo van used to transport supplies over shorter distances (**b**)

second in Warsaw-Połczyńska, in the western part of Warsaw. Next were Warsaw-Białołęka (in 2005) and Warsaw-Wawer (in 2016) respectively in the north and south-east of the city, while the Warsaw-Piaseczno warehouse located between Warsaw and Piaseczno opened in 2017.[4]

At first glance, MAKRO and Selgros operate in the same way, notably in the location of their warehouses. However, a deeper analysis reveals some differences. Selgros is more oriented towards supplying restaurant and bar owners. This can be seen in a separate section supplying goods to the catering sector. Secondly, the car parks are smaller as are parking spaces (designed for cars). This is due to the fact that customers often buy goods in packages of not more than 5 kg. Other differences are seen in the size of the trolleys that customers use to transport their purchases. At Selgros, they are similar to those used at retail hypermarkets (Auchan, TESCO, Carrefour, etc.), while MAKRO provides robust platforms that can carry loads of over 30–50 kg. To summarize, the main difference between the two companies is that Selgros is more oriented towards serving the catering sector, while most of MAKRO's customers are owners of small food stores. On the other hand, although in both cases, customers must prove that are business owners, the majority of items are available in retail-size packaging. Thus, both Selgros and MAKRO are, on the one hand, wholesalers and, on the other hand, retail outlets (Fig. 13.2).

---

[4]*Selgros*, https://pl.wikipedia.org/wiki/Selgros (Accessed 10 Feb 2018).

13.3 A Case Study from Warsaw (Poland): Menu Express Sp. z o. o.

Most restaurant suppliers in big cities have their headquarters and warehouses in that city or its immediate vicinity. The location of these warehouses is determined by many factors, notably:

- ownership structure,
- square metre costs of storage,
- warehouse surface,
- location of the warehouse within the city,
- accessibility,
- technical infrastructure,
- safety and security.

In the restaurant sector, all of these factors (and others) influence not only the location of the supplier, but also its activities and relations with customers.

Menu Express sp. z o. o. is a Polish company operating in Warsaw, which provides supplies to catering establishments. Here it is used as a case study to illustrate the problem presented above. Its clients are typically independent restaurants and bars, and less frequently establishments in hotels or cafeterias in office buildings. The company was established in 2015. Its headquarters are in Warsaw, as are its warehouses. The company can be classified as an intermediary between the wholesale trade and the gastronomy sector. Although some of its goods are purchased from producers, most are purchased from wholesalers. This position is reflected in its customers, who encompass both restaurant operators (owners, managers, chefs) and catering companies, and wholesalers, or other intermediaries in the wholesale–retail chain.

Its headquarters and warehouse are located in the northern part of Warsaw, in the former industrial and warehouse district of Żerań Wschodni (Żerań East). This district was built in the 1950–1960s as one of the industrial zones of the city. Four other, similar, zones were established in Warsaw at that time: Służewiec Przemysłowy (Służewiec Industrial), Okęcie Przemysłowe (Okęcie Industrial), Targówek Przemysłowy (Targówek Industrial) and Odolany. All are located in the vicinity of railway lines at the periphery of the city, and between the early 1950s and the late 1980s they have been the location for industrial plants (for the food, chemical, engineering and building materials industries) and large warehouses. Beginning in 1989, government-owned factories and wholesalers began to be transformed. Following political and economic reform, and the move to a capitalist economy, most factories were liquidated. Although warehouses remained they were privatized, and the new owners began to rent them to other entrepreneurs. This included cold storage facilities, formerly used by the meat industry plants in the district of Żerań Wschodni. Some of these refrigerated warehouses are rented by Menu Express sp. z o. o., which initially mainly supplied fish and seafood to restaurants and bars (mainly shrimps and shushi products) as well as meat and

Table 13.4 Geographical distribution of Menu Express sp. z o. o. suppliers 2015–2017

Area	Year					
	2015		2016		2017	
	Number	%	Number	%	Number	%
Warsaw city	26	36.6	26	34.2	30	37.9
Warsaw metropolitan area	16	22.5	16	21.1	13	16.5
Mazovian region	5	7.0	3	3.9	5	6.3
Other regions in Poland	24	33.8	29	38.2	29	36.7
Other countries	–	–	2	2.6	2	2.6
Total	71	100	76	100	79	100

Source: Andrzej Kowalczyk based on data from Menu Express sp. z o. o.

poultry. Since then, the scope of its activity has expanded and it now provides almost everything (except for furniture and kitchen appliances) needed by restaurant owners and chefs. This was a natural extension, as the company's owners had previously worked in the largest Polish company specializing in the distribution of food products from countries outside Europe (mainly typical Asian, Middle Eastern and Mexican products).

In 2017, 30.8% of the money spent on total delivery to all the suppliers of Menu Express sp. z o. o. were paid to suppliers from the Warsaw metropolitan area (18.13% from Warsaw alone, and 12.67% from nearby towns). Foreign suppliers earned 5.58% of the money spent by Menu Express sp. z o.o. on delivery, suppliers from the Mazovian Region—1.52%, and as much as 62.1% were paid to the suppliers from other regions of Poland, mainly the Wielkopolska and Pomerania Regions.

Table 13.4 shows that the number of companies supplying Menu Express sp. z o. o. increased slightly. Beginning in 2016, two foreign suppliers (in Germany and Belgium) were added. At the same time, the geographical distribution of Polish suppliers has not changed much. Over half are based in Warsaw or its surroundings. Other suppliers of fish and seafood are located in the cities of Gdynia, Gdańsk and Szczecin on the Baltic Sea coast, while berries, mushrooms, fruit and meat, milk and dairy products are sourced from small towns. In some cases, these suppliers' headquarters are located far from Warsaw, but they have warehouses or wholesalers in the city.

The map presented in Fig. 13.3 shows that most suppliers are located outside the city centre. Most—generally larger warehouses and wholesale traders—are in parts of the city that have already been mentioned—Żerań Wschodni (in the north), Targówek Przemysłowy (in the east) and Okęcie Przemysłowe and Służewiec Przemysłowy (in the south-west, near the international airport). There is a concentration in the industrial and warehouse district of Żerań Wschodni, which is the location of the headquarters of Menu Express sp. z o. o. Other companies act as both suppliers and buyers. Some Menu Express sp. z o. o. suppliers are located in

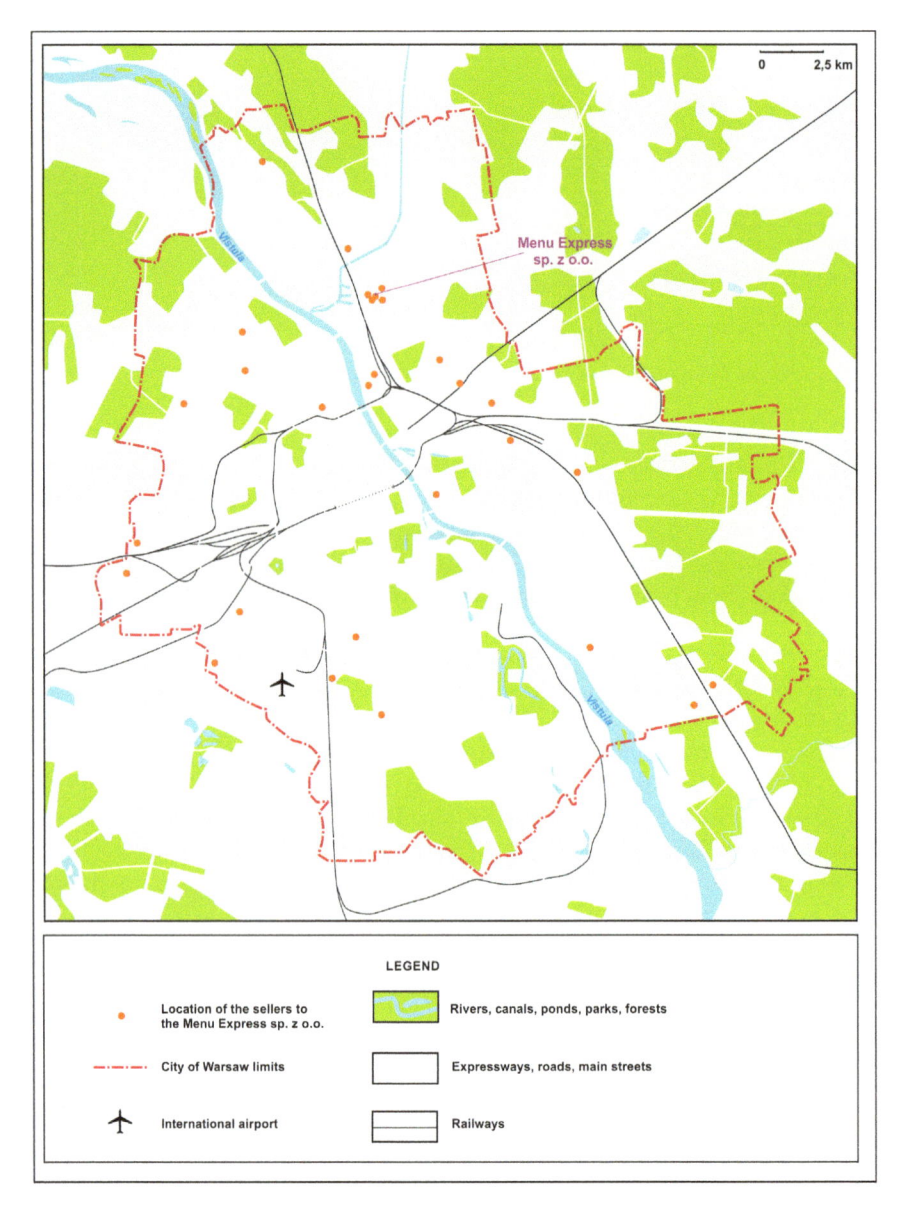

Fig. 13.3 Geographical distribution within Warsaw (Poland) of Menu Express sp. z o. o. suppliers in 2017. Source: Author based on data from Menu Express sp. z o. o.

Table 13.5 Geographical distribution of Menu Express sp. z o. o. customers (buyers) 2015–2017

Area	Year					
	2015		2016		2017	
	Number	%	Number	%	Number	%
Warsaw city	75	69.4	72	73.5	91	72.2
Warsaw metropolitan area	27	25.0	21	21.4	26	20.6
Mazovian region	3	2.8	3	3.1	3	2.4
Other regions in Poland	3	2.8	2	2.0	6	4.8
Other countries	–	–	–	–	–	–
Total	108	100.0	98	100.0	126	100.0

Source: Andrzej Kowalczyk based on data from Menu Express sp. z o. o.

areas formerly (pre-1990s) occupied by the largest Polish car factory (FSO Żerań) in the area of the city known as Żerań Zachodni.

Table 13.5 presents the geographical distribution of Menu Express sp. z o. o. customers or buyers. They are mainly restaurants, bars and other companies operated in the gastronomy sector; however, some are wholesalers and companies with a profile that is similar to Menu Express sp. z o. o. Some are both suppliers and customers for the company. The table shows that from the time Menu Express sp. z o. o. launched its activities, 90–95% of its customers have been in Warsaw and the surrounding towns.

The spatial distribution of Menu Express sp. z o. o. customers and suppliers are very different. Figure 13.4 shows that the vast majority of customers are in the central part of Warsaw, either in the city centre or, especially, in the outer city. These zones correspond to new financial and business districts to the south and west of the centre of Warsaw, and a new and rapidly growing middle-class residential district known as Miasteczko Wilanów to the south near the Vistula river. This neighbourhood is mainly inhabited by young professionals working in banking, insurance, and for international corporations, and there is a considerable demand for eating out.

To conclude this discussion, Fig. 13.5 highlights that Menu Express sp. z o. o. supplies no restaurants or bars located in the historical part of Warsaw or the prestigious Saska Kępa district, which was established in the 1930s on the eastern Vistula river. There may be several reasons for this. First, the historical part of the city (especially the Old Town) is dominated by foreign tourists and restaurants offer traditional Polish dishes. As Menu Express sp. z o. o. mainly supplies non-local products, clearly Polish restaurants are not among its customers. The second relates to logistics. Traffic restrictions in the Old Town make access difficult for Menu Express sp. z o. o. delivery vans. The third reason may be related to the strategies of restaurant owners in the Old Town and Saska Kępa, which are typically luxury dining establishments, often with long traditions. Such businesses have long-standing relationships with their suppliers, while Menu Express sp. z o. o. only started to operate in 2015. A final observation is that some Menu Express sp. z o. o.

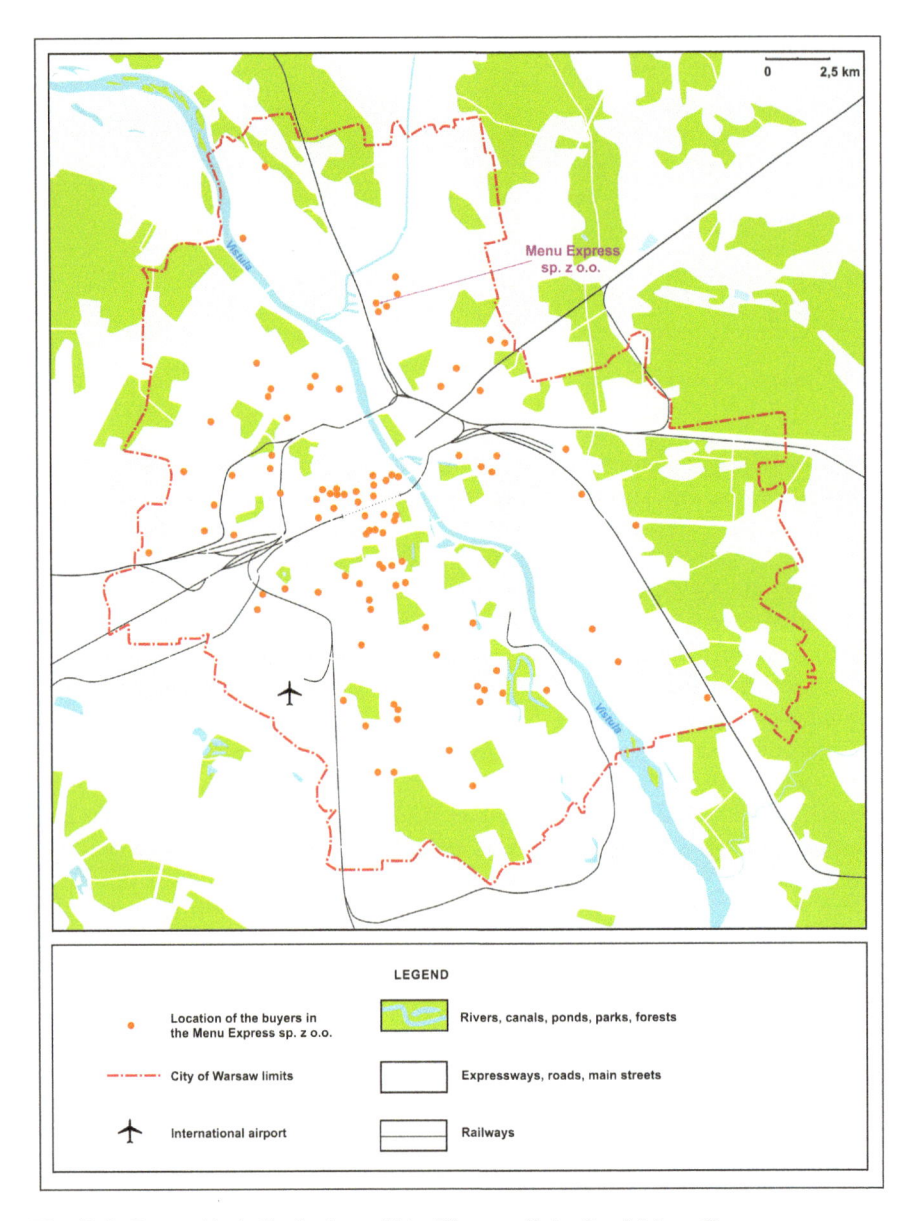

Fig. 13.4 Geographical distribution within Warsaw (Poland) of Menu Express sp. z o. o. customers in 2017. Source: Andrzej Kowalczyk based on data from Menu Express sp. z o. o.

customers are food chains. This is the case for one family-owned company, which was founded in 1991 and owns seven boutiques in Warsaw and one at the suburban spa-town of Konstancin-Jeziorna. This company delivers goods from suppliers (among them Menu Express sp. z o. o.) of cakes, ice cream and other desserts, to

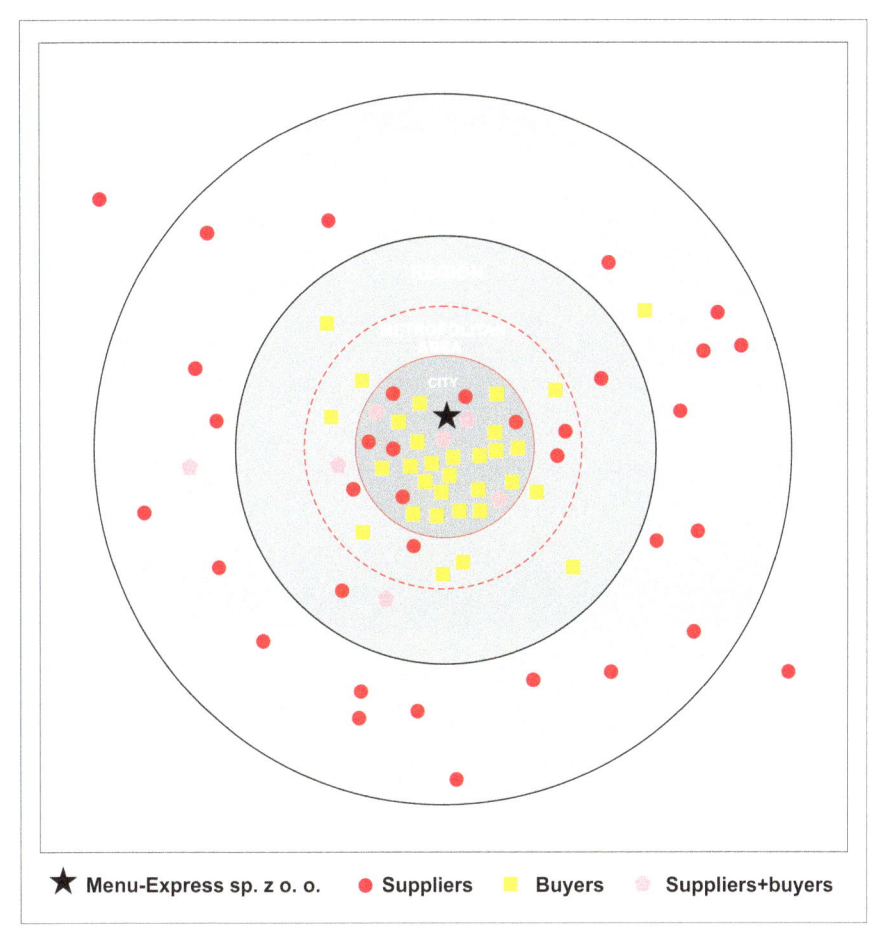

Fig. 13.5 Distribution of contractors in the supply chain for Menu Express sp. z o. o. in Warsaw (Poland)

boutiques located in various parts of Warsaw. The same is true for a chain of sushi bars. This company was established in 2005 and manages four bars in Warsaw (one in the city of Szczecin and other in Wrocław); here, dishes are prepared in one place and delivered to other bars.

13.4 Conclusions

The situation presented in the above text confirms the thesis of Doreen Massey (1999), that in the modern world the 'place' is determined not only by local but also global conditions. Thus, on the example of spatial relationships between suppliers and recipients of the Menu Express sp. z o. o., the concept of glocalization can be positively verified (Robertson 1994). At this point, it can be stated that in terms of financial flows (which due to trade secrets have been omitted in this text) Menu Express sp. z o. o.'s links with suppliers from other parts of Poland and from abroad are much stronger than with suppliers from Warsaw and its surrounding area.

The research also proved that in many cases the location of suppliers is not accidental, as it is determined by the appropriate infrastructure (in the case of location of industrial cold storages), which is not present in the urban space with the same intensity. This peculiar 'geographical determinism' can be also applied to the location of customers in the urban space. In general, they have restaurants and bars in the central part of Warsaw. This is not a surprise, because, drawing on the models of spatial structure of cities, the largest concentration of gastronomic establishments is precisely in the downtown of large cities.

References

Elektromis. https://pl.wikipedia.org/wiki/Elektromis. Accessed 10 Feb 2018
Eurocash. https://pl.wikipedia.org/wiki/Eurocash. Accessed 10 Feb 2018
Frontczak J (2017) Rynek rośnie a wraz z nim dostawcy. Rynek dostawców HoReCa w Polsce. Raport 2017, Wydanie Specjalne Czasopism Nowości Gastronomiczne, Świat Hoteli, Sweets Coffee, Styczeń 2017, p 4–7. https://issuu.com/brogmarketing/docs/rynek_dostawc__w_horeca_w_polsce_-_. Accessed 25 Feb 2018
Hellier P (2019) Sustainable restaurant system and gastronomy. In: Dixit, S.K. (ed.), 2019, The routledge handbook of gastronomic tourism, Routledge, Oxon–New York, pp 280–288
Historia. http://eurocash.pl/grupa-eurocash/historia.html. Accessed 10 Feb 2018
Historia MAKRO w Polsce. https://www.makro.pl/o-makro/historia-makro-w-polsce. Accessed 10 Feb 2018
Kock MG (2013) The development of an eco-gastronomic tourism (EGT) supply chain-analyzing linkages between farmers, restaurants and tourists in Aruba, University of Central Florida, Department of Hospitality in Rosen College of Hospitality, Orlando, FL, Ph.D. thesis
Massey D (1999) Power-geometries and the politics of space-time, Hettner-Lecture 1998, Hettner-Lectures 2, Department of geography, University of Heidelberg, Heidelberg
Mentzer JT, DeWitt W, Keebler JS, Min S, Nix NW, Smith CD, Zacharia ZG (2001) Defining supply chain management. J Bus Logist 22(2), (Autumn), pp 1–25
Murphy J, Smith S (2009) Chefs and suppliers: an exploratory look at supply chain issues in an upscale restaurant alliance. Int J Hosp Manag 28(2), pp 212–220
Newman L, Ling C, Peters K (2013) Between field and table: environmental implications of local food distribution. Int J Sustain Soc 5(1), pp 11–23
Robertson R (1994) Globalisation or glocalization? J Int Commun 1(1), pp 33–52
Selgros. https://pl.wikipedia.org/wiki/Selgros. Accessed 10 Feb 2018
Steel C (2013) Hungry city: how food shapes our lives, Random house
Szot M (2018) Jakość najważniejszym kryterium wyboru, Rynek dostawców HoReCa w Polsce. Raport 2018, Wydanie Specjalne Czasopism Nowości Gastronomiczne, Świat Hoteli, Sweets Coffee, Styczeń 2018, pp 8–11. https://issuu.com/brogmarketing/docs/rynek_dostawc_w_horeca_raport_2018. Accessed 10 May 2018
Telfer DJ, Wall G (2000) Strengthening backward economic linkages: local food purchasing by three Indonesian hotels. Tour Geogr 2(4), 421–447

Part III
Challenges

Chapter 14
New Trends in Gastronomy in the Context of the Urban Space—Introduction to Part III

Andrzej Kowalczyk

Abstract In contemporary gastronomy, there are several trends that can be considered as challenges for the current model of restaurant business operation. This short section is an introduction to the part of the book paying attention to this problem. Examples of the challenges include: (1) the development of eating services in shopping centres, (2) the growing role of street food, (3) the emergence of a dining function in new public spaces (parks, riverside areas) and (4) the increasing of the popularity of delivering meals to the home (which is greatly influenced by new technologies in the IT sector).

Keywords Gastronomy · Urban space · Challenges

In the closing decades of the twentieth century, new trends appeared in city gastronomy, which continued to develop in the following century. These trends have both spatial and non-spatial dimensions. According to Park, '…there are a variety of possible locations for restaurants including a freestanding unit, located in shopping mall outlet, food court, or a multiple concept unit within an existing facility designed for another business such as a gas station and convenience store' (Park 2002, p. 2). While this statement may be true of any place, it is not true of any time: Shopping malls and gas stations are relatively new phenomena which, in many cases, are exceptions to the traditional locations described by Park as freestanding units or convenience stores (in the context of traditional restaurants and bars). So what does this mean?

In many cases, new ways of offering food are providing serious competition for traditional forms of gastronomy. Selling coffee from a vending machine at a gas station may mean that the potential customer will not drink coffee at the bar. Offering kebabs and pizzas in the form of street food may reduce the number of customers visiting Middle Eastern-style bars and pizzerias. Providing free beer or

---

A. Kowalczyk (✉)
Department of Tourism Geography and Recreation, Faculty of Geography and Regional Studies, University of Warsaw, ul. Krakowskie Przedmieście 30, 00-927 Warsaw, Poland
e-mail: akowalczyk@uw.edu.pl

© Springer Nature Switzerland AG 2020
A. Kowalczyk and M. Derek (eds.), *Gastronomy and Urban Space*,
The Urban Book Series, https://doi.org/10.1007/978-3-030-34492-4_14

wine at a culinary festival may be reflected in a drop in the number of customers visiting the beer and wine bars. Ordering food at home means fewer guests in restaurants and bars, which may not be a threat to cooks, but could lead to waiters losing their jobs.

Maybe in the context of gastronomy it sounds awkward, but in the language of modern economics, the phenomenon of such internal competition is called *cannibalisation*. Regardless of the terminology that is used, the phenomenon presents a new and serious challenge for the gastronomy business, and for individual entrepreneurs (e.g. restaurant owners) in particular. In the urban context, the first phenomenon with a spatial dimension is the growing appearance of eating establishments in shopping centres. Initially, this concerned mainly North America, then Western Europe, but after political transformations in Central and Eastern Europe it has also surfaced in these parts of the world. It should be noted, however, that the presence of dining outlets in places where trade is concentrated is not something new. It was already common in the Middle Ages, when many of the inns and meal stands were located in marketplaces and covered halls. This trend developed and strengthened until the turn of the nineteenth and twentieth centuries. The difference between then and now, however, is that unlike the nineteenth century, the majority of eating outlets in modern shopping centres are part of restaurant chains and offer similar food in Europe, North America and Asia.

The development of gastronomy in shopping centres has not only led to the harmonisation of the dishes on offer. The presence of transnational restaurant chains is a threat to independent bar owners, who often lose out in a competitive fight with the big players, for many reasons. One of the main ones is that the dining/catering establishments of global restaurant chains (e.g. McDonald's, KFC or Pizza Hut) very often have a shared consumption space in shopping centres. This means that they share the costs of renting the space, and that the area is in use almost all of the time. In the case of independent operators, this is not the case: Their restaurants and bars often have breaks in receiving guests and their rented consumer areas are not always occupied. This, in turn, increases relative operating costs and reduces their competitiveness. Although this phenomenon mainly applies to dining establishments located inside the malls, restaurants and bars around them may also be less competitive, particularly if the dishes they offer do not stand out. This eviction of independent gastronomy in the vicinity of shopping centres may be dangerous, potentially leading to gastronomic deserts around malls. Although this is not yet commonplace, the frequency with which old restaurants and bars close, only to be replaced by new ones suggests that creating a successful business can be very difficult for independent operators.

Restaurants in shopping galleries are not the only new challenge for urban gastronomy. As Park (2002) also noted, there is a rising presence of bars at gas stations, which can also be problematic. A dozen or so years ago, bars were mostly seen at gas stations located outside large cities, but now they are almost everywhere, including city centres. At the same time, more and more convenience stores have stands selling pizza or sandwiches, coffee, etc. This means that the gastronomic function now extends beyond its traditional boundaries and is moving into new areas of the urban space.

Another new trend in gastronomy in urban areas is greater concern for the health of residents. This applies not only to healthy eating, but also to healthy lifestyles. It is also related to the concept of 'food deserts', which assumes access to healthy food, understood as '…areas of relative exclusion where people experience physical and economic barriers to accessing healthy food' (Shaw 2006, p. 231). This phenomenon is a consequence of pro-environmental attitudes, which have become increasingly popular since they emerged in the 1960s and 1970s. The movement for healthy eating grew stronger in the mid-1980s, when the Slow Food initiative was founded in Italy 1986 by Carlo Petrini. It has subsequently spread worldwide, promoting not only local food and traditional cooking, but also changes in lifestyle. It was established as an alternative to fast food, although it forms part of the broader, so-called Slow movement. Slow Food's goals, including the promotion of ecological food and local small businesses, are paralleled by an ideology directed against the globalisation of agricultural products—see Parham (2008, 2015a), Pink (2008) and Pothukuchi (2009) for more details.

According to Jang et al. (2011, p. 803) 'Consumer demand for environmentally friendly food has risen quickly over recent decades. (…) Increasing consumer interest in eco-friendly food has brought a number of changes to the restaurant industry'. These words ring true, in particular, in relation to the behaviour of consumers born in the 1980s and 1990s. It is no wonder that these new trends have been noticed by city authorities. In October 2015, the local authorities of Milan (Italy) together with the consortium 'Food Smart Cities for Development' launched the Milan Urban Food Policy Pact. This international protocol engages cities to use more sustainable and fair urban development methods that leverage food policies. It is worth noting that the protocol was signed by more than 130 cities, including Barcelona and Bilbao (Spain), Bruges and Gent (Belgium), Utrecht (The Netherlands) and Turin (Italy), as well as Dakar (Senegal) and Accra (Ghana) (*Food smart cities for development. Recommendations and good practices* 2017). There is also a group in Oakland, California in the United States, which has an active policy of implementing solutions that are consistent with the Slow Food ideology (Unger and Wooten 2006). These trends also apply to the restaurant sector. A recent American survey showed that organic produce was one of the most important menu trends, albeit mainly in fine dining restaurants (Poulston and Yiu 2011, pp. 186–187).

One of the most common manifestations of the Slow Food ideology in the urban space is the so-called farmers markets. These are an example of a specific type of food retailing known as direct marketing, where producers sell directly to consumers. It creates a direct link between the people who grow food and those who eat it. This phenomenon is not new in the history of cities: It existed in ancient times and in the Middle Ages. In the modern capitalist economy, however, the food supply chain from the producer to the consumer has become very elongated and nowadays consumers often receive highly processed agricultural products, examples of which include fast food. Until recently, this phenomenon concerned cities in

Fig. 14.1 New dining hotspots in the urban space. In Wrocław (Poland), the former railway station (Dworzec Świebodzki, which closed in 1991) has been a popular meeting place for foodies for several years (**a**). In some urban agglomerations people eat and meet friends in parks and other green areas within the city, such as the Parc de la Ciutadella in downtown Barcelona (Spain) (**b**). Source: Anna Kowalczyk 2017, 2015

North America and Europe, but towards the end of the twentieth century, it increasingly appeared in South American, Asian and even African cities. Farmers markets are not the only opportunity for consumers to buy food directly from the producer: Often customers can meet chefs preparing meals with fresh products. In many cases farmers markets include meal stands, as occured in European cities during the Middle Ages, and as currently happens in cities in South Asia, Latin America and Africa. Farmers markets are located in different places in different cities: Sometimes they have a permanent home; sometimes their location changes. Often, they are in former market halls, on city squares, on river banks,[1] in green areas, or former industrial or railway areas (Fig. 14.1). The growing popularity of farmers markets and similar initiatives (food fairs, culinary festivals, etc.) means that they are beginning to compete with restaurants and bars in today's capitalist economies.

> A lot of attention to city marketplaces can be found in the works of Susan Parham. The problem of traditional food markets has been extensively described in her PhD dissertation on three districts of London: Borough Market in Southwark, Broadway Market in East London and Exmouth Market in Islington (Parham 2008). An important element of her deliberations, which is already signalled in the title, is the idea of the so-called *food quarter*. The term goes on in many of Susan Parham's works, which defines the *food quarter* as '… a "fuzzy edged" food-centred area of an urban

---

[1]For example, in the summer of 2016 in Warsaw (Poland), there were 31 eating establishments on the banks of the Vistula river (Jankowska 2016).

settlement, predicated on human scaled, highly mixed, walkable and fine grained urbanism that reflects the European City Model (Clos 2005). The food quarter thus conceptualised is generally located in traditional urban fabric but its elements are capable of being retrofitted into more sprawling locations and built into new areas too. The market at the centre of the food quarter is not necessarily a farmers' market—and its operation challenges the easy stereotypes that situate traditional markets and farmers' markets as a clear cut duality—nor is it only servicing elite consumption needs' (Parham 2012. p. 3). The mentioned above Parham studies in London have shown that an important role in maintaining the eating function in a given part of the city is the character of urban structure. This means that historical conditions are important for the current use of space. The research also showed that at first the urbanisation processes after World War II—above all suburbanization— caused that the role of marketplaces decreased. However, later the actions of the authorities and the support of local communities contributed to the 're-birth' of the trade fair function. The proper urban planning was of great importance for the success of these activities (Parham 2008, p. 139).

The next challenge for the restaurant industry in urban areas is the growing popularity of street food. Although this has been a way to satisfy food needs in many areas of the world, it became less significant in large cities in Europe and North America in the mid-twentieth century. It has seen something of a resurgence, however, since it came to be regarded as a fashionable lifestyle choice, rather than a necessity. Eating favourite dishes bought from hawkers or permanent stalls has always been possible, but food wagons—which have recently become a permanent element of the urban space—are a relatively new phenomenon. The basic feature of food wagons is that, by definition, they do not have to have a fixed parking place. Although they sometimes stay in one location for extended periods, they are usually moving, sometimes within a given city, but often to other cities or towns. So-called 'food truck festivals' bring together trucks from different cities and regions. Because their dishes are increasingly popular among foodies in big cities, they have become a threat to fixed restaurants and bars, notably because their operating costs are usually much lower.

The last, but by no means least, important spatial challenge for city gastronomy is home delivery. Providing food to the customer's place of residence is not new. For centuries, milk and cheese, meat, potatoes, vegetables, etc. were delivered to the homes of (mainly richer) customers. However, the scale of this phenomenon in the twentieth century weakened, largely due to the rapid development of the retail sector and the growing popularity of the automobile. Customers preferred the convenience of buying many products in one place from one seller (for instance a supermarket) rather than having to buy products separately from many places.

Initially, home food delivery concerned uncooked or semi-processed food products, rather than ready-made meals. Over time, it has triggered a significant

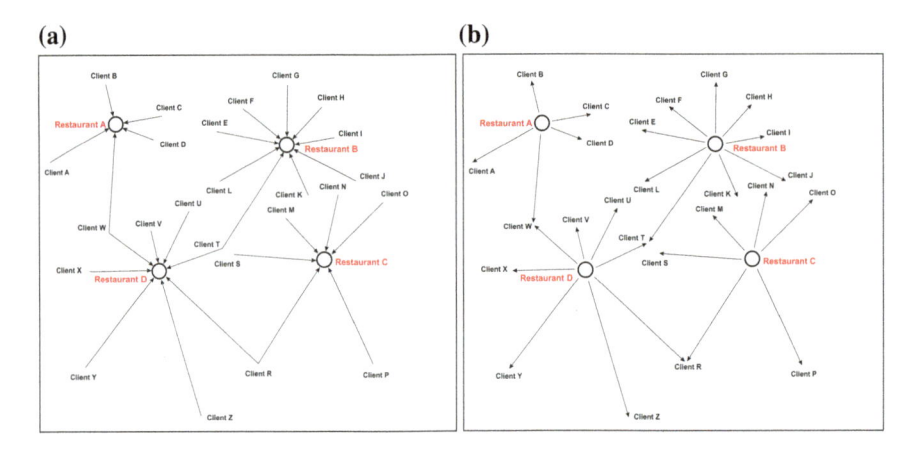

Fig. 14.2 Two types of spatial relations between restaurants and their clients. Type A (the old model): customers search for a restaurant (**a**). Type B (the new model): restaurants search for customers (**b**). Type B represents the process described as home food delivery

change in research into spatial relations. Until now, the main problem has been the spatial behaviour of restaurant customers. In the context of home food delivery, the problem is how to apply spatial analysis to the behaviour of the restaurant sector. In other words, according to the old paradigm the *consumer* was looking for a *restaurant*, while now *restaurants* are looking for *customers* (Fig. 14.2).

The growth in home food delivery may lead to some restaurants and bars turning into de facto catering companies. Although this will bring financial returns, they will have to change their employment structure. Waiters and other staff who serve customers will become unnecessary, while the demand for kitchen staff, drivers, people taking orders, etc. will increase. This may mean that restaurants and bars disappear from their current locations (for example, along principal streets or squares) and the entire spatial pattern of gastronomy in urban space may change. The growing popularity of home food delivery may also lead to significant non-spatial changes. People will still meet for lunch and dinner with friends, but in private apartments, rather than in restaurants. The social (integrating) role of gastronomy will remain, but it will move to private rather than public spaces. Moreover, the spatial aspect of gastronomy in cities will begin to take on a different dimension. Until now, geographical analyses have primarily focused on the location of eating establishments and spatial relations between their locations and the place of residence of the people using them. These spatial relations will continue to be explored, but in the context of the distribution of demand (customers) rather than supply (restaurants). This means that research will be less focused on the spatial availability of restaurants and bars, and more on the optimal delivery of meals to customers.

Finally, it should be noted that the literature on gastronomy in cities focuses on many other problems, which are often very important for social reasons. One of the major, unfortunate, challenges is the development of new eating habits that have led to obesity and serious illnesses. An American study covering California, Michigan, New Jersey and Texas showed a significant effect of the proximity to fast-food restaurants on the risk of obesity: the presence of a fast-food restaurant within 0.1 mile of a school was associated with at least a 5.2% increase in the obesity rate in that school (Currie et al. 2010, p. 60). A similar picture was detected in a study of the relationship between obesity and the location of eating places offering takeaway food in Boston, Massachusetts in New England. Many studies of the relationship between the location of catering establishments and obesity in schools have been carried out in the United States, including the work of Frank et al. (2006) who focused on the quality of the available food. As Tian et al. (2005, p. 5) noted, '… areas with a higher density of take out restaurants have a higher percentage of obesity, except for those neighborhoods that reside in the central downtown area (…). The disparity for these downtown areas may be due to the fact that restaurants in these areas are catering to visitors, students, or business people who do not typically reside in that neighbourhood'. Although this subject mainly interests medical sciences (Brennan and Carpenter 2009), the problem of obesity and related diseases has forced authorities in some cities to introduce regulations governing the location of fast-food restaurants in the urban space (Mair et al. 2005). It is worth noting that similar results were found in a study of 537 fast-food restaurants in the state of Victoria (Australia). As Thornton (2016, p. 8) noted, 'Understanding the distributional patterns of fast food restaurants may assist authorities to target appropriate potential policy mechanisms, such as planning regulations, where they are most needed'.

Another problem that has been present from ancient Roman times, through the Middle Ages (as already mentioned in Sect. 1.1.), and still exists today is the relationship between gastronomy and crime. These relationships have a spatial dimension and it is unsurprising that research into drug crime in Cleveland, Ohio (Roncek and Maier 1991) and in the Tower Hamlets neighbourhood in London (Tarkhanyan 2013) showed that bars are hotspots.

The big challenge—perhaps the biggest—for the restaurant sector in cities is a phenomenon that does not have a spatial dimension: cooking at home with friends. Although the practice meets social needs, it is dangerous for the restaurant industry. While it has been encouraged by television programmes, the internet and in lifestyle magazines, it can reduce visits to restaurants. In combination with home food delivery, cooking with friends at home can be a serious blow to restaurants and bars operating under the current rules. It is possible that restaurants will begin to look more and more like catering companies, rather than places where people meet to socialise.

References

Brennan D, Carpenter Ch (2009) Proximity of fast-food restaurants to schools and adolescent obesity. Am J Public Health 99(3):505–510

Clos J (2005) Towards a European city model. In London: Europe's Global City? Urban Age Conference Newspaper, London, November 2005

Currie J, Della Vigna S, Moretti E, Pathania V (2010) The effect of fast food restaurants on obesity and weight gain. Am Econ J: Econ Policy 2, 3(August):32–63. https://www.aeaweb.org/articles?id=10.1257/pol.2.3.32. Accessed 10 June 2017

Food smart cities for development. Recommendations and good practices (2017) Milan Urban Food Policy Pact, Milano. http://www.milanurbanfoodpolicypact.org/wp-content/uploads/2017/02/FSC4D-Recommendation-and-good-practices.pdf. Accessed 30 Jan 2018

Frank L, Glanz K, McCarron M, Sallis J, Sealens B, Chapman J (2006) The spatial distribution of food outlet type and quality around schools in differing built environment and demographic contexts. Berkeley Plan J 19(1):79–95

Jang YJ, Kim WG, Bonn MA (2011) Generation Y consumers' selection attributes and behavioral intentions concerning green restaurants. Int J Hosp Manag 30(4):803–811

Jankowska E (2016) Lista wszystkich nadwiślańskich knajp [MAPA] i cena piwa w każdej z nich, metrowarszawa.pl, 29.07.2016. http://metrowarszawa.gazeta.pl/metrowarszawa/1,141635,20430426,mapa-knajp-nad-wisla-ile-zaplacisz-za-piwo.html. Accessed 30 July 2016

Mair JS, Pierce MW, Teret SP (2005) The use of zoning to restrict fast food outlets: a potential strategy to combat obesity. https://www.jhsph.edu/research/centers-and-institutes/center-for-law-and-the-publics-health/research/ZoningFastFoodOutlets.pdf. Accessed 31 Dec 2018

Parham S (2008) Exploring London's food quarters: urban design and social process in three food-centred spaces. A thesis submitted to the Department of Sociology of the London School of Economics, London, December 2008

Parham S (2012) *Market Place: Food Quarters, Design and Urban Renewal in London*, Cambridge Scholars Publishing, Newcastle upon Tyne

Park K (2002) Identification of site selection factors in the U.S. franchise restaurant industry: an exploratory study, Faculty of the Virginia Polytechnic Institute and State University, January 10, 2002, Blacksburg, Virginia, MSc thesis. https://theses.lib.vt.edu/theses/available/etd-01112002-135621. Accessed 10 June 2017

Pink S (2008) Mobilising visual ethnography: Making routes, making place and making images. Forum Qualitative Sozialforschung/Forum: Qualitative Social Research, 9, 3, 36. http://nbn-resolving.de/urn:nbn:de:0114-fqs0803362. Accessed 18 Feb 2019

Pothukuchi K (2009) Community and regional food planning: building institutional support in the United States. Int Plan Stud 14(4):349–367. https://doi.org/10.1080/13563471003642902

Poulston J, Yiu AYK (2011) Profit or principles: why do restaurants serve organic food? Int J Hosp Manag 30(1):184–191

Roncek D, Maier P (1991) Bar, blocks, and crime revisited: linking theory of routine activities to the empiricism of "hot spots". Criminology 29, 4(November):725–753

Shaw HJ (2006) Food deserts: towards the development of a classification. Geogr Ann: Ser B, Hum Geogr 88(2):231–247. https://doi.org/10.1111/j.0435-3684.2006.00217.x

Tarkhanyan L (2013) Drug crime and the urban mosaic: the locational choice of drug crime in relation to high streets, bars, schools and hospitals. In: Kim YO, Park HT, Seo KW (eds) Proceedings of the Ninth International Space Syntex Symposium, 31 October-3 November 2013. Sejon University, Seoul, 089, pp 0101:1–101:13. http://www.sss9.or.kr/paperpdf/ussecp/sss9_2013_ref101_p.pdf. Accessed 20 June 2017

Thornton LE, Lamb KE, Ball K (2016) *Fast food restaurant locations according to socioeconomic disadvantage, urban–regional locality, and schools within Victoria, Australia.* Popul Health 2:1–9

Tian YQ, Troppy TS, Dripps W, Chalfoiux D (2005) *An exploratory spatial analysis to identify the appropriate scale and potential risk factors for understanding obesity distributions.* http://www.geog.leeds.ac.uk/groups/geocomp/2005/Tian.pdf. Accessed 10 March 2018

Unger S, Wooten H (2006) A food systems assessment for Oakland, CA: toward a sustainable food plan. Oakland Mayor's Office of Sustainability and University of California, Berkeley, Department of City and Regional Planning June 21, 2006, Oakland. http://oaklandfoodsystem.pbworks.com/f/Oakland%20FSA_6.13.pdf. Accessed 04 March 2017

Chapter 15
New Gastronomic Hotspots in the Urban Space. Food Courts in Poland

Andrzej Kowalczyk, Magdalena Kubal-Czerwińska, Katarzyna Duda-Gromada and Aleksandra Korpysz

Abstract Shopping centres occupy an increasingly important position in urban society and visits to them have become part of residents' lifestyles. Most malls contain not only shops, but also restaurants, bars and cafés. This chapter looks at the development of gastronomy services in these centres in terms of both the external context (the shopping centre in urban space) and the internal context (how establishments are distributed within centres). Taking the example of Warsaw, changes in the number of eating establishments in the biggest shopping centres are shown between 2001 and 2017. The chapter also considers the negative impact on restaurants and bars located outside malls, which lose business because consumers visit them less frequently.

Keywords Gastronomy · Shopping centres · Food courts · Warsaw

15.1 The Beginning and Evolution of Shopping Centres

The origins of what we now call shopping centres can be traced back to the Middle Ages, when market halls began to appear in the cities of Europe, and roofed bazaars appeared in cities such as Damascus and Istanbul in the Middle East. In the modern

---

A. Kowalczyk (✉) · K. Duda-Gromada · A. Korpysz
Department of Tourism Geography and Recreation, Faculty of Geography and Regional Studies, University of Warsaw, ul. Krakowskie Przedmieście 30, 00-927 Warsaw, Poland
e-mail: akowalczyk@uw.edu.pl

K. Duda-Gromada
e-mail: kduda@uw.edu.pl

A. Korpysz
e-mail: aleksandra.korpysz@uw.edu.pl

M. Kubal-Czerwińska
Department of Tourism and Health Resort Management, Jagiellonian University in Kraków, Institute of Geography and Spatial Management, ul. Gronostajowa 7, 30-387 Kraków, Poland
e-mail: magdalena.kubal@uj.edu.pl

era, one of the earliest precursors of today's shopping centres was the Great Gostiny Dvor (Большой Гостиный Двор in Russian) in Saint Petersburg. Built in 1785, it had more than 100 outlets. Later, similar buildings were built in London, Paris, Milan, Moscow, Berlin and many other large cities in Europe and North America (Reikli 2012, p. 22).

The concept of the modern shopping centre, which can be traced back to the beginning of the 20th century, grew in subsequent decades. It developed rapidly in the 1950s and 1960s, expanding the presence of restaurants and bars.[1] One of the earliest in Europe was the Kaufhaus des Westens (KaDeWe) in Berlin (Germany), which was built in 1907 and expanded in the 1930s. It was rebuilt after its destruction during World War II, reopened in 1950 and subsequently expanded in 1956 and 1996.[2] From the very beginning, the KaDeWe included restaurants and bars (in 2017 there were 21[3]) as well as shops. Located in central Berlin, it has always been popular with Berlin residents and others from outside the city.

But the idea of shopping centres has gained most popularity in the United States. This process began in the 1910s and 1920s. Among the first were Market Square in Lake Forest, Illinois (in 1916), and the Country Club Plaza in Kansas City, Missouri, which was opened in 1923 and was the first shopping centre on the outskirts of the city designed for customers using automobiles (Reikli 2012, p. 22).

According to Cohen the growth of shopping centres accompanied the suburbanization of residential life. These new suburban residents, '…who had themselves grown up in urban neighborhoods walking to corner stores and taking public transportation to shop downtown were now contending with changed conditions' (Cohen 1996, p. 1051). Some developers responded by placing the shopping district at the core of the residential community, while others made no effort to provide for residents' needs. Suburbanites were expected to fend for themselves by driving to existing market towns (Cohen 1996, pp. 1051–1052). In the early 1950s some traders aggressively reached out to these new communities whose buying power exceeded their numbers. Initially they built stores along new highways, in commercial strips that consumers could easily reach by car. But by the mid-1950s, developers had constructed a new type of marketplace: the regional shopping centre, which was located at highway intersections or along the busiest expressways. As Cohen notes, after a time these shopping centres became '…the ideal core for a settlement that grew by adding residential nodes off of major roadways rather than concentric rings from downtown, as in cities and earlier suburban communities' (Cohen 1996, p. 1053). It should be noted that while these shopping centres were designed by many architects, Victor Gruen (in the 1950s and 1960s), James Rouse (1970s and 1980s) and Jon Jerde (1980s and 1990s) were the most

---

[1]We use the terms *shopping centre, shopping gallery and shopping mall* interchangeably.

[2]*The KaDeWe—history of our story*, https://www.kadewe.de/en/das-kadewe-die-geschichte (Accessed 08 March 2018).

[3]*Bars & restaurants*, https://www.kadewe.de/bars-und-restaurants-im-kadewe-berlin (Accessed 08 March 2018).

significant. They paid attention not only to technical issues, but also to social goals and '…believed in the possibility of a new humanistic city' (Wall 2003, p. 106). Well-designed malls were intended to be an important part of this idea.

From the geographical perspective shopping centres and galleries can be divided into inner city- and outer city-oriented, including those located in the suburban zone. Shopping centres and galleries in the central part of the city are often opened in commercial districts and near railway stations. Shopping centres in the outer zones, however, are usually located at highway junctions and adjacent to housing estates. Shopping centres in suburban areas are distinguished by their location at the intersections of roads and adjacent to business parks. Another way of categorising shopping centres is by separating them into those that are integrated into their surroundings (e.g. in former industrial zones, port warehouses, railway stations) and those that are not (e.g. a new shopping gallery in the old part of the city, a large shopping centre in the vicinity of bungalows or villas).

15.2 Shopping Centres and Gastronomy—General Outlook

In recent years, shopping centres have occupied an increasingly important position in urban society and visits to them have become part of residents' lifestyles. Most shopping malls contain not only shops, but also restaurants, bars and cafés (as well as entertainment facilities, beauty salons, fitness clubs and many other services). This explains why some authors put gastronomy in second place (after shopping) when determining the social role of shopping centres (Frączkiewicz 2013, p. 340). Research in the early 1990s found that nearly 50% of clients visiting three malls in suburban areas of the United States consumed some form of snack, while about 30% ate a full meal (Bloch et al. 1994, p. 30).

One of the first researchers to draw attention to the role of other factors (which he called 'leisure') that motivated visits to shopping centres was Jackson (1991, p. 283). He based his conclusions on work carried out at West Edmonton Mall (Edmonton, Canada), which was analysed by many other authors. Shopping centres have become places that meet both the entertainment and shopping needs of contemporary urban people. They therefore incorporate recreational elements, such as cinemas, play areas, exhibitions, places to chat, places to listen to music, food and beverage outlets, etc. (Haseki 2013, p. 41). Some authors, such as Goss, have drawn attention to the important sociocultural role of food courts in shopping centres, which '…have become an absolute necessity, in part because of the increased role of food as a marker of social taste, in part also because the presentation of diverse culinary experiences enhances the sense of elsewhere (food courts now typically present a range "ethnic" cuisines), and because it provides a vantage point for watching others display their commodified lifestyles' (Goss 1993, p. 34). Goss also emphasises that food courts are the main attraction for office workers, who often spend 10–15 min there (Goss 1993, pp. 34–35).

The growing role of gastronomy in shopping centres has been noted outside Canada and the United States. The introduction of food courts was observed in the United Kingdom in the mid-1980s (Sirpal and Peng 1995, p. 13), and the process has subsequently intensified. In Nuremberg (Germany) the proportion of weekday (Monday–Friday) visitors who said the main purpose of their visit was related to gastronomy and culture, rose from 8% in 1988, to 13% in 2005. A similar picture emerged for Saturday visitors, rising from 6 to 14% over the same period (Monheim 2008, p. 16). This growing demand for eating services in cities has led to the introduction of the term *Gastrofizierung*, which means the trend of opening restaurants and bars in cities, into the German language. There are two reasons for this. The first is the transfer of consumers from downtown shops to large suburban shopping centres. The second is related to the ageing of urban residents, who are less and less interested in shopping and prefer to socialise in restaurants or cafés (Hengst and Steinbach 2012, p. 516).

In 2012, Dublin City Council (Ireland) commissioned a study to examine gastronomy trends. This found that 66% more people visited restaurants than pubs and this trend has continued to grow. In their next survey in 2014, they found that 50% of people visited the downtown area to go to a restaurant, while only 29% went to a pub (it should be noted, however, that many Dublin pubs also serve food). Of those who said that they visited the city centre to go shopping, 73% said that they did so because of the restaurants, coffee shops, pubs, etc. (*Café & restaurants sector report* 2016, p. 5). In a recent survey of about 22,000 people across several European countries, 41% said they looked to shopping centres when thinking about eating and drinking, and only 10% chose restaurants and cafés in downtown areas. What is particularly interesting, though, is that over 30% of people said that they visited shopping centres only to eat and drink (*Eat the mall. Dining, enjoying, relaxing, and shopping. Centre operators are completely revamping their menus. The new food experience* 2017).

Some analysts say that shopping centres have evolved from places to shop into places to eat and, hence, places to be. About 60% of visitors to ECE (a German company that owns shopping centres in many European countries) shopping centres take advantage of onsite restaurants and snack bars. In fact, 40% chose a particular shopping centre on the basis of its food offerings (*Food & beverages in shopping centres* 2016, pp. 2–3). This trend is supported by the figures: From 2010 to 2015, food and beverage turnover at German ECE shopping centres grew by 54, and 89% of visitors to food outlets stayed there for more than 15 min (*Food & beverages in shopping centres* 2016, pp. 4–5).

This shift does not only apply to Europe and North America. In Southeast Asian shopping centres the area occupied by food and beverage enterprises has risen from 15 to 25% (Haseki 2013, p. 43). This process, which has been particularly noticeable in Singapore, is connected to growing demand from office workers at lunchtime. The availability of food courts in shopping centres has helped to attract more customers and has improved the performance of other businesses in the centre (Sirpal and Peng 1995, p. 13).

So, who are the customers of these restaurants and bars in shopping centres? A survey of visitors to the M1 Merkez Adana Shopping Centre in Adana (Turkey) showed that 35.3% (n = 680) were married, 59.7% were single and 5% were either divorced, widowed or separately living (Haseki 2013, p. 49). The main reasons people gave for visiting a particular restaurant were: eating and/or drinking (42.8%) and resting and relaxing after shopping (26.9%). Fewer respondents indicated a desire to eat a healthy meal (12.9%), meeting friends (12.2%) or passing time (11.8%) (Haseki 2013, p. 50). Many of those interviewed (40.1%) said they visited restaurants in shopping centres once a month, 24.3% said they went once every two weeks, and 18.4% said they visited regularly (at least once a week). Usually they spent between 30 and 120 min; only 17.4% spent more than three hours. About a third of respondents said that they '…generally use the same restaurant all the time', the remainder said that they '…do not always use the same restaurant' (Haseki 2013, p. 51). The main reasons for preferring a particular restaurant were the high quality of products (29.6%) and brand loyalty (19%). The proximity of the restaurant to their home was regarded as very important for some (12.5%), while 10% said the most important thing was a restaurant '…offering different flavors/tastes' (Haseki 2013, p. 52).

An analysis of preferences of United States residents (n = 1737) showed that the most popular configuration of mall food service combined a moderate-size food court with several casual and fast-casual restaurants. Malls with only table service restaurants and no food court were least popular (Taylor and Verma 2010, p. 10). Different demographic groups exhibited different preferences for food court configurations, cuisines and restaurant brands. The youngest respondents visited most often, and were most likely to eat at fast-food outlets. On average, 27.6% used quick-service restaurants, although this was higher for people under 25 (40.4%) and much lower for people over 55 (16.8%) (Taylor and Verma 2010, p. 11). Older people also preferred table service brands (e.g. Red Lobster and Outback Steakhouse), while younger people preferred casual, fast-food brands such as Burger King, Taco Bell, and KFC (Taylor and Verma 2010, p. 13).The most popular cuisines were American (85.3%), Chinese (64.3%), Mexican (60.5%), and Italian (54.3%) (Taylor and Verma 2010, p. 12). In Germany, the most popular meals offered by food courts in shopping centres owned by ECE fall into the category of fast food (burgers, currywurst, chips, etc.), which are preferred by 53% of diners. On average 37% of customers preferred Asian food, although this varied with age group. Within the group of people aged 18 to 29, for example, 53% favoured Asian food, and 26% pasta and pizza (*Food & beverages in shopping centres* 2016, pp. 9–10).

An American survey published in late 2017 (conducted in March 2017) highlighted that 66% of mall visitors regarded the range of food and beverages on offer as important when choosing a centre to visit. Furthermore 55% of respondents said they visited these establishments when they were in shopping centres (*The successful integration of food & beverage within retail real estate* 2017, pp. 8–9) (Table 15.1).

Table 15.1 Key indicators of food services in shopping centres by category

Food service metaphor (examples)	Typical trading periods	Typical purchase	Typical length of stay
Impulse (sweet treats)	All day—anytime	Single item of core product	Takeaway
Refuel and relax (coffee and cake)	Breakfast and snack times	One regular coffee and pastry/bagel	Takeaway, but sometimes 1 h (average 20 min)
Speed eating (fast food)	Lunch, dinner, rarely snack times	Meal deal including side and drink	10–20 min
Fast-casual	Mostly lunchtime and dinner	Meal deal including side and drink	15–30 min
Casual dining	Mostly lunchtime and dinner	Two-course meal with/without alcohol	30–60 min
Finer dining	Lunchtime and dinner	Two- three-course meal with alcohol	60–180 min
Social drinking	Mostly dinner and evening snack	Drink and snack or meal	60–90 min
Gourmet food	All day—day time	Artisan food products	19–15 min

Source: *The successful integration of food & beverage within retail real estate* 2017, p. 13, Table 2 with slight modifications

The authors suggest that those who fall into the *impulse* category may be anyone who happens to be passing a shopping centre, but in practice are '…predominantly retail visitors, often younger and female'. The second category (*refuel and relax*) describes people who use the café or bar as a kind of pit stop. This type of eating establishment is often visited by office workers (as a meeting place) and by people on their way to work. The *speed eating* category is quite broad and covers office workers and blue-collar workers, as well as families and shoppers. The *social drinking* category comprises people meeting after work, as well as ladies who lunch, more affluent couples without children, young adults, groups celebrating special occasions and leisure users. *Fast-casual* restaurants and bars are typically used by office workers, as well as young adults, so-called hipsters, foodies and some of the richer retail visitors. The latter also frequent *casual dining* restaurants, which are often visited by affluent people without children, but also by families celebrating a special occasion. As the name suggests, *fine dining* restaurants and bars are not usually associated with a visit to the mall for shopping purposes. Fine diners tend to be '…destination diners, corporate lunchers, special occasion users'. Last but not least, there is *gourmet food*. The people who seek this out include shoppers (browsers and buyers), office workers and foodies (*The successful integration of food & beverage within retail real estate* 2017, p. 14).

15.3 Gastronomy in Shopping Centres in the Urban Context

15.3.1 The External Context

As noted above, the locations of shopping centres can be divided into: (a) edge-of-centre, (b) out-of centre and (c) out-of-town (Guy 1998, p. 262). The first category operates in the inner city and usually charges higher prices because of factors such as higher rents and levels of investment (Kunc et al. 2012, p. 40). In the context of gastronomy in urban space, we can divide shopping centres into two main types:

- large, destination centres that not only serve the immediate population, but people from elsewhere in the city (office workers) as well as visitors, both from other cities and abroad (tourists);
- smaller neighbourhood shopping centres that are less focused on providing a destination for dining, and more concerned about servicing the needs of local residents.

There are also mixed-use centres that are somewhat similar to large shopping centres. These cater to captive customers, such as train passengers, hotel guests, office workers, etc. They are often located near (or at) railway stations. One example is the new main railway station in Berlin (Berliner Hauptbahnhof), which opened in 2006 and has more than 15 eating establishments. There are similar examples in some larger Polish cities—Galeria Krakowska in Kraków (built in 2006 with 40 eating establishments[4]); Galeria Katowicka in Katowice (opened in 2013, with 26[5]) and Avenida Poznań (formerly Poznań City Centre, built in 2013, with 29[6]).

The data indicate that the area dedicated to gastronomy within shopping centres is increasing globally: from 5% a decade ago, to 10–15% nowadays in Europe, and 8–9% in the United States. The trend continues to grow, and there are already newly built shopping centres where 25% of the area is occupied by gastronomy. The Hungarian capital, Budapest is one example. Table 15.2 shows that the share of commercial space dedicated to eating is under 10% in only two cases. In most cases it is between 10 and 20%, while in two cases it exceeds 20%. The table below also gives an index of the number of food establishments per 100 square metres of store space. In general, this varies between 0.020 and 0.050. However, in one case it is only 0.011 (MC Ferihegy close to the international airport), and in two it is over 0.100, with the highest value being 0.208.

---

[4]http://www.galeriakrakowska.pl/pl/sklepy/restauracje,18,cat,9; http://www.galeriakrakowska.pl/pl/sklepy/kawiarnie-i-cukiernie,1,cat,9 (Accessed 08 March 2018).
[5]http://galeriakatowicka.eu/kawiarnie-i-restauracje (Accessed 08 March 2018).
[6]http://avenidapoznan.com/gastronomia (Accessed 08 March 2018).

The position of gastronomy as a reason to visit a shopping centre is confirmed by research conducted in Prague, while the centre's function can differ. In the case of the Palladium shopping gallery,[7] 24% of people reported visiting a restaurant as one of their goals, while this was only 12% for the Arkády Pankrác Praha centre.[8] This difference is largely due to their location: The Palladium is in the immediate vicinity of the Old Town and is often visited by tourists, while Arkády Pankrác Praha is located in the outer zone of Prague. The difference in the function of the two centres is clearly visible. Only 19% of visitors to the Palladium reported 'shopping for food' as the purpose of their visit, while in Arkády Pankrác Praha that figure was 58% (Sikos 2013, p. 365).

The growing share of catering space in shopping centres is particularly noticeable in Asia. New, 'mega malls' are being developed, where the area allocated to food services is as high as 30–40%. According to analysts, this is in response to the tradition of eating out in Asian society (*The successful integration of food & beverage within retail real estate* 2017, p. 9). One example, which opened in 2012, is the Parkview Green FangCaoDi in Beijing. This shopping centre has about 90 stores, and almost 35% of the retail space is allocated to restaurants and bars. They not only draw visitors, but also help to keep its 8, 000 office workers onsite for longer (*The future of the shopping center industry. Report from the ICSC Board of Trustees* 2016, p. 37).

The trend is also seen in Poland. The latest shopping centre in Warsaw, the Galeria Handlowa Rondo Wiatraczna opened in February 2018; out of the 35 outlets, 11 (31.4%) are eating establishments, notably cafés and pastry shops. As with other large, urban centres in Europe and North America, the Polish capital, Warsaw built many shopping centres after 1989, which became the target of visits by city residents (and visitors) for both dining and commercial purposes. The role of eating services in malls was confirmed by research carried out in Warsaw in 2009, which showed that 9.6% of the tenants in shopping centres were related to gastronomy (second only to clothing stores) (Celińska-Janowicz 2013, p. 295).

The first shopping centre in Warsaw opened in 1993. The Panorama gallery (number 31 on the map presented in Fig. 15.1 and Table 15.3), now known as Galeria Panorama, is in the southern part of the inner city (Celińska-Janowicz 2013, p. 262). This was followed in 1996 by King Cross Praga (originally Géant King Cross) (number 19 in Fig. 15.1 and in Table 15.3) in the eastern part of the city, and Centrum Handlowe Auchan Piasczno in the city of Piaseczno (number 18 in

---

[7]The centre opened in 2007. It is in a historical building, which from the 1780s served as army barracks. By the second half of the twentieth century the property had deteriorated and renovation started in 2005. It has 213 outlets and 15% (32 outlets) are eating facilities (https://www.palladiumpraha.cz/obchody-sluzby; https://www.palladiumpraha.cz/restaurace-obcerstveni (Accessed 08 March 2018).

[8]The Arkády Pankrác Praha, which is owned by ECE, is a modern building that opened in 2008. It houses 136 retail outlets, while the restaurants, bars, etc. make up 12.5% (17 outlets) (http://www.arkady-pankrac.cz/cz/obchody, accessed 08 March 2018).

Table 15.2 Characteristic of the main shopping centres in Budapest (Hungary) 2008–2011

Shopping centre	Gross leasable area (m^2)	Total number of tenants	Food outlets Number	%	Per 100 m^2
MC Ferihegy	44, 000	29	5	17.2	0.011
Savoya Park	30, 000	74	7	9.5	0.023
Rózsakert	7, 700	68	16	23.5	0.208
EuroPark	24,700	64	6	9.4	0.024
CsepelPlaza	13, 654	60	8	13.3	0.059
Campona	40, 000	145	15	10.3	0.038
MOM Park	30,000	86	18	20.9	0.060
Duna Plaza	47,000	123	16	13.0	0.034
Mammut	57,000	318	43	13.5	0.075
Aréna	67, 000	181	27	14.9	0.040
Árkád	45,000	155	17	11.0	0.038
Lurdy Ház	33,000	115	17	14.8	0.052
Allee	47,000	138	22	15.9	0.047
WestEnd	44,800	403	46	11.4	0.103
Polus Centre	56,000	209	25	12.0	0.045
Corvin	34,600	61	10	16.4	0.029
KÖKI	59,000	134	24	17.9	0.041

Source: Adapted from Reikli (2012, p. 153, Table 24)

Fig. 15.1 and in Table 15.3).[9] In both cases, the galleries were mainly funded by French capital, which was very visible in the initial period of development. In 1997, King Cross Ursynów was created (currently Ursynów Centre, number 26 on the map and in the table), followed in 1998 by Auchan Modlińska (number 30) and Atrium Targówek (with its Carrefour hypermarket, number 7). However, the real expansion of shopping galleries started in 1999–2001, when five malls were created in Warsaw, including Galeria Mokotów (number 3) with a commercial floor space of 65,000 square metres (Jarosz 2002, p. 262).

Table 15.3 shows that for most of the shopping centres that existed in 2001, most either maintained or increased the numbers of eating establishments. The biggest increases were in Galeria Mokotów (from 22 to 35), Atrium Targówek (13 to 21), Atrium Promenada (8 to 16) and Atrium Reduta (7 to 14). Everywhere, the increase was related to expansion. It should be noted that Galeria Mokotów is situated in the centre of the main office district in Warsaw. Atrium Reduta is in a similar location, but it also serves inhabitants of the western suburbs. Atrium

---

[9]The hypermarket HIT (currently Pasaż Tesco Górczewska) opened in 1994 on the western outskirts of the city. It was only officially recognised as a shopping gallery after its expansion in 2001 (Jarosz 2002, p. 259).

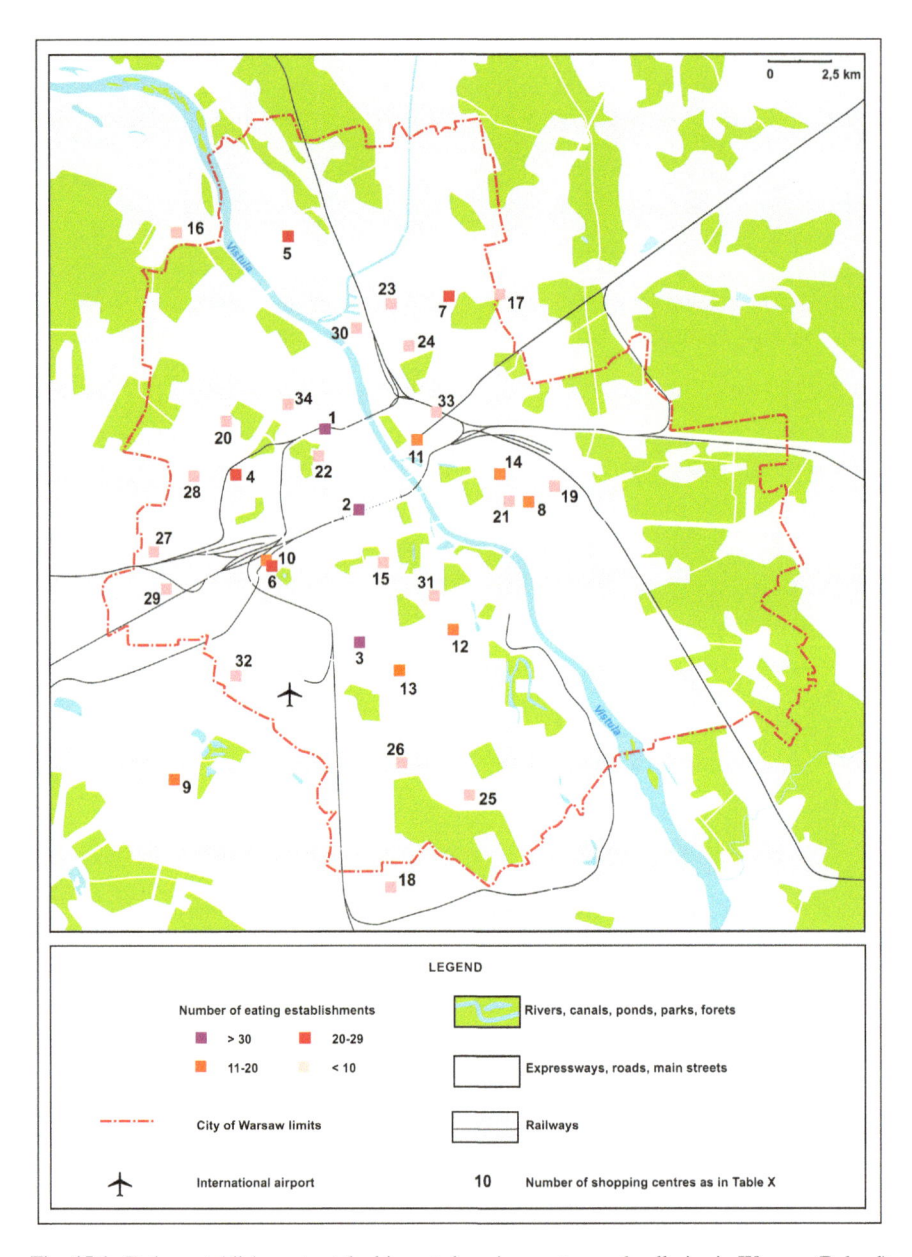

Fig. 15.1 Eating establishments at the biggest shopping centres and galleries in Warsaw (Poland) in 2017. Source: Authors' own research

Targówek and Atrium Promenada, however, are situated in the vicinity of large new housing estates on the eastern bank of the Vistula River. As with Atrium Reduta, both of these centres are important for residents of north-eastern and south-eastern

Table 15.3 Changes in the number of eating establishments in the biggest shopping centres and galleries in Warsaw (Poland) between 2001 and 2017

Number (see Fig. 15.1)	Name (in 2017)	Number of facilities 2001	Number of facilities 2017	Growth index (2001 = 100)
3	Galeria Mokotów	22	35	159
7	Atrium Targówek	13	21	161
8	Atrium Promenada	8	16	200
9	Centrum Janki	12	16	133
10	Atrium Reduta	7	14	200
11	Galeria Wileńska	13	13	100
12	Sadyba Best Mall	8	13	162
13	Land Centrum Handlowe	6	12	200
17	M1 Marki	2	8	400
18	Centrum Handlowe Auchan Piaseczno	5	7	140
19	King Cross Praga	12	7	58
20	Galeria Bemowo	7	12	171
22	Dom Mody Klif	6	6	100
25	Pasaż Tesco Kabaty	5	4	80
26	Centrum Ursynów	2	3	150
27	Pasaż Tesco Połczyńska	4	3	75
28	Pasaż Tesco Górczewska	3	3	100
30	Centrum Handlowe Auchan Modlińska	3	2	67
31	Galeria Panorama	3	2	67
33	Pasaż Tesco Stalowa	1	1	100
34	Galeria Żoliborz	3	1	33

Source: Adapted from Sosnowska (2002) and survey conducted by Katarzyna Duda-Gromada

suburbs. The table also shows that in some centres the number of catering establishments has decreased, in most cases only by one, but in King Cross Praga the number reduced by five. This can be explained by the fact that this mall—as previously mentioned, one of the first in Warsaw—came up against very serious competition from the nearby Atrium Promenada, which was built as a new-generation facility with better organised food courts. The latter also includes several branded shops, which means that it is visited by richer customers who often dine at its restaurants, bars and cafés.

It should also be noted that between 2001 and 2017 two shopping malls disappeared. The first of these, Galeria Centrum (which included three eateries in 2002) was located in the downtown district, in a building dating back to the mid-1960s. It ceased to be a shopping centre when it was divided into independent stores. The second, King Cross Fort Wola (in the western part of Warsaw), had six dining outlets in 2001, although it only had four when it closed in 2016. Although it was scheduled to be modernised, no construction works had been undertaken by summer 2018. Before its closure, the centre was in trouble, when the establishment of nearby shopping galleries led to some of its stores and restaurants going into liquidation.

Comparing the list of shopping centres in Table 15.3 with that in Table 15.4, it can be seen that the number in 2017 was much higher. In 2001 (just after the survey by Sosnowska), Wola Park opened, followed in 2002 by Factory Ursus (which was expanded in 2006)—these were the main competitors for the King Cross Fort Wola gallery. Two years later, in 2004, Blue City and Arkadia (the largest mall in Poland) both opened, and more have opened in subsequent years. The largest of these are Złote Tarasy, which opened in 2007, Galeria Remblielińska in 2008 (and renamed Galeria Renova in 2013), Centrum Krakowska in 2010, Plac Unii City Shopping and Factory Annopol in 2013, and Galeria Północna, which opened in 2017. Most of these newly built malls are new-generation centres with a more extensive dining and entertainment zone.

Analysing the distribution of eating establishments in urban spaces in Warsaw, most are in two shopping galleries located on the inner-city border (Arkadia, number 1 in Fig. 15.1 and Table 15.3 and Galeria Mokotów, number 3), and the Złote Tarasy mall (number 2) in the downtown area. The Złote Tarasy mall is frequently used for shopping purposes by foreign and domestic tourists, as well as suburban passengers waiting for trains. A common feature of all of these shopping centres is that they are among the largest in terms of the number of stores (Arkadia has 115,000 square metres of retail space). It is also worth noting that Złote Tarasy, one of the biggest malls in Warsaw, is located in the immediate vicinity of the main train station (they are connected, but constitute separate buildings) and forms an office building (26 floors) with a total area of 225,000 square metres. It therefore differs from shopping centres at the main train stations in Katowice or Poznań (Fig. 15.2).

Table 15.4 shows the number of eating establishments relative to the total number of commercial and service outlets in shopping centres located in Warsaw. The number of restaurants, bars, etc. covered by the analysis ranged from 1 to 40. Most centres have between 5 and 20 dining outlets, while three have over 30 (restaurants, bars, cafés, tea rooms, ice cream parlours, pastry shops, etc.).

The excessive concentration of shopping centres in selected places (clusters) is often considered a negative influence on the spatial development of commerce and services. This problem is increasingly related to gastronomy, especially in smaller cities and tourist destinations. At Costa Calma in the Canary Islands (Spain), for example, by 2013 nearly two-thirds (64.7%) of restaurants, bars and cafés (outside hotels) were located in shopping centres (Fig. 15.3).

It is worth highlighting at this point that architects and restaurant and bar owners often aim to make their location *the* place to be. This does not refer to prestigious locations (in the sociocultural sense), but to the uniqueness of the broader

Table 15.4 Eating establishments located at the biggest shopping centres and galleries in Warsaw (Poland) in 2017

Number (see Fig. 15.1)	Name	Total number of enterprises	Gastronomy Number of facilities	Per cent
1	Arkadia	256	40	15.6
2	Złote Tarasy	222	39	17.6
3	Galeria Mokotów	274	36	13.1
4	Wola Park	212	27	12.7
5	Galeria Północna	206	26	12.6
6	Blue City	204	23	11.3
7	Atrium Targówek	169	21	12.4
8	Atrium Promenada	142	16	11.3
9	Centrum Janki	112	16	14.3
10	Atrium Reduta	135	14	14.4
11	Galeria Wileńska	102	13	12.7
12	Sadyba Best Mall	107	13	12.1
13	Land Centrum Handlowe	205	12	5.9
14	Galeria Handlowa Rondo Wiatraczna	35	11	31.4
15	Plac Unii City Shopping	84	10	11.9
16	Galeria Łomianki	88	9	10.2
17	M1 Marki	69	8	11.6
18	Centrum Handlowe Auchan Piaseczno	49	7	14.3
19	King Cross Praga	86	7	8.1
20	Galeria Bemowo	78	7	9.0
21	Pasaż Tesco Fieldorfa	42	7	16.7
22	Dom Mody Klif	106	6	5.7
23	Factory Annopol	110	6	5.5
24	Galeria Renova	49	5	10.2
25	Pasaż Tesco Kabaty	24	4	16.7
26	Centrum Ursynów	48	3	6.3
27	Pasaż Tesco Połczyńska	25	3	12.0
28	Pasaż Tesco Górczewska	41	3	7.3
29	Factory Ursus	110	3	2.7
30	Centrum Handlowe Auchan Modlińska	19	2	10.5
31	Galeria Panorama	83	2	2.4
32	Centrum Krakowska	12	2	16.7
33	Pasaż Tesco Stalowa	13	1	7.7
34	Galeria Żoliborz	34	1	2.9

Source: Surveys conducted in 2017 and 2018 by Katarzyna Duda-Gromada, Andrzej Kowalczyk and Aleksandra Korpysz

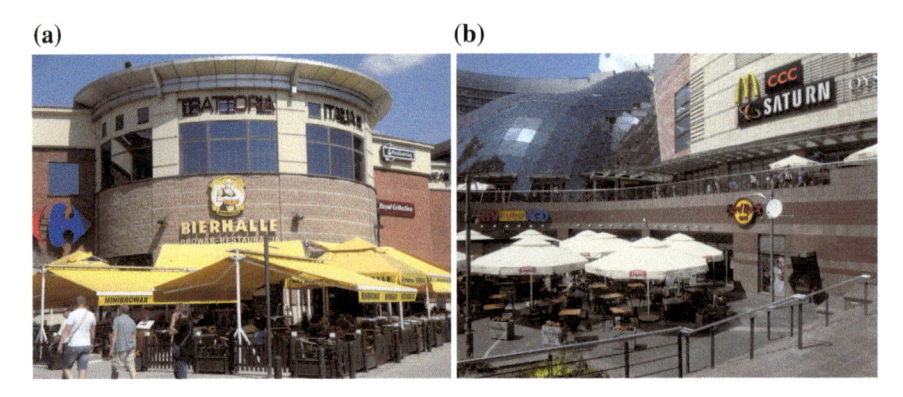

Fig. 15.2 Dining establishments at shopping centres in Warsaw (Poland). The Bierhalle chain restaurant in the Arkadia mall (**a**) and Złote Tarasy mall (**b**)

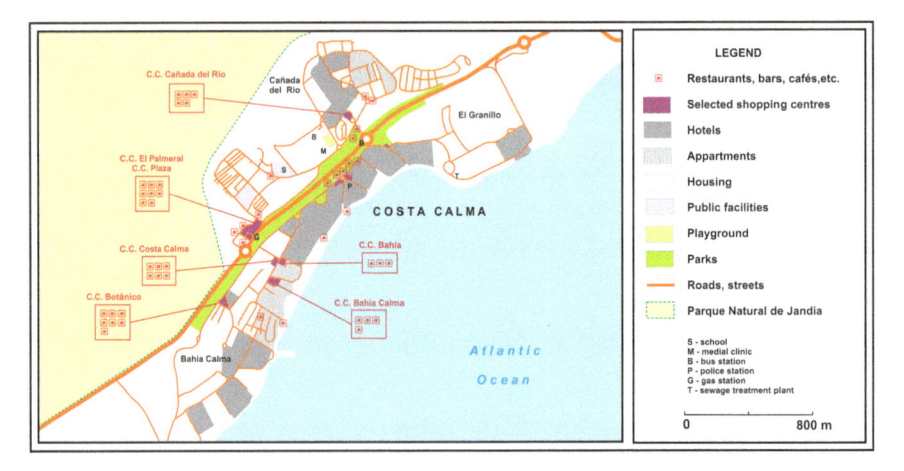

Fig. 15.3 Clusters of eating establishments at shopping centres in Costa Calma, Fuerteventura Island (Spain) in 2013

geographical location. Consequently, in many cities there are restaurants on hilltops (The Peak Lookout restaurant in Victoria, Hong Kong), television towers (Restaurant Sphere im Berliner Fernsehturm near the Alexanderplatz in Berlin, Germany), pylons of bridges (UFO Bar & Restaurant at the top of the Most Slovenského národného povstania in Bratislava, Slovakia), and on the highest floors of skyscrapers, which are often hotels (The Sun Dial Restaurant, Bar & View at top of The Westin Peachtree Plaza in Atlanta, Georgia, United States). In many large cities, restaurants are increasingly being located on the roofs of buildings. One such example, which opened in 2013, is the ALEX Frankfurt Skyline Plaza in Frankfurt am Main (Germany). This restaurant is adjacent to the Skyline Garden, which is located on the roof of the Skyline Plaza shopping centre from where there is a spectacular view of the Europaviertel's skyscrapers. In addition to the restaurant, the Skyline Plaza also offers a market hall with 16 food service concepts (*Skyline Plaza: spectacular gastro experience in the roof garden* 2017).

15.3.2 The Internal Context

From the point of view of a spatial (or geographical) analysis, it is not only who uses particular types of eating establishments, but also how these establishments are distributed within shopping centres. There are two key criteria when it comes to defining restaurant operators' choice of location. The first concerns the number of food establishments in the immediate vicinity. The second concerns the type and category of these food services. One of the more common solutions is the clustering of gastronomy in so-called food courts. But too many outlets offering the same or similar type of cuisine (such as Indian and Pakistani restaurants) or product duplication (fast-food joints) within the same area can lead to what is referred to as *cannibalisation*. Although this clearly depends on the size and configuration of the mall, analysts argue that at least 20% of food service facilities should be distributed around the shopping centre (*impulse* and *refuel and relax* outlets), while the remaining 80% should be clustered and focused on other forms of dining (*speed eating*, *fast-casual*, *casual dining* or combinations of these). In smaller shopping centres this clustering means that there is one food court, but in larger malls there may be more such hubs (Table 15.5).

The location of food services in shopping centres depends on several factors. These include considerations such as:

- whether the mall is tourism-oriented or transit-oriented, which determines how long people will spend in the centre;
- the location of the mall in the urban space. If it is in the downtown area of the city, for example, its weekday customers may include office workers who spend their lunchtimes there, but may have different types of visitors at evenings and weekends;
- the cost of investing in the food establishment.

Inside shopping centres, analysts categorise how catering establishments can be organised as follows:

- food courts, which are food zones built around a common seating area surrounded by several food vendors;
- food halls, which offer a wide variety of eating-related services, including ethnic restaurants and bars, artisanal food vendors, luxury restaurants, as well as fast-food bars;
- restaurant clusters, made up of several restaurants and bars in the same area of a shopping centre that do not share a common dining area; and

Table 15.5 Optimal location of eating establishments within shopping centres by food service category

Food service metaphor (examples)	Location	Optimal unit size (in m^2)
Impulse (sweet treats)	Dispersed within shopping centres, but at key mall intersections and in corridors with high traffic (internal spaces preferred by visitors).	20–30
Refuel and relax (coffee and cake)	Not clustered. In different parts of the malls with seating. With access from mall and/or outside.	15–250
Speed eating (fast food)	Clustered in a food court. Sometimes with outside space.	200–400
Fast-casual	Clustered in a food court. Sometimes with outside space.	200–400
Casual dining	Clustered in a food court. Outside space required. Sometimes with access from the mall and/or outside. Sometimes opening hours are longer than in other outlets within the shopping centre.	300–600
Finer dining	Not clustered. With access from mall and/or outside. Sometimes opening hours are longer than in other outlets within the shopping centre.	Over 400
Social drinking	Clustered in a food/drink court. Outside space required.	350 (with food)
Gourmet food	Not clustered in a food court, but can work in a gourmet food court. In-line retail on the mall.	50–150

Source: Adapted from *The successful integration of food & beverage within retail real estate* 2017, p. 15, Tables 4 and 5

- stand-alone establishments scattered throughout the centre (*The successful integration of food & beverage within retail real estate* 2017, pp. 14–15).

Figure 15.4 presents a more elaborate proposal for the classification of the distribution of dining establishments in shopping centres, based on this categorisation.

In general, the distribution of eating outlets in shopping centres can be divided into two categories: single and multiple. In the first case, there is only one restaurant or bar. Where there are multiple restaurants or bars, they may be organised into one or more clusters (consolidated patterns) or they can be scattered across different areas. They may be distributed randomly or regularly. Whether restaurants and bars are concentrated in one (mono-nucleus pattern) or more (multi-nucleus pattern) clusters, they may follow either a node or linear pattern.

Single restaurants and bars mostly occur in smaller shopping centres, which may include a hypermarket and only a few other retail or service outlets. There are usually many eating establishments in the larger shopping galleries, and restaurants

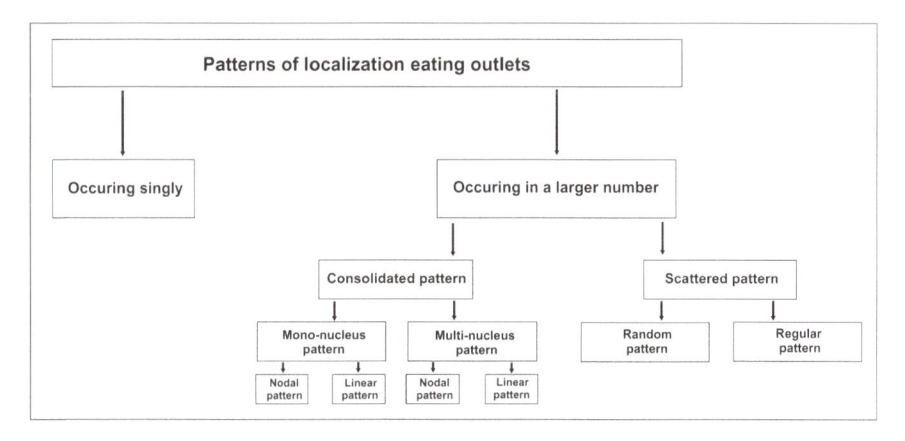

15 New Gastronomic Hotspots in the Urban Space … 289

Fig. 15.4 Classification of the spatial distribution of eating outlets within shopping centres

and bars may be distributed according to a scattered pattern or a consolidated pattern. In medium-sized centres, there is generally a mono-nucleus, consolidated pattern. In the largest galleries, however, restaurants, bars and cafés often form a multi-nucleus pattern that is designed to enhance accessibility.

The largest shopping centres often have several restaurants and bars on each floor, with different spatial distributions. Figure 15.5 shows the location of dining establishments in Galeria Mokotów, one of the largest shopping galleries in Warsaw (Poland). There are a total of 36 catering establishments. The figure shows that on level 0, the eight restaurants and bars appear in several places, with concentrations near the main gallery entrances. On level 1 there are only three eating establishments and they are scattered across the floor. On level 2, gastronomy occupies almost half of all commercial space (the rest is a multiplex cinema, gym and a large shop with sports accessories). In this part of Galeria Mokotów there are

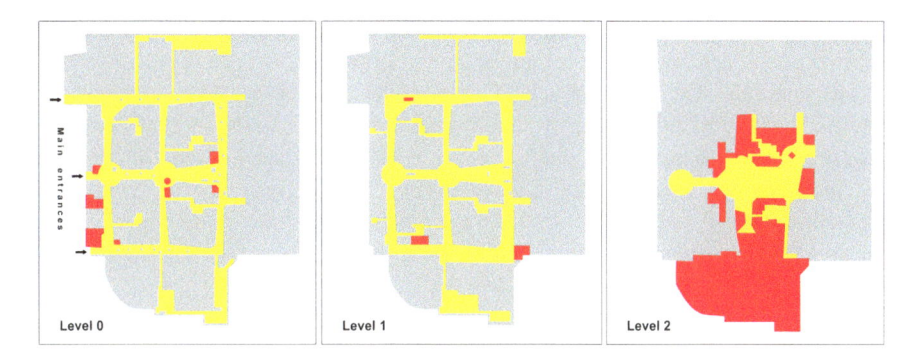

Fig. 15.5 a–c The distribution of eating establishments (shown in red) and corridors (yellow) in the Galeria Mokotów shopping centre in Warsaw (Poland) in 2017

25 restaurants, bars, cafés, ice cream shops, etc. This is due to the centre's three main functions. Firstly, the mall is located in the immediate vicinity of the largest concentration of corporations, banks, insurance companies, etc. in Warsaw. This is especially important for casual restaurants on level 2. Second, on two sides of the centre, there are large, middle-class housing estates. Third, the mall is only about 50–100 metres from a major street intersection and is less than 3.5 kilometres from the international airport.

Research conducted by ECE showed that whether a particular food court in a particular shopping centre succeeds or not depends on:

- the catchment area (for instance, a high percentage of visitors with more than 30 min of travel time, a large number of office workers);
- the tenant mix (diversity of cuisines, categories of restaurants, etc.);
- the atmosphere (interior,[10] noise level, music);
- the location;
- the quality of the dining experience (*Destination food court. Facts. Success factors. Insights* 2017, p. 4).

Finally, it should be noted that some customers are attracted to a particular restaurant or bar in a mall by its name. When considering the organisation and symbolism of the space inside the Mall of America in Bloomington, Minnesota, Goss noted that one of the metaphors used in the mall is *primitiveness*. This applies not only to terms used to describe specific dishes, but also to the names of restaurants and bars, such as Rainforest Cafe (Goss 1999, p. 65).

15.4 Conclusions

The importance of eating establishments located in shopping centres, although it has been growing in recent decades, is not something new in itself. This trend, which is not so important in the period of Fordism (twentieth century), reminds us to some degree of what happened in the cities of ancient Greece or Rome, as well as in mediaeval cities. At this point, there are also analogies between shopping galleries or malls and bazaars in contemporary Middle East and South Asian cities.

The research described above, which was carried out in Warsaw, confirmed the results of research undertaken in other agglomerations. It shows that shopping galleries are increasingly visited not in order to make purchases of clothing or food, but in order to meet friends in a restaurant, celebrate a child's birthday at McDonald's, drink coffee after a visit to the cinema, etc. Many malls often become places where traditional food fairs, culinary competitions, cooking schools, etc. are

---

[10]According to an ECE customer survey carried out in 2015 exploring food court use (n = 2598), around 63% of those interviewed said they valued the idea of a special architectural design (*Destination food court. Facts. Success factors. Insights* 2017, p. 4).

organised. Although Marc Augé (1995) treats the supermarkets as a 'non-places', for many visitors (especially for younger people) they are friendly and far from 'inauthentic' (Relph 1976). This applies in particular to new-generation malls, where it is not a commercial function which attracts visitors, but gastronomy and entertainment. It should also be mentioned that new shopping centres (in contrast to the first generation malls) are often created in connection with the revitalization of the central parts of cities and refer to the concepts of festival market place (Gravari-Barbas 1998) or city as an entertainment machine (Lloyd and Clark 2001).

References

Augé M (1995) Non-places. Introduction to an anthropology of supermodernity, Verso, London— New York, English translation by J. Howe. *Bars & restaurants*. https://www.kadewe.de/bars-und-restaurants-im-kadewe-berlin. Accessed 08 March 2018

Bars & restaurants. https://www.kadewe.de/bars-und-restaurants-im-kadewe-berlin. Accessed 08 March 2018

Bloch PH, Ridgway NM, Dawson SA (1994) The shopping mall as consumer habitat. J Retail 70 (1):23–42

Café & restaurants sector report (2016) Report to the Economic Development and Enterprise SPC, 22nd November 2016, Item Number 3, Dublin City Council, Economic Development & Enterprise, Dublin. https://www.dublincity.ie/councilmeetings/documents/s7137/Cafe%20and%20Restaurants%20Report%20to%20Economic%20Development%20Enterprise%20SPC%2022%20Nov%202016.pdf. Accessed 08 March 2018

Celińska-Janowicz D (2013) Przestrzenne aspekty funkcjonowania warszawskich centrów handlowych, Wydział Geografii i Studiów Regionalnych, Uniwersytet Warszawski, Warszawa, Ph.D. thesis

Cohen L (1996) From town center to shopping center: the reconfiguration of community marketplaces in postwar America. Am Hist Rev 101, 4(October):1050–1081

Destination food court. Facts. Success factors. Insights (2017). https://www.ece.com/fileadmin/PDF_englisch/Studies/ECE_Destination_Food_Court_eng.pdf. Accessed 08 March 2018

Eat the mall. Dining, enjoying, relaxing, and shopping. Center operators are completely revamping their menus. The new food experience. ACROSS. The European Retail Real Estate Magazine. November 13, 2017. https://www.across-magazine.com/eat-the-mall. Accessed 08 March 2018

Food & beverages in shopping centres. Destination Food. ECE Market Research", 2016, 1, ECE Projektmanagement G.m.b.H. & Co., Hamburg. https://www.ece.com/fileadmin/_processed_/csm_ECE_Destination_Food_eng_ad389eeff4.png. Accessed 08 March 2018

Frączkiewicz M (2013) The cultural role of the malls, "Zeszyty Naukowe Uniwersytetu Jagiellońskiego" MCCCXXIX, "Prace Etnograficzne" 41(4):335–342

Gastronomia. http://avenidapoznan.com/gastronomia. Accessed 08 March 2018

Goss J (1993) The 'magic of the mall': an analysis of form, function and meaning in the retail built environment. Ann Assoc Am Geogr 83(1):18–47

Goss J (1999) Once-upon-a-time in the commodity world: an unoffcial guide to the Mall of America. Ann Assoc Am Geogr 89(1):45–75

Gravari-Barbas M (1998) La "festival market place" ou le tourisme sur le front d'eau. Un modèle urbain américain à exporteur. Norois 45, 178(April–June):261–278

Guy CM (1998) Classifications of retail stores and shopping centres: some methodological issues. GeoJournal 45(4):255–264

Haseki MI (2013) Customer expectactions in mall restaurants: a case study. Int J Bus Soc 14 (1):41–60

Hengst M, Steinebach G (2012) Multi-Channel-Konzepte als Chance für eine nachhaltige und zukunftsfähige Entwicklung der Innenstädte? In: Schrenk M, Popovich VV, Zeile P, Elisei P (eds) Proceedings REAL CORP Tagungsband 14–16 May 2012, Schwechat, pp 513–522. http://www.corp.at. Accessed 08 March 2018

Jackson EL (1991) Shopping and leisure: implications of West Edmonton Mall for leisure and for leisure research. Can Geogr 35(3):280–287

Jarosz A (2002) Miejsce hipermarketów w przestrzeni miejskiej aglomeracji. In: Węcławowicz G (ed) *Warszawa jako przedmiot badań w geografii społeczno-ekonomicznej*, „Prace Geograficzne", 184, Instytut Geografii i Przestrzennego Zagospodarowania Polskiej Akademii Nauk, Warszawa, pp 253–264

Kawiarnie i cukiernie. http://www.galeriakrakowska.pl/pl/sklepy/kawiarnie-i-cukiernie,1,cat,9. Accessed 08 March 2018

Kawiarnie i restauracje. http://galeriakatowicka.eu/kawiarnie-i-restauracje. Accessed 08 March 2018

Kunc J, Tonev P, Szczyrba Z, Frantál B (2012) Shopping centres and selected aspects of shopping behaviour (Brno, The Czech Republic). Geogr Tech 7(2):39–51

Lloyd R, Clark TN (2001) The city as an entertainment machine. Research Project no. 454, Paper prepared for presentation at the annual meeting of the American Sociological Association, 2000. http://www.sociologia.unimib.it/DATA/Insegnamenti/4_2996/materiale/the city as an entertainment machine.doc. Accessed 10 July 2017

Monheim R (2008) Zunehmende Vielfalt der Aktivitäten beim Innenstadtbesuch als Ausdruck der Entwicklung von Lebensstil und Stadtstruktur—das Beispiel der Nürnberger Innenstadt. In: Spannungsfeld Innenstadt—zwischen Shopping-Center und Fußgängerzone?! Berichte des Arbeitskreises Geographische Handelsforschung 24(Dezember):15–20

Obchody & služby. https://www.palladiumpraha.cz/obchody-sluzby. Accessed 08 March 2018

Obchody. http://www.arkady-pankrac.cz/cz/obchody. Accessed 08 March 2018

Reikli M (2012) The key of success in shopping centers. Composing elements of shopping centers and their strategic fit, Corvinus University of Budapest, School of Business Administration, Budapest, Ph.D. dissertation. phd.lib.uni-corvinus.hu/742/6/Reikli_Melinda_den.pdf. Accessed 08 March 2018

Relph E (1976) Place and placelessness. Pion Limited, London

Restauracje. http://www.galeriakrakowska.pl/pl/sklepy/restauracje,18,cat,9. Accessed 08 March 2018

Restaurace, kavárny & bary. https://www.palladiumpraha.cz/restaurace-obcerstveni. Accessed 08 March 2018

Sikos TT (2013) Prague, the city of traditions and shopping centres. A case study. Hung Geogr Bull 62(4):351–372

Sirpal R, Peng O (1995) Impact of food courts and other factors on tenants' businesses for a major shopping centre in Singapore. Prop Manag 13(4):13–20

Skyline Plaza: spectacular gastro experience in the roof garden. ACROSS. The European Retail Real Estate Magazine, September 26, 2017, p 7. https://www.across-magazine.com/new-ece-food-experience-2. Accessed 08 March 2018

Sosnowska A (2002) Usługi gastronomiczne w wielkopowierzchniowych obiektach handlowych na terenie aglomeracji warszawskiej. Wydział Geografii i Studiów Regionalnych, Uniwersytet Warszawski, Warszawa, MA thesis

Taylor WJ, Verma R (2010) Customer preferences for restaurant brands, cuisine, and food court configurations in shopping centers. Cornell Hosp Rep 10(3):6–19

The future of the shopping center industry. Report from the ICSC Board of Trustees (2016) International Council of Shopping Centers, New York, NY. https://www.icsc.org/uploads/default/Envision-2020-Report.pdf. Accessed 03 Aug 2018

The KaDeWe—history of our story. https://www.kadewe.de/en/das-kadewe-die-geschichte. Accessed 08 March 2018

The successful integration of food & beverage within retail real estate (2017) Report from the ICSC Board of Trustees, 2016, International Council of Shopping Centers, New York, NY. https://www.icsc.org/uploads/.../Food__Beverage_Study_US.pdf. Accessed 08 March 2018

Top 10: Galerie handlowe Warszawa. Które centrum handlowe jest największe? FOTOGALERIA, 06.02.2018. https://www.muratorplus.pl/inwestycje/inwestycje-komercyjne/top-10-galerie-handlowe-warszawa-ktora-najwieksza-aa-rLuK-26F4-nZm5.html. Accessed 08 March 2018

Wall A (2003) Victor Gruen: the transformation of the American cityscape and landscape. In: Shopping_Center_Stadt. Urbane Strategien für eine nachhaltige Entwicklung, StadtBauKultur NRW, Weimar, pp 106–112

Chapter 16
Challenges to Urban Gastronomy: Green and Blue Spaces

Sylwia Kulczyk, Monika Kordowska and Katarzyna Duda-Gromada

Abstract Eating in urban parks or by water bodies is a popular activity in warm weather, as it makes it possible to combine time spent in nature and socialising. Although managing green and blue spaces in a way that enhances both ecological and human needs is one of the principles of contemporary urban policy and planning, it remains a challenge. This chapter discusses the problem using several examples from Warsaw, including gastronomy in the parks of Warsaw's central borough, the Breakfast Market, and the banks of the Vistula River. Fluctuating numbers of visitors, and accommodating the different needs of visitors, are only two of the most important challenges for the development of gastronomic services in green and blue urban spaces.

Keywords Urban parks · River banks · Social trends · Environmental threats · Warsaw · The breakfast market

16.1 Green and Blue Urban Spaces as Recreational Areas

Green or *blue* space refers to natural and semi-natural areas in the urban system. Among the various types of green spaces, urban parks—areas dominated by vegetation and water, designed for public use (Konindejndijk et al. 2013)—are significant because of their multiple uses. Their ecological functions (microclimate, air quality and noise reduction), combined with their social functions (opportunities for interaction, active and passive recreation, strengthening of identity and cognitive devel-

---

S. Kulczyk (✉) · M. Kordowska · K. Duda-Gromada
Department of Tourism Geography and Recreation, University of Warsaw,
ul. Krakowskie Przedmieście 30, 00-927 Warsaw, Poland
e-mail: skulczyk@uw.edu.pl

M. Kordowska
e-mail: monika_kordowska11@gazeta.pl

K. Duda-Gromada
e-mail: kduda@uw.edu.pl

© Springer Nature Switzerland AG 2020
A. Kowalczyk and M. Derek (eds.), *Gastronomy and Urban Space*,
The Urban Book Series, https://doi.org/10.1007/978-3-030-34492-4_16

opment) influence the quality of life and human well-being (Giedych and Maksymiuk 2017). Like urban parks, blue spaces, rivers and their banks also support both the urban ecosystem and society. Managing green and blue spaces in a way that enhance their multifunctionality and provides both ecological and social (and, to a lesser extent, monetary) benefits is one of the principles of contemporary urban policy (Stepniewska and Sobczak 2017). However, combining the different ecological and social functions remains a challenge. To avoid both environmental threats and social conflicts it is essential to carefully plan their composition, including the spatial distribution of greenery, as well as recreational facilities and other infrastructure (Voigt et al. 2014).

Eating in a natural setting not only supports the basic life functions of visitors, but also allows them to combine enjoying nature with social interactions. The tradition of organising outdoor meals and social gatherings has developed over the centuries, following the evolution of social habits. The demand for spending time in the natural environment followed the development of the urban lifestyle, as green areas offer the possibility of balancing the rush of everyday life with tranquillity, aesthetic qualities and a suitable microclimate. Outdoor eating has become a preferred way to spend time, combining a meal, the cultivation of social relationships and contact with nature. Originally feasts in urban parks and gardens were reserved for the highest social classes, because they were organised in private green areas. The lifestyle of the upper classes was subsequently adopted by the wider public.

16.2 From King's Fetes to Beer Bars

In Warsaw, this trend was started by the Polish king, Stanisław II Augustus who ruled in 1764–1795. Meetings in the king's residence of Łazienki, called *Thursday dinners*, were renowned as intellectual meetings, where invited guests, apart from enjoying their meals, engaged in storytelling and exchanging opinions. The beautiful gardens surrounding the palace provided a special atmosphere. Apart from these private meetings, the king also organised public feasts and ordered that the traditional plays that were organised in urban markets should be transferred to the suburban wood of Bielany, where the first Pentecostal feast took place on 19 May 1766. Picnicking in Bielany became an annual activity for most Warsaw citizens and its form evolved throughout the nineteenth century. The programme, composed of parades and public plays, required new spatial arrangements (the creation of alleys and squares) as well as the construction of temporary facilities. Beyond the consumption of their own meals, visitors could use tented restaurants offering different qualities and price levels (Waszkiewicz 2013). At the same time, gastronomic outlets began to emerge in areas that surrounded the city. The most popular destination for these suburban trips was the Wiejska Kawa (Rural Café) situated on the route to the residences of the nobility, south of the city (Herbaczyński 2005).

Saski Garden was the first public park in Warsaw. The part of the royal residence in the centre of Warsaw opened to the public in 1727. In the second half of the nineteenth century, the Garden was *the* place to visit. Its 14.4 hectare landscaped park housed various facilities, including gastronomic establishments, some of which promoted a healthy lifestyle. The Institute of Mineral Waters offered water treatments, in the form of drinks as well as baths. The dairy was located in an adjacent building. Fresh milk was available along with a chance to observe cows in the barn. Cafés were another type of gastronomic establishment that attracted visitors by offering them the chance to relax with a cup of coffee accompanied by pastries. Two were situated at the borders of the park: one in the Institute of Mineral Waters and the other at the opposite corner in a private building that opened onto the green area. Other, smaller establishments selling fruits and soda water were situated in the central part of the park (Marcinkowski 2003).

There is also a long tradition of using the Polish capital's blue spaces for recreational purposes. Warsaw lies along the course of the Vistula, the longest river in Poland. The 28-kilometre-long stretch of the Vistula that lies within Warsaw's administrative boundaries retains its highly natural character despite the presence of the large city. Within Warsaw, the Vistula valley is asymmetrical: the right bank is low, while the left bank is high. In the early eighteenth century the Polish king, Augustus III (1733–1763), organised fetes in the meadows on the right bank. The first forms of recreation were boat excursions and baths, followed by the creation of the Warsaw Rowing Society in the mid-nineteenth century (Stefanowska 2012). In the 1960s, an average of 2,000 people visited the beaches on weekdays and 24,000 on holidays (Król 1969). Increasing water pollution and the opening of Zegrze Reservoir—an artificial lake 30 km from Warsaw—meant that the popularity of the capital's beaches and sport clubs fell significantly in the 1980s.

Along with a long tradition of recreation, the shores of the Vistula have been important for the gastronomy sector. Although a few snack kiosks existed in sport clubs and on beaches, they only supported the very basic needs of visitors. Beer bars, created in the early 1960s, were an important element in the development of catering on the left bank. In addition to beer, they also sold snacks and simple, warm dishes. While the food was little more than basic, some of these places were important social hotspots. They were visited by a diverse clientele, which included workers from nearby industrial plants, university students, anglers warming up after hours spent by the river, and families on their Sunday walks. In the early 2000s the municipality decided to modernise the Vistula boulevards. Bars were gradually closed down, and the area cleaned up and prepared for renovation.

In the next section, we present the contemporary state of gastronomy in the green and blue spaces of central Warsaw and describe the main challenges faced by its development.

16.3 Parks in the Central Borough of Warsaw

To some extent, the development of contemporary gastronomy in the parks of Warsaw's central borough (Śródmieście) followed patterns established in the eighteenth and nineteenth centuries. Green areas in the central district vary in both

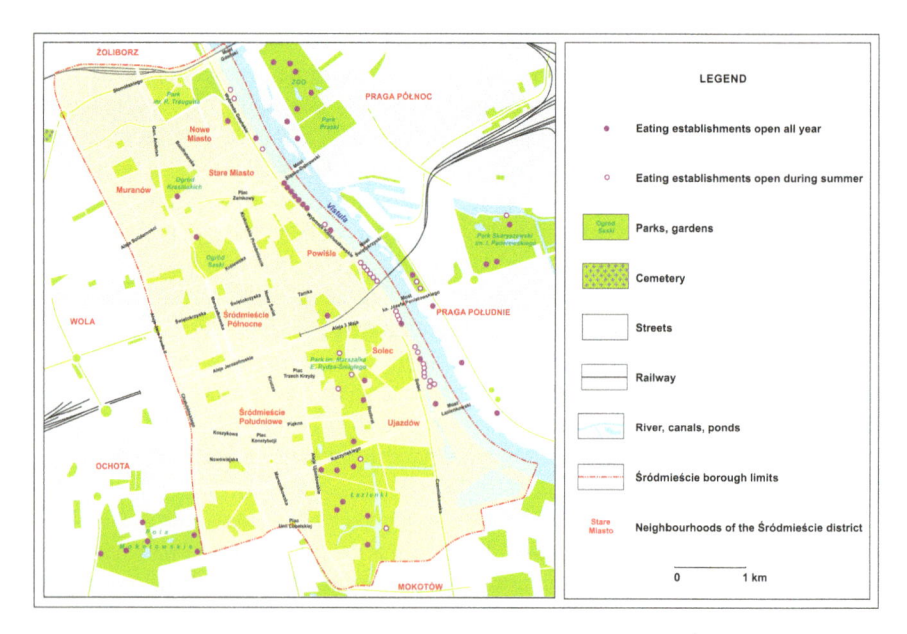

Fig. 16.1 Distribution of eating establishments concentrated in the Śródmieście borough (Warsaw) parks and along the Vistula river in 2017. Source: Authors' own research, underlay map by Andrzej Kowalczyk

size and character, with 18 parks over one hectare and 20 smaller, green squares. The biggest park is the 76-hectare *Łazienki* (officially *Łazienki Królewskie*, or the Royal Łazienki Park) which connects to other parks and forms part of the green corridor that continues along the slope of the Vistula river valley (Fig. 16.1). Within this area, there are four parks that are designated as historical monuments and attract significant number of tourists and city dwellers. The Park is one of Warsaw's most famous tourist attractions: there were 2.1 million visitors, including 240,000 to its museums (Derek 2018).

The eating and drinking points located within these green areas, or adjacent to them include restaurants, cafés and bars. An inventory conducted in the summer of 2017 showed that there were both seasonal and year-round facilities. In the summer period there are five temporary establishments (four bars and one café), mobile ice cream and coffee vendors, which supplement establishments that are open all year round (seven restaurants, two cafés, one tea house and seven bars) (Fig. 16.1). All of these establishments consider their setting as special and explicitly refer to their surrounding natural environment (particularly its aesthetical and relaxing properties) in their marketing materials. For more upmarket restaurants it is the food that draws in visitors, and the park setting is secondary. This is particularly true of Atelier Amaro, a 1 star Michelin restaurant, that has an experimental Polish kitchen, and is set in a modest building dating from the 1970s, in Agrykola park. Like Atelier Amaro, most of the gastronomic establishments have introduced extra functionality to existing places. For example, in the Łazienki Royal Park, a restaurant and two cafés are located in historic monuments. However, some parks are protected historical monuments, and the introduction of any new infrastructure raises significant legal problems. Consequently, the location of gastronomy in existing buildings is the exception rather than the rule.

Furthermore, such locations are usually popular with users, who can combine sightseeing with an eat or drink stop. The location may be so specific that it influences the entire character of the gastronomic establishment and is used as a brand. This is the case of the Belvedere restaurant in the Royal Łazienki Park, which is situated in the orangery. Built in 1861, the building still houses tropical plants, which give the eating space its unique character. In areas that have recently been restored new eateries have been established, reflecting the general interest of the authorities in implementing a gastronomy function in green areas. On the one hand, this approach follows the rising trend of eating outside; on the other, the income from renting out facilities helps in managing green areas and maintaining their high quality.

Modern establishments tend to be located on the outskirts of green areas. Some visitors are attracted by the opportunity to eat and drink; such customers may not be interested in long walks. In addition, places that are not formally part of green areas may profit from their attractiveness. If a restaurant or a bar borders an unfenced park, the owners may arrange a terrace, verandah or sunloungers in the adjacent space. These are usually temporary facilities that respond to the changing seasons. In general, the number of visitors to Warsaw's green areas rises significantly at the start of spring (typically around mid-April) and falls in the second half of September.

There are also differences between weekdays and weekends. On weekdays the majority of park visitors are small children with their carers, and pensioners, together with a few office workers. At weekends whole families and groups of friends arrive and the demographic profile is much broader.

Apart from a few restaurants that offer different and, in some cases, very unique cuisines, the food on offer is fairly homogenous, and includes typical international and Polish dishes (grilled dishes, pizza, dumplings), coffee, sweets, beer and soft drinks. Field observations have identified several factors other than food that attract visitors to particular establishments:

- setting: places with nice views (for example a panorama over parks and the Vistula river valley) and those located close to the main walking and cycling trails are among the most popular; a convenient or beautiful setting can attract random visitors, i.e. those, who had not intended to visit a café, but decided to stay because they like the place;
- additional facilities: sunloungers and children's playgrounds draw visitors on sunny days. Such facilities may attract repeat visits from families with children;
- events: workshops, live music and yoga sessions can help to transform a gastronomy point from an average café to a special, socially meaningful place. For some of café owners it is a way to both expand and profile their clientele. On the other hand, societal hotspots can be formed by customers themselves and

usually appear at places that are located in parks but are easily reachable by customers from other places such as office blocks or sport facilities. For example, one of the café bars in Agrykola Park is a traditional meeting place for supporters of Legia, Warsaw's biggest football team, whose stadium is located within walking distance.

Urban parks cater for many different groups of visitors, and the spread of gastronomic facilities and accompanying events is not always popular among people looking for a natural setting unspoiled by human influence. One solution may be to limit the timing of gastronomy and leisure events, or moving them to smaller, less frequented green areas. One example of such an initiative is the Breakfast Market.

16.4 The Breakfast Market—a Short-Term Green Space Gastronomy

The Breakfast Market is held at weekends and blends elements of gastronomy, farmers markets and cultural events, while creating a green space for resting. The original concept refers to the tradition of picnics in green urban areas but enhances it with a variety of other experiences and activities, from shopping to dancing. The event is organised by a private entrepreneur who has trademarked the name. It is therefore carefully managed. This makes it different from, for example, the Thai Wiese in Berliner Preussen Park, which started as a spontaneous meeting of Asian migrants and then gained fame as a gastronomy attraction. All of the data presented here were gathered by Monika Bartman for her Bachelor's degree thesis (Bartman 2017). The authors are grateful for the opportunity to present her findings here.

The idea of the Breakfast Market focuses on enjoying a diversity of high-quality products while spending time outdoors. Both aspects are considered as elements of a healthy lifestyle and important aspects of the well-being of urban citizens, who warmly welcomed the initiative when it was launched in 2013. The outdoor version of the Market takes place at the weekend from April to October at various points in Warsaw. In 2017, there were two main events on Saturdays and Sundays, supported by two smaller Saturday markets. All events are located in central Warsaw away from the tourist/business centre, as one of the organiser's aims is to attract and integrate local communities. It is typically located in a green square or small park close to a bigger green (or blue) space. The location has to be big enough to host more than 50 stands, tables, and places for side activities and—particularly important—sufficient lawn space to accommodate several hundred people with their blankets, bicycles, children's buggies, etc. (Figure 16.2). The ideal location is easily accessible for both delivery services and visitors. The large numbers of visitors who attend can pose a problem for those looking for a more peaceful environment and who may feel that they are losing what they consider to be their green oasis. The location therefore also has to be somewhat isolated from the main body of green areas.

Fig. 16.2 The Breakfast Market in Żoliborz borough in Warsaw (Poland) in 2017. Picknique at the Breakfast Market. Note the effects of trampling (**a**). More than a breakfast—Brazilian food at the Breakfast Market (**b**). Source: Monika Bartman 2017

Up to 70 stands form a circle that surrounds dining tables, while a picnic space lies outside the circle. Blankets and sunloungers can be borrowed from the organisers. Three groups of services are offered: the sale of food; the sale of other articles; and cultural activities.

The sale of food is the main pillar of the event. In 2017, 80 vendors offered dishes and drinks for onsite consumption, and grocery products that could be used to prepare meals at home. The organisers select vendors carefully, looking both for high-quality products and diversity. Thanks to this policy, the gastronomy of the Breakfast Market is rich and unique, as a variety of culinary trends can be offered in one place. In 2017, 17 ethnic cuisines could be sampled, including Thai, Vietnamese, Brazilian (Fig. 16.2), Italian, French, Georgian and Hungarian dishes. Some sellers represent regular eating establishments (e.g. Rong Vang, a well-known Vietnamese restaurant in Warsaw), but menus usually differ from those offered by permanent locations. This may be due to technical limitations imposed by temporary stands, but it can also be a business solution, because the portions offered are usually smaller. Apart from stands, the Market includes up to a maximum of four food trucks. This regulation aims to keep the event as unique as possible and, although most of the food trucks offer original food concepts, they are mobile and consequently available all over the city. Another consideration is to preserve green areas and avoid leaving tyre tracks.

A significant group of sellers offer both ready-to-eat and takeaway products. This is typical of bakery and pastry stands, which constitute around 20% of the market and attract many customers. In the summer, drink stands are the most popular. Lemonade stands that use original receipts attracted so many customers that the organisers decided to withdraw the uniqueness criterion and allowed multiple stands.

The Market runs from 9 am until 4 pm and offers a mix of snacks and lunch dishes. Although typical breakfast food (bread and cheese) is available, it can be only purchased to take away. After some clients complained on the Market's Facebook page, saying that they would rather buy slices of delicious bread and cheese to consume immediately, than carry items around with them all day, a stand offering typical Polish breakfast food (simple sandwiches and egg dishes) was opened and immediately became popular.

Around 20 stands offer takeaway food, such as baked goods, fish, dairy products and organic fruit and vegetables. Most of these goods can only be purchased at the Breakfast Market or, in some cases, online. In their marketing materials, vendors stress the high quality of their products and their environmentally friendly nature, creating an impression of a connection between a particular product and the land. This, however, comes at a cost: prices are around 20% higher than similar products in the city's other farmers markets and organic food stores.

There are 17 stands offering non-food items, mainly handicrafts (e.g. bags and hats) and natural cosmetics. These stands are grouped together, separate from the food stalls.

Cultural offerings are also diverse. Families (or children) can participate in games and plays, others can test cosmetics, take part in do-it-yourself workshops, or join literary meetings. In 2017, themed versions of the Breakfast Market were launched, including Brazilian and Mexican. During these special events, ethnic cuisine was complemented by concerts, dance lessons and other activities that allowed people to experience an exotic culture. This change from the standard weekend routine was warmly welcomed by visitors.

The idea of the Breakfast Market is an innovative, social response to demand for a healthy lifestyle. The event attracts different groups of visitors. Before 11 am, most are older people from local neighbourhoods. They are mainly interested in buying high-quality food products, rather than prepared food. Younger visitors arrive around noon. These are usually friends or family groups in their twenties and thirties, many of them accompanied by small children. They spend several hours picnicking, eating and playing together. Many visitors arrive by bike. The peak time is from noon to 2 pm and, on sunny days, it can be hard to find a free patch of grass.

In just three years the Breakfast Market has become a trusted brand. It has appeared in the foreign press and on television as one of Warsaw's attractions. In summer 2017, around 15% of visitors were foreigners, although the market is still largely a local event. The majority of visitors live in the same borough or nearby. Because there are several versions of the event each weekend, people can choose the one that is closest to where they live. Participating in the Market is very different to walking in a green area, or relaxing alone on a park bench. Despite being advertised as a place to make contact with nature, the event is rather noisy and crowded. Moreover, sitting and eating is not particularly healthy. The fact that visitors frequently arrive by bike, however, and the location of the markets (close to bigger green areas) makes it possible to combine a visit with other, more directly nature-based activities.

This regular mass event poses some environmental threats. As noted above, they are kept to a minimum thanks to appropriate localization. In previous years, however, there have been complaints about the amount of rubbish left behind. The problem has been resolved by putting more effort into clearing up after the event.

Trampling the grass, however, is hard to avoid and it is particularly difficult to keep it healthy during extended dry spells.

Unlike the Breakfast Market, the banks of the Vistula River offer a long strip of gastronomy space. Here, relaxing and chilling out are deemed to be more important to visitors than eating.

16.5 Blue Space Gastronomy—the Banks of the Vistula River

After a period of neglect by city authorities and local inhabitants, the banks of the Vistula have undergone a renaissance. Social and promotional activities have been developed, aimed at bringing Warsaw back to the Vistula. In response to the need for holistic action, an informal 'Borough Vistula' has been created. The project, implemented by the City of Warsaw, aims to promote riverbank areas. A website (http://www.dzielnicawisla.um.warszawa.pl) has been created, with up-to-date information about events, riverboat cruises, a photo gallery, and more detailed information on the development of the area. Cultural events, such as *Wianki* (Wreaths, an open-air event on St. John's Night) are held and there is an outdoor cinema.

In the summer of 2017, catering services on the riverbanks in the city centre (between Łazienkowski and Gdański bridges) consisted of 40 facilities, most in a section of the left bank (34 facilities). This can be divided into three parts. The north section, located in front of the Old Town, offers a limited number of eating places. Except for one café, they are all temporary outlets, mostly located on moored boats. These places cater mainly for tourists who visit the Old Town and do not serve the best Polish and European dishes. City authorities have imposed limitations on gastronomy facilities in this section for two main reasons. Firstly, the Old Town is part of a protected UNESCO site and its panorama is also preserved. Second, parking is limited in the area. The earlier development of gastronomy in this section of the river (around 15 years ago) resulted in frequent and sometimes dangerous traffic violations. Therefore, the modernization of boulevards included shifting bars and restaurants further south.

Between the Śląsko-Dąbrowski and Świętokrzyski bridges, gastronomic offerings are more diverse. Recently renovated, this section of the boulevards provides year-round eating facilities (Fig. 16.3a). Bars and restaurants in glass-walled pavilions serve a variety of world cuisines. The most dynamic and diversified, but also the most frequented and energetic part of the left bank is the section between Świętokrzyski and Łazienkowski bridges. The majority of establishments here are temporary, only operating between spring and autumn. Moored boats (Fig. 16.3b) and other outlets are supplemented by mobile facilities, such as food trucks and bike cafés. It is worth noting that the most readily accessible part of the boulevard (near the Poniatowski Bridge) is designated for food trucks and is known as the

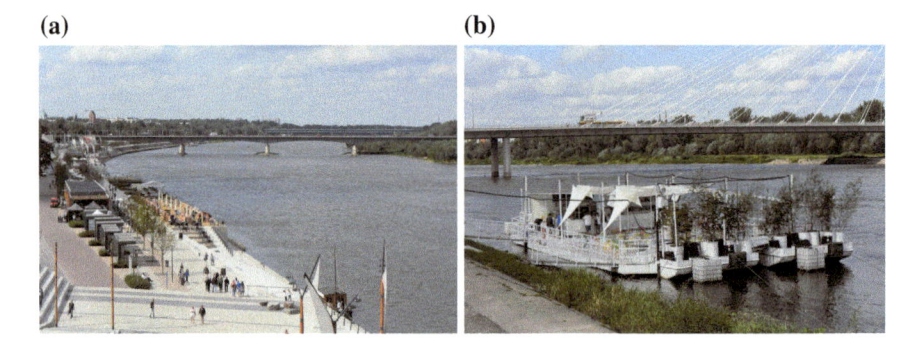

Fig. 16.3 Vistula River in the centre of Warsaw (Poland) in 2017. Recently renovated left bank with a number of eating establishments faces the 'wild' right bank of the river (**a**). Floating bars and restaurants allow to stay as close to the river as possible (**b**). Source: Monika Kordowska 2017

Food Port. The southernmost part of this section is a combination of blue and green space. In addition to the shade provided by the trees, there is an open-air gym, kayak rental and other leisure facilities. Here, the river is furtherest from adjacent streets and as more space is available, gastronomy establishments are more like leisure resorts, in that they combine drinking and eating with playground or sports facilities. During the day, various workshops attract families, and in the evening concerts and dance events are staged. Menus vary, following current trends. It should be noted that some popular city restaurants open a summer outpost by the river, usually with a smaller and simpler selection of dishes. The availability of exotic (mainly Asian) food and Caribbean-style drinks clearly underlines the leisure character of the space.

The right bank of the Vistula is flat, and sand sedimentation has determined the location of the city's beaches. The tradition of beach recreation dates back to the period before World War II. Nowadays, three public beaches are open on the right bank, managed by Warsaw City Hall. These areas are leased from the Regional Water Management Authority and therefore cannot be utilised for commercial purposes (Dobrowolska 2014). For this reason, catering facilities are located some distance from the river, close to the street. Consequently, they cater for more than just those seeking recreation in the riverbank area. Beaches, however, are equipped with self-service grills for use by visitors.

Catering services on the two riverbanks differ significantly in terms of their number and character. This is determined by the natural environment, as well as the history of the development of these areas and legal regulations. Several offer additional services that are of particular importance in the context of open-air recreation, making contact with nature, and spending time with family and friends. These are: sunloungers (14 facilities), sunloungers and tables on the pier (19 facilities), and tables that are directly situated on the boulevard (eight facilities on the left bank). For a city riverbank this is essential, because it allows visitors to take advantage of the river and its landscape. Moreover, some catering facilities are

Table 16.1 Additional attractions at eating facilities on the Vistula River in Warsaw

Attraction	Number of facilities
Playground for children	4
Open-air fitness club	1
Beach volleyball court	3
Other amenities	6
Concerts	16
Workshops	17
Theme events, discos	19

Source: Authors' own research

equipped with sport and recreational amenities, such as playgrounds for children (four facilities), which attracts families and people with younger children. These customers use the catering services during the day, and less frequently in the evening. There are also three beach volleyball courts (two of them at bars on the right bank). Other amenities include an open-air fitness centre, a scenic viewpoint, a skimboarding course, and a karting course for children.

An important additional feature is organised events. Concerts have been held in six of these facilities, and 17 have organised workshops and games. Workshops included jewellery making, others relate to sport and recreation (fitness, group exercise, golf classes, skimboarding classes, dance classes, etc.). Themed events are also organised, such as film shows, live music, dances and karaoke (Table 16.1).

The fact that many eating facilities enhance their traditional food offer with other elements, such as services and events, reflect the widespread current trend for combining catering services with other functions, particularly recreation, sport and entertainment.

Inhabitants of large cities regard making contact with nature as particularly important and desirable. Riverbank areas in Warsaw are becoming exciting, new public spaces, where the population and visitors can spend time. In this context, the development of catering services makes their stay there more attractive. The shape of the river valley makes it possible to spend an entire day and a good part of a night exploring a variety of bars. Sunloungers are available, and visitors can walk or bike further afield in search of entertainment. Small ferries allow people to easily cross from the more formal left bank to the beaches and urban wilderness on the right shore. This pattern of use is, however, limited by the weather. The area, which is full from spring to autumn (on most days), remains almost deserted in winter. While some eating establishments remain open all year round, it is clear that dining opportunities are not enough to attract people to the area throughout the year: leisure activities, some of which are not available during the colder months, are the main draw.

16.6 Conclusions: Challenges for Gastronomy in Blue and Green Spaces

The paradox of modern society is that a general interest in following an environmentally friendly lifestyle is accompanied by a loss of contact with the natural world. Urban parks are attractive to visitors not only because they are close and easily accessible, but also because they are constructed and managed in a way that supports comfort and safety. Losing contact with nature means, however, that while people may recognise the importance of visiting green areas, they do not know how to organise their time when they are there. Although they do not know how to make contact with nature, they are familiar with the facilities and enjoy using them. While this has provided a green light for further development of gastronomic services in green and blue urban spaces, there are still some challenges that need to be overcome.

The main barrier is the fluctuating number of visitors. Differences between weekday and weekend numbers, as well as seasonal variations create busy bars at weekends and on sunny days, but empty bars on rainy weekdays in cold months. Coping with this requires flexibility, both in terms of space for visitors and cooking capacity. Usage patterns can be balanced in parks that are tourist attractions, where a drop in the number of local users may be offset by the number of tourists. Other bars and cafés try to attract more regular users by offering food combined with regular activities such as yoga or art workshops.

Another problem is how to accommodate the different needs of park visitors. Although gastronomy attracts many people, it also raises objections from those who are more interested in making direct contact with nature. This is especially true in small parks, where even one gastronomy unit can totally change the character of the green area and turn a peaceful hideaway into a fashionable (and noisy) hangout. Finally, the development of gastronomy may significantly affect the ecological balance in green and blue areas as it may require the construction of new buildings. This means not only loss of vegetation, but also significant environmental hazards during the construction process. Deliveries to these new outlets may result in additional traffic. Temporary facilities such as summer gardens and terraces also have a negative environmental impact, as the trampling of the ground results in a significant loss of biodiversity (Sikorski et al. 2013). As for the local wildlife, the ready availability of food can lead to changes in the feeding habits of birds and small mammals (e.g. squirrels) who forage in dustbins or even at tables. Furthermore, the light and noise generated by gastronomy establishments after nightfall can disturb the natural biorhythms of animals.

Facing these challenges requires not only the proper management of existing facilities, but also a desire to seek out new locations and new forms of gastronomy that offer not just food and drink, but also the opportunity to make contact with nature.

References

Bartman M (2017) *Targ śniadaniowy jako miejsce spędzania wolnego czasu w Warszawie*, Wydział Geografii i Studiów Regionalnych, Uniwersytet Warszawski, Warszawa, BA thesis

Derek M (2018) Spatial structure of tourism in a city after transition: the case of Warsaw, Poland. In: Müller DK, Więckowski M (eds), Tourism in transitions, Springer, Cham, pp 157–171

Dobrowolska W (2014) *Plaże miejskie. Nowe (?) zjawisko w przestrzeni Warszawy*, Wydział Geografii i Studiów Regionalnych, Uniwersytet Warszawski, Warszawa, BA thesis

Dzielnica Wisła. http://www.dzielnicawisla.um.warszawa.pl. Accessed 24 Oct 2017

Giedych R, Maksymiuk G (2017) Specific features of parks and their impact on regulation and cultural ecosystem services provision in Warsaw, Poland. Sustainability 9(792). http://www.mdpi.com/2071-1050/9/5/792/pdf. Accessed 20 May 2018

Herbaczyński W (2005) W dawnych cukierniach i kawiarniach warszawskich. Veda, Warszawa

Konijnendijk CC, Annerstedt M, Nielsen AB, Maruthaveeran S (2013) Benefits of urban parks. A systematic review. A report for IFPRA, Copenhagen & Alnarp

Król B (1969) Wypoczynek Warszawiaków. In: Nowakowski S (ed) Warszawa: socjologiczne problemy stolicy i aglomeracji, Książka i Wiedza, Warszawa

Marcinkowski R (2003) Ilustroway Atlas Dawnej Warszawy. Pangea, Warszawa

Sikorski P, Szumacher I, Sikorska D, Kozak M, Wierzba M (2013) Effects of visitor pressure on understory vegetation in Warsaw forested parks (Poland). Environ Monit Assess 185(7):5823–5836

Stefanowska A (2012) Zagospodarowanie rekreacyjne Wisły na jej warszawskim odcinku—historia i perspektywy. Problemy Turystyki i Rekreacji 1:69–79

Stepniewska M, Sobczak, U (2017) *Assessing the synergies and trade-offs between ecosystem services provided by urban floodplains: the case of the Warta River Valley in Poznań, Poland*. Land Use Policy 69:238–246

Voigt A, Kabisch N, Wurster D, Haase D, Breuste J (2014) Structural diversity: a multi-dimensional approach to assess recreational services in urban parks. Ambio 43(4):480–491

Waszkiewicz K (2013) *Piknik w kulturze polskiej*, Wydział Filologii Polskiej, Uniwersytet Warszawski, Warszawa, MA thesis

Chapter 17
Street Food and Food Trucks: Old and New Trends in Urban Gastronomy

Andrzej Kowalczyk and Magdalena Kubal-Czerwińska

Abstract Street food refers to ready-to-eat food sold in a street or other public place (markets, bazaars, etc.) often prepared at a temporary facility. Originally, it was offered by vendors as they traversed a city; nowadays, dedicated portable food carts in the shape of modified bikes, motorbikes, etc., as well as cars known as food trucks have become increasingly popular. Although the history of the way in which ready-to-eat food is offered on the street stretches back to ancient times, it is especially popular nowadays not only in the cities in developing countries, but in the European and North American agglomerations too. In Poland, the development of food truck gastronomy has been encouraged by a rise in the number of public events, such as outdoor concerts, historical reconstructions, and sports and recreation events. Drawing on the examples of Warsaw and Kraków, this chapter seeks to explore places where food tracks congregate during the week (business and corporate districts in particular) and at the weekend (freetime districts).

Keywords Urbanization · Street food · Food trucks · Warsaw · Kraków

17.1 Street Food—An Old or a New Form of Urban Gastronomy?

One of the more serious competitors for restaurants and bars is the phenomenon referred to as street food. The supply and consumption of food in the streets and squares of cities and towns existed in ancient civilizations and has survived the

---

A. Kowalczyk (✉)
Department of Tourism Geography and Recreation, Faculty of Geography and Regional Studies, University of Warsaw, ul. Krakowskie Przedmieście 30, 00-927 Warsaw, Poland
e-mail: akowalczyk@uw.edu.pl

M. Kubal-Czerwińska
Department of Tourism and Health Resort Management, Jagiellonian University, Institute of Geography and Spatial Management in Kraków, ul. Gronostajowa 7, 30-387 Kraków, Poland
e-mail: magdalena.kubal@uj.edu.pl

© Springer Nature Switzerland AG 2020
A. Kowalczyk and M. Derek (eds.), *Gastronomy and Urban Space*,
The Urban Book Series, https://doi.org/10.1007/978-3-030-34492-4_17

passage of time to the modern day. Street food refers to ready-to-eat food sold in a street or other public place (markets, bazaars, etc.) often prepared at a temporary facility. Most street foods are classified as fast food or finger food, and are cheaper on average than meals from restaurants, bars, or other eateries.[1] Originally, it was offered by vendors as they traversed a city; nowadays, dedicated portable food carts in the shape of modified bikes, motorbikes, etc., as well as cars known as food trucks have become increasingly popular (Kowalczyk 2014, p. 146). Street food is a heterogeneous category, both in terms of what is available (drinks, meals and snacks) and the ingredients from which they are made. Processing ranges from preparation in the street in relatively heterogeneous and unregulated conditions, to the central processing of ready-to-eat foods, such as snacks, by the wider food industry that are distributed and retailed via street food vendors (Draper 1996, p. 1). The official definition (accepted by Food and Agricultural Organization in 1986) is: 'Street foods are ready-to-eat foods and beverages prepared and/or sold by vendors and hawkers especially in streets and other similar public places' (Fellows and Hilmi 2011, p. 2). This definition means that the crucial characteristic of street foods is their retail location, i.e. on the street. Henderson (2019) makes a distinction between vending and hawking, as the latter refers strictly to sellers who are mobile, but the terms are usually used interchangeably.

Although the history of the way in which ready-to-eat food is offered on the street stretches back to ancient times, it is especially popular nowadays in African, Asian and Latin American cities, where it is a common way of obtaining a (usually) warm meal for many residents. As Henderson (2019) points out, in Asia food vendors are an integral part of life and an important tourism resource. Nevertheless, this does not mean that street food does not exist in European and Northern American cities, although its importance began to weaken in the nineteenth century with the development of modern capitalist cities. It is only in recent decades—due to the arrival, among others, of immigrants from the Middle East, Northern Africa and parts of Asia —that the consumption of street food has started to rise again. In large North American cities on the Atlantic coast and in the Great Lakes the revival of street food, which had been available in the seventeenth and eighteenth centuries, coincided with the arrival of immigrants from Central and Southern Europe (Taylor et al. 2000, pp. 26–27).

17.2 The Growing Importance of Street Food

Draper (1996, p. 1) attributed the expansion of street food trade in the global South (Third World countries) to urbanization and the need of urban populations for both employment and food. As Naidoo et al. (2017, p. 2) observed, '...many Asian, Latin American and African cities have a long tradition of street food vendors and hawkers'. This statement was supported by a report by the Food and Agricultural Organization.

---

[1] *Street food*, http://pl.wikipedia.org/wiki/Street_food (Accessed 02 July 2014).

According to it street foods are consumed by 2.5 billion people on a daily basis, and may contribute as much as 40% of the daily calorific intake in some Asian cities such as Bangkok. Like many other Southeast Asian countries, Singapore has a long tradition of street vendors and hawkers (Fellows and Hilmi 2011, p. 7). According to recent data there are more than 100 hawker centres, with between 30 and 50 food stalls that provide Singaporeans with a variety of traditional ethnic dishes (Naidoo et al. 2017, p. 2). It is worth noting that they are considered one of Singapore's most important tourist attractions (Hakam and Leong 1989; Henderson et al. 2012).

Research carried out in the 1980s and 1990s in various cities showed that in Zinguinchor (Senegal) and Manikganj (Bangladesh) street food trade concerned 6% of the total labour force, in Iloilo (Philippines) 15%, and in Bogor (Indonesia) 25% (Draper 1996, p. 6). The same study found that vendors usually operated from selected strategic locations, including bus and train stations, markets and shopping areas, commercial districts, near schools and hospitals, in residential suburbs, close to factories and construction sites, etc. The general perception is that street food vendors and hawkers tend to concentrate in downtown commercial areas. But in Nigeria 23% were located in residential areas of cities (Draper 1996, p. 8).

Draper's study also looked at the products and dishes offered to customers. For example, research in Bogor showed that nearly 300 items were on offer, including numerous varieties of rice-based meals, fried snacks, traditional cakes, soups and porridges, drinks and fruit. Raw products included meat, poultry, fish, seafood, eggs, cereal products, soya products, fruits and vegetables. These ingredients were fried, roasted, boiled, baked, steamed or eaten raw (Draper 1996, p. 11). Generally, the number of vendors correlated closely with the total urban population, which supports the assumption that there is a direct association between the number of street food enterprises and city size. The number of inhabitants per vendor ranged from 34 to 69; in Bogor there were only 14, while in Minia it was 255 (Draper 1996, pp. 12–13)

When analysing the importance of street food in large cities in South Asia, Etzold (2014) noted that it is important for urban food security, because it is cheap, readily available and nutritious. Street vending is also one of the most important employment providers for poorer residents (Etzold 2014, p. 1). In Bangkok, for example, it is perceived as a tool in the city's poverty reduction strategies (Henderson 2019). Several forms of street food were identified in Dhaka, including: '…mobile hawkers who walk around with a basket, tray or flask (20% of interviewed vendors); semi-mobile push-carts and rickshaws that are occasionally moved to reach consumers at different places (36%); semi-permanent vending units like tables that are set up for the day (13%); to more permanent, but not consolidated, food stalls that are built illegally (21%); and permanent consolidated huts, which cannot be considered as street vending units and served as comparison group in this study (11%)' (Etzold 2014, p. 3). Despite this, the role of street food in meeting the nutritional needs of urban populations has received little official attention. Governments and local authorities have paid more attention to the potential dangers arising from the consumption of street foods than to any benefits they might offer. Misunderstandings include:

- street food trade is a marginal and transitory economic activity, which is a hangover from traditional market activities and will, over time, disappear;
- it is dominated by women;
- it is focused in the main commercial areas of urban centres;
- street foods are 'dirty' and 'dangerous'[2];
- only the lower classes (poor) eat them;
- they do not make an important contribution to dietary intake (Draper 1996, p. 2).

Singapore has a long history of regulating street food (Sirpal and Peng 1995, p. 13). The results of a survey (n = 554 respondents) undertaken by Hou and Swinyard (1986) showed that the traditional strengths of hawker centres in Singapore were being eroded. They were no longer perceived as providing the best-quality food, and their popularity had declined compared to fast-food chains. The study concluded that it was time for the bigger and more established hawker centres to become organized. Client preferences for clean, friendly, speedy and comfortable places to eat was a trend that could not be ignored.

> A survey carried out in 2010–2011 showed that Singapore (and probably several other Southeast Asian urban areas) was undergoing a socio-economic transition: home cooking was perceived as less attractive and eating out was becoming more common. At the same time, traditional street food retailers (hawkers) were identified as important food providers whose role must be considered in policies that seek to address changes in eating habits and rising rates of cardio-metabolic diseases. As a result of regulations, Singapore was hailed as a model of good practice regarding food hawker management, and some cities (e.g. Ho Chi Ming) have spoken of emulating Singapore. The challenge is how to promote home cooking; Naidoo et al. (2017, p. 10) suggest that this new policy should be adopted by other Asian countries (Malaysia and Hong Kong in particular) with similar models. It should be remembered, however, that Singapore is prosperous and renowned for its competent and corruption-free government. Moreover, visitors may prefer the disorder and energy which prevails in many other Asian cities (Henderson 2019).

---

[2]Some of these statements are confirmed by research. The street foods available in Singapore are generally high in sodium and saturated fat and whole-grain options are limited (Naidoo et al. 2017, p. 2). Studies carried out in the Philippines have shown that the food offered on the street contain high levels of bacteria, as it is prepared and sold in unhygienic conditions (Canini et al. 2013, pp. 116–117).

Table 17.1 Cities with the world's best street food 2012–2017

Heelan (2014)	Forbes magazine (2012)	Fisher (2014)	New York Daily News magazine (2013)	Bhide (2013)	CNN (2016)	Forbes magazine (2017)
Bangkok	Bangkok	Tokyo	Hong Kong	Bangkok	Bangkok	Bangkok
Tel Aviv	Singapore	Tel Aviv	Rio de Janeiro	Istanbul	Tokyo	Singapore
Istanbul	Penang	Taipei	Paris	New York	Honolulu	Penang
Paris	Marrakesh	Singapore	Boston	Marrakesh	Durban	Marrakech
Mexico	Palermo	Seoul	Istanbul	Ho Chi Minh	New Orleans	Palermo
Hong Kong	Ho Chi Minh	San Juan	Mexico	Palermo	Istanbul	Ho Chi Minh
Kuala Lumpur	Istanbul	San Francisco	Ottawa	Rio de Janeiro	Hong Kong	Instanbul
Mumbai	Mexico	Rio de Janeiro	Marrakesh	Paris	Paris	Mexico City
Tokyo	Brussels	Portland	Berlin	Hong Kong	Mexico	Brussels
Singapore	Ambergris Caye	Penang	Fukuoka	London	Cairo	Ambergris Caye

Source: Adapted from Heelan (2014), Fisher (2014), Bhide (2013) and Henderson (2019)

Table 17.1 shows that the most attractive cities for street food are in Asia. The cities that are ranked most highly, based on snacks bought from street vendors (and usually eaten in the same place) are mostly in Southeast Asia (particularly Bangkok, Singapore, Ho Chi Minh) and East Asia (Tokyo and Hong Kong). However, Istanbul, Tel Aviv, and Paris were all ranked highly; street food in North America was ranked significantly lower. Bangkok, the most famous destination for street food according many assessments (Table 17.1).

The popularity of typical snacks and meals classed as street food is usually due to their low price and the fact that they offer a quick way of satisfying hunger. These two factors are crucial for both local inhabitants and tourists. Some street food is typical of the national or local cuisine, but many snacks and meals have been exported to other parts of the world. For example, hot dogs, hamburgers, pizzas, chips, kebabs, spring rolls, sushi and many other snacks were originally symbols of particular culinary traditions. Some of these meals have regional varieties. For example, in Turkey or Iran there is a pizza that is prepared according to local preferences (in Turkey called *Turkish pizza*). In Poland, for similar reasons, Chinese, Vietnamese and Indian meals are often less spicy than in their country of origin because they are too spicy for the Polish palate (Figs. 17.1, 17.2 and 17.3).

In theory, street food could be offered in every street. In practice, however, it is usually offered in places that are more likely to be visited by local inhabitants and

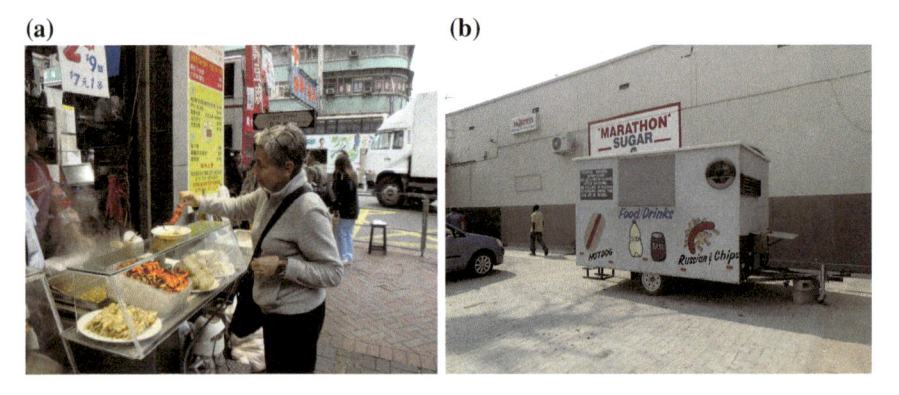

Fig. 17.1 Street food stalls offering traditional and non-traditional meals. Permanent food stall with Chinese traditional delicacies in the Mong Kok district of Kowloon (Hong Kong Special Administrative Region of the People's Republic of China) (**a**). Semi-mobile food stall with hotdogs, drinks and chips and sausage in Divundu (Namibia) (**b**)

Fig. 17.2 Ways of offering street food. Mobile hawker in Lahore (Pakistan) (**a**). Semi-mobile hawker with clams at Ölüdeniz-Belceğiz (Turkey) (**b**). Semi-mobile push-cart with fresh juices in Marrakesh (Morocco) (**c**). Source: Anna Kowalczyk 2017

tourists, notably the principal streets and squares frequently located near railway and bus stations or ports, as well as areas with highly developed trade functions (e.g. Arab bazaars). The term 'street food' covers a wide range of snacks, ready-to-eat meals and beverages. Most are linked with local culinary traditions and

Fig. 17.3 Locations of stalls. Outskirts of Rehoboth (Namibia) (**a**) and the old part of the Havana (Cuba) (**b**). Source: Katarzyna Kowalczyk 2015

Fig. 17.4 Permanent food stalls. Food stall selling *bak kwa* in Macao (Macao Special Administrative Region of the People's Republic of China) (**a**). Food stall with dumplings at Tai O (Hong Kong Special Administrative Region of the People's Republic of China) (**b**). Source: Anna Kowalczyk 2011

attract tourists interested in discovering other cultures and foods. For example, one of the most commonly offered snacks in the streets in Hong Kong, and particularly the old Portuguese colony of Macao is fried meat known as *bak kwa*.[3] It is one of

---

[3]*Bak kwa* (in the *hokkien* dialect of the South China Sea coast, or *rou gan* in Mandarin) is a fried, flat and usually square portion of pork popular in Hong Kong, Macao and nearby southern regions of China. In addition to migrating from Fujian and Guangdong provinces to Southeast Asia, it has become one of the most popular snacks in places like Malaysia and Singapore (especially during the celebration of the Chinese New Year). Bee Cheng Hiang, Lim Chee Guan, Fragrance Foodstuff and Kim Joo Guan are also known for *bak kwa* (http://eresources.nlb.gov.sg/infopedia/articles/SIP_1746_2010-12-30.html, accessed 02 July 2014).

Fig. 17.5 Ways of offering street food. Semi-mobile push-carts in Cienfuegos (Cuba) (**a**) and at Stanley (Hong Kong Special Administrative Region of the People's Republic of China) (**b**). Source: Katarzyna Kowalczyk 2015

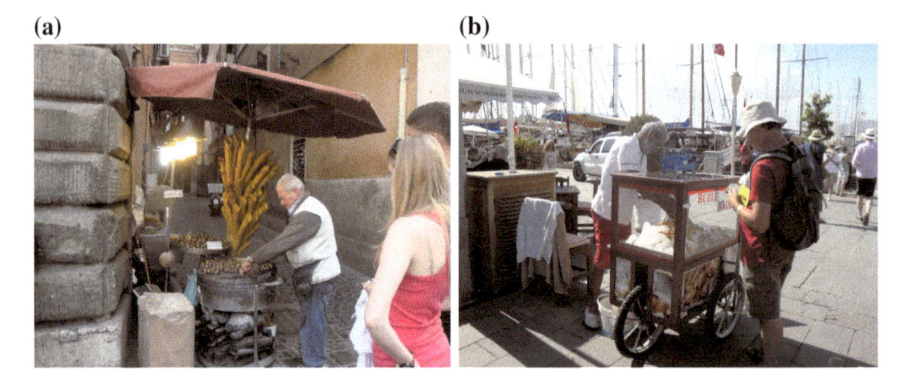

Fig. 17.6 Ways of offering street food. Semi-permanent vending table with chestnuts at Piazza di Spagna in Rome (Italy) (**a**). Semi-mobile push-cart with iced almonds at Bodrum harbour (Turkey) (**b**). Source: Anna Kowalczyk 2013

the most popular snacks, alongside almond biscuits. The first shop selling *bak kwa* is called Heong Kei Iok Kon.[4] It was founded in 1969 in Rua de Cinco de Outubro and has a floor area of only 50 square metres (Figs. 17.4, 17.5 and 17.6).

---

[4]For more details of Kei Iok Kon jerky see http://www.heongkei.com/en/content/about-carnes-secas-heong-kei-iok-kon-jerky (Accessed 02 July 2014).

17.3 Food Trucks—A Revolution in Street Food

One type of street food that has developed in recent years is the food truck (Hawk 2013). They have started to appear in larger Polish cities, offering a range of dishes (sometimes pre-prepared). As well as becoming popular, they are also considered very fashionable. Food trucks owe much of their popularity to serendipity: they do not have a permanent location, unlike fixed (bricks-and-mortar) cafés and restaurants. They depend on the availability of parking spaces on different days of the week, and at different times of the day. Information about their changing location is therefore usually available through social media channels, such as Twitter and Facebook. Their owners are very active web users, regularly updating social channels with their latest location and daily offerings. Sometimes, customers rely on luck to track down a favourite truck in a specific location. Owners use social media to build networks of supporters, promote the visual appearance of their trucks and keep people aware of their latest daily offering, all of which has helped to foster their expansion. These technologies have helped to change the overall diversity of urban consumption (Ananberg and Kung 2015).The dynamic nature of social media means that trucks have an advantage over the local competition, which explains the special emphasis on promoting their services via the Internet.

The trend for *small gastronomy* in the form of food trucks has developed hand in hand with the healthy eating trend. This phenomenon can be traced by looking at social media (Instagram, Facebook, Snapchat), in particular, where customers share photos of food supplied by trucks. Dishes include: meat, vegetarian, those from different parts of the world, dishes prepared using local ingredients, dishes with regional modifications and those that faithfully reproduce the flavours of different continents. This trend can be directly linked to the Slow Food movement: people celebrating and savouring food, rather than hurried consumption. Owners also often use biodegradable napkins and disposable organic dishes, in line with the high ethical and moral standards that are considered values in the era of Slow Food.

But the role of food trucks in urban space is not only limited to satisfying the food needs of residents in their leisure time. Nowadays, food trucks are an increasingly common sight in car parks at business centres, serving corporate employees. They are not only visible on the street: their routes can be followed online and in the press. These media channels (and word of mouth) are used to disseminate information to residents of larger cities across the globe about when and where the next Street Food Festival will take place or where the next trucker will park. They are guided by the motto: *Eat healthy, fresh and local*, no matter where you are. The fashion for eating out and adopting a healthier lifestyle—healthy eating and being fit—plus the desire to support local products have all contributed to their success. These changes in consumption, including healthy nutrition and the culture of eating and preparing meals have led to an increase in eating awareness. Healthy food and careful meal preparation is now more appreciated. It is not only the type and taste of prepared dishes, but also their presentation and ready availability (including during working hours) that have helped to popularize them.

The largest concentrations of trucks were created in urban spaces with developed business and service areas, in cities like New York, Los Angeles and San Francisco, where they became an integral part of gastronomic culture. They served healthy food to people working in the areas, prepared on the spot, fresh every day, at prices that compete with food served in cafés and restaurants. The development of the urban food truck network has made them more competitive with chains such as McDonald's and KFC. Furthermore, the price of their food is unaffected by the increasing cost of renting bricks-and-mortar premises. While they are particularly widespread in the United States, they are also gaining popularity in Europe. In Poland, the development of food truck gastronomy has been encouraged by a rise in the number of public events, such as outdoor concerts, historical reconstructions, and sports and recreation events. These events typically attract thousands of people and are generally staged in places that do not have any permanent catering establishments. Nevertheless, they are mainly found in the larger cities. According to a census carried out in July 2014 (n = 150), there were 64 food trucks in Warsaw, 34 in the seaside region of Gdańsk-Sopot-Gdynia, 11 in Poznań, nine in Łódź, eight in Kraków and six in Wrocław. All of these areas have over 500,000 inhabitants (*Polskie Food Trucki w liczbach* 2014). They play an important role in propagating the cuisine of other nations and countries. The average Polish consumer now has the opportunity to learn new flavours, which may also lead to a change in culinary preferences.[5]

One such example is the La Chica Sandwicheria food truck in Warsaw, which specializes in simple dishes that are typical of the cuisine of Cuban immigrants living in the state of Florida in the United States. Having spent a few years in North America, the owners succumbed to the Cuban fascination for street food and, on returning to Poland, decided to develop their own food truck to sell such products. The second example is the Carnitas Food Truck, also in Warsaw, which specializes in typical Mexican street food dishes: tacos and burritos made using corn or wheat tortilla, with additional sauces (salsa), habanero chilli, chipotle and guacamole. Both the range of dishes and the design of the vehicle reflect close ties with Mexican culinary traditions. A slightly different example comes in the shape of another Warsaw food truck, Pepe Crepe, which offers pancakes. Although the intention is to promote Japanese cuisine, as evidenced by the Polish–Japanese inscription, it sells contemporary, rather than classical food. The close links with modern Japanese culture are also apparent on the truck's side, which depicts a girl drawn in the anime tradition. Pepe Crepe's pancakes are extremely popular (like the anime subculture) not only in Japan, but also in the United States, and it is difficult to say whether they company is really propagating traditional Japanese cuisine, or a globalized version.

---

[5]It should be noted that it is not the case that new nutritional trends are only transferred to Europe or North America from countries on other continents. The process goes both ways, confirmed by the popularity of a dish called kebab or *kebap* in Germany or in Poland, as well as dishes referred to as *Turkish pizza* in Turkey (similar observations apply to Iran and Pakistan).

One interesting example of a mobile street food stall in Warsaw in the classic food truck tradition is Tuk Tuk Thai Street Food, which looks exactly like a Thai taxi. It belongs to the restaurant of the same name in the city centre and is very popular among foodies due to the quality and price of its dishes. It has also benefited from the recent Polish fashion for Thai food, which stems from the fact that Thailand is a very popular destination for Polish tourists in Asia.

Very often the same food trucks can be found in different parts of the city on different days. This applies not only to North American cities, but also to large cities in Poland, most notably Warsaw and Kraków. At the start of the summer in 2015, for example, La Chica Sandwicheria was present at the Slow Weekend event. The following week it was on Jutrzenki Street in the Włochy borough on the western outskirts of the city: first in front of COMP S.A. (an IT company) and then near the headquarters of the Lionbridge Poland corporation. Then it moved to Powsińska Street (next to the headquarters of TVN, one of the biggest private TV companies in Poland) in the southern part of Warsaw. The following weekend it was at the Breakfast Market (Targi Śniadaniowe) in the Powiśle neighbourhood (Śródmieście borough) and in the borough of Mokotów.

Food trucks often congregate in particular regions and can be in one place during the week and another at the weekend. For example, in Warsaw, from Monday to Friday, many congregate near to office buildings in the southern (Służewiec) and western (Śródmieście Zachodnie) parts of the inner city, as well as along Aleje Jerozolimskie Street in the western part of the outer city. On Saturdays and Sundays, however, they mostly congregate in areas along the Vistula River, in green areas in the inner city, and in revitalized, so-called brownfield sites (Figs. 17.7 and 17.8). The Soho Factory is the former site of a car and motorcycle factory (after World War I it was an ammunition factory) in the Kamionek neighbourhood on the east bank of the Vistula River, and the Asian Street Food Fest was organized on recreational areas adjacent to the Vistula close to the city centre. Food trucks only began appearing in Warsaw in about 2013. According to estimates, there were about 30 at that time. Their number doubled in the next year, and in 2015 had risen to about 250 (Staniak 2017, p. 27). The main clusters are shown in Fig. 17.9, which highlights that they are concentrated not only in downtown areas but also in the surrounding neighbourhoods. This particularly applies to the areas south and west of the city centre.

During the week, trucks mainly serve people working in the area and the main clusters can be found at the following places (Fig. 17.9):

- office parks in the Służewiec neighbourhood in the borough of Mokotów (A),
- office parks along Aleje Jerozolimskie Street in Włochy borough (B),
- office parks in the Śródmieście Zachodnie neighbourhood in Wola borough (C),
- the downtown area in the vicinity of the main railway centre (D),
- office parks along Powsińska Street in the Sadyba neighbourhood of Wilanów borough (H).

At weekends, they move to the following places (Fig. 17.9):

Fig. 17.7 Food trucks during the Slow Weekend festival organized in Warsaw at Soho Factory (Mińska Street) in June 2015. La Chica Sandwicheria (**a**), Carnitas Food Truck (**b**) and Pepe Crepe (**c**). Source: Anna Kowalczyk 2015

Fig. 17.8 The Tuk Tuk Thai Street Food truck at the Asian Street Food Fest in Warsaw on 27 June 2015 taken at 12.20 am (**a**) and 1.17 pm (**b**). Source: Anna Kowalczyk 2015

- the waterfront along the Vistula River (E1 and E2),
- the Soho Factory in the revitalized post-industrial area in Praga Południe borough (F),

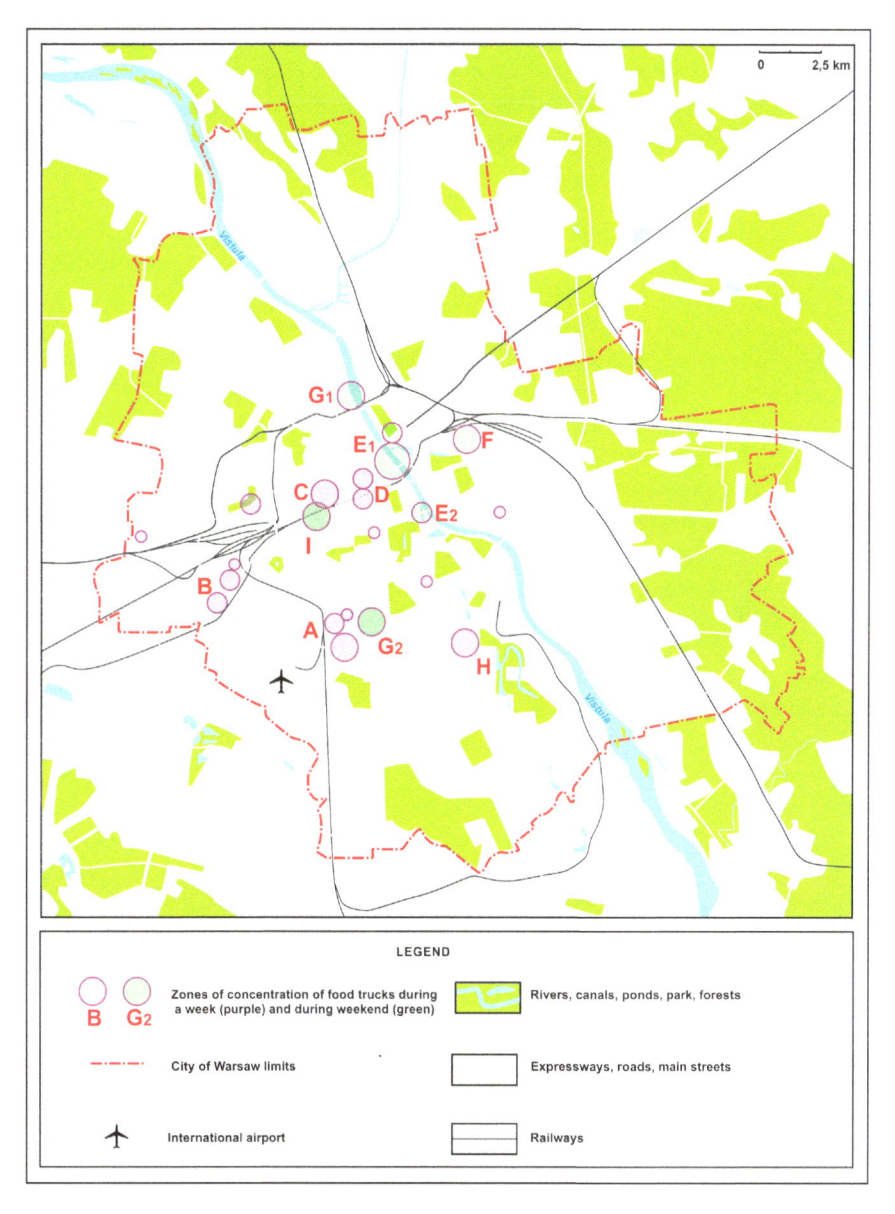

Fig. 17.9 The main clusters of food trucks in Warsaw in 2015–2018. Source: Adapted from Staniak (2017, p. 36) and a survey conducted by Andrzej Kowalczyk

- the sites of the Breakfast Market in Żoliborz (G1) and Mokotów (G2) boroughs,
- the Night Market in the Śródmieście Zachodnie neighbourhood of Wola borough (I), where they gather on the platforms of the closed railway station.

There has also been an increase in the concentration of food trucks in some parts of Kraków. These zones can be divided into two types: daytime in the business and corporate districts of the city; and freetime where people spend time after work and seek out nightlife. Daytime clusters are found in business districts where office complexes are rented by domestic and international corporations. In Kraków, these districts are: Zabłocie (business complexes including the Orange Office Park and the Zabłocie Business Park); Wola Duchacka (e.g. Bonarka for Business); Kazimierz (e.g. the Kazimierz Office Centre); Czyżyny (office complexes including Avia Offices and the Podium Park); and Osiedle Europejskie near Ruczaj (which includes Krakowski Park Technologiczny Sp. z o. o., Shell Energy Campus, Ericsson Sp. z o. o., Nokia Solutions and Networks Ltd., and Motorola Solutions Systems Polska Sp. z o. o.). Employees, who are often Baby Boomers or Millennials, frequent food trucks stationed in designated parking spaces on weekdays. They know exactly where and when particular trucks will appear as the constant online presence of their owners on social media means that their current schedules are immediately available. Consequently, customers never show up at a designated catering area only to find that their favourite food truck is not there. This regular, ongoing contact also serves to tie customers to particular brands. Many trucks visit the Kraków Business Park in Zabierzów, where many international companies have their headquarters. Although it is outside the city centre, the park functions well, thanks to a convenient train connection and a large supply of qualified employees from Kraków and its surrounding areas. Freetime districts are where consumers, especially Millennials and other social groups spend their time after work and seek out nightlife. The largest concentration of trucks is in the district of Kazimierz.

In 2015, a new location for mobile catering facilities appeared on the culinary map of Kraków: the square at Dajwór Street 21. This was replaced in 2017 by a bar with a food truck park, called Truckarnia. This covered, heated area welcomes people who can indulge their interest in food and spend time eating. Truckarnia also has a new outlet to easily order and eat food. Its founders are crew members of the food truck *Pogromcy Głodu* who decided to overhaul the square and invited renowned Kraków brands to cooperate. Trucks from other locations soon appeared, serving fresh products and dishes. The food truck park in the area of Isaac Square, at the intersection of Isaac Street, Jacob Street and Ciemna Street, near Nowy Square emerged as a bottom-up initiative. Its opening led to the opening of the truck, 'the Kitchen for the Brave', whose owner serves offal and original Polish sausages. In addition to regional sausages, the trucks located there offer biscuits, potatoes, Tex-Mex cuisine, Brazilian drinks, Indian curry specialties and Thai ice cream. Another place on the street food map of Kraków is the Bezogródek Food Truck Park. It is located on Piastowska Street and occupies 2,000 square metres of outdoor park area in the vicinity of Błonia, Cichy Kacik, Jordana Park and student dormitories. This seasonal recreation area is open from April to October, and provides Kraków residents with music and eating opportunities. Facilities include sunloungers, a picnic area and the family-friendly zone (with appropriate levels of

Fig. 17.10 A food truck park in Kraków (Poland). Food trucks at Isaac Square on the plot between the old buildings of Kazimierz, which was an independent city inhabited by the Jewish community and is now a tourist attraction. Source: Magdalena Kubal-Czerwińska 2017

security) also allows pets. There are 14 stands, representing the influence of almost 30 cuisines from around the world, including Thai, Japanese, Hungarian, Polish, Mexican and American.

Culinary festivals are also organized periodically. The range of food truck offerings in Kraków is very diverse.[6] Some trucks specialize in beverages, serving a range of smoothies, coffees, teas or alcoholic drinks. Almost half offer several cuisines, and menus are not always limited to specific regional dishes. Most of the trucks that serve meat or vegetarian dishes also offer vegan dishes. Many specialize in serving several variants of a particular type of dish, such as desserts, coffees, soups, fish, pancakes or potato dishes. There are usually 75 food trucks stationed in Kraków. This number increases in summer, when they arrive for street food and catering events. Many have also opened up small, permanent restaurants, which operate all year round. Examples include CurryUp, which has its own premises under the same name on Krakowska Street in Kazimierz and Calavera Mexican Grill, which has a restaurant at Biskupia Street in Śródmieście (Fig. 17.10).

> The food truck industry in Poland, especially in Kraków and Warsaw, is particularly attractive to young people, notably young graduates. These people have already gained several years of professional experience in gastronomy, including time spent abroad. Most food truck owners are aged 25–45 (figures are based on average data collected during conversations with owners). Many do not have a formal catering education, but are passionate cooks. This hobby often translates into a visible and significant commitment

---

[6]It includes vegetarian, vegan, Indian (including curry), Japanese (sushi, ramen), Mexican, Hungarian, Italian (pizza), Georgian (kebabs, khachaputi, chinkali), Asian, Belgian (fries, pancakes), Turkish (kebabs, falafel), Iraniane, American (burgers, grilled dishes, fries, steaks, hotdogs), fusion and regional (*maczanka krakowska* or Krakovian pulled pork sandwich, casseroles and sandwiches) cuisines.

to the business. It should be emphasized that the passion for cooking and sharing the pleasure of eating is very visible during street food festivals.

Over half of Kraków's food truck owners started up after the age of 30, either alone or with friends. Almost 60% run microenterprises, employing up to 10 people. These young Polish business people form an extremely creative group that recognizes the needs of the market and is ready to take risks. Many have a passion for cooking and creating meals and were already considering starting their business during their studies. A combination of a poor job market, a passion for cooking and eating, and the willingness to work independently are the main motives for starting a gastronomy business. A lack of practical knowledge, combined with a lack of financing, in particular from non-bank sources, are major barriers for those wanting to start their own small restaurant or diner. The new 'kitchens on wheels' respond to the growing popularity of small, Slow Food-style street food and overcome these problems.

Each truck is an original, individual entity; however, they share a love of food, and the desire to prepare and serve meals to others. Owners are free to create their business plan. Nearly all owners use information technology to promote their business. Access to the Internet is very important as owners are permanently online and can instantly share information about their current offers and location. More and more trucks use the Internet to support online payments by credit card, PayPal and Apple Pays, while the equipment they use to access the web can be quite basic.

Food trucks compete with the traditional model of restaurants and bars. However, in many cases, notably when it belongs to the owner of a restaurant, they also represent an opportunity as their mobility makes it easier to acquire new customers. One of the best examples (from Kraków) is Gruzja na kółkach, a food truck serving Georgian cuisine. Its owners, who have roots in Georgia, also run Tbilisuri, which is a Georgian restaurant in the district of Kazimierz. Although street food has existed in cities and towns since ancient times, its importance has waned with the advent of industrialization and modernization. While it has survived in the cities of the global South, in the urban areas of the global North it has become less significant. The situation has changed over recent decades, and street food has once again become popular (and a challenge to fixed-location restaurants and bars) as a result of factors such as South–North migration, tourist trips to exotic regions and a return to traditional food.

17.4 Conclusions

As shown in the above text, street food is not a new phenomenon and one can assume that it was common in the cities of ancient Egypt, Greece, Rome, etc. (similarly to the cities of Asia, Africa or the America in pre-Columbian era). However, at the end of the twentieth century in Europe and North America we have observed the 'renaissance' of street food, largely due to the influx of immigrants who bring their current eating habits to new places of residence. The popularity of street food is also related to the way of living in big cities, where residents do not have time to spend 1 or 2 h in a restaurant. Observing the growth of the phenomenon in recent years, we can also recognize that the popularity of eating on the street is the result of fashion or lifestyle.

Also, the food truck phenomenon, closely related to street food, is not entirely new. In the past, it was a man who was carrying a meal (which, for example, had slings with baskets of fruit), or it was sold from carts pulled by donkeys, camels, horses or mules. In some cities, dishes were (and still are, e.g. in Bangkok) sold to customers from a boat. In the twentieth century animals have been replaced by a car, a motorcycle or a bicycle. More and more often food trucks have begun to be created especially for this purpose (small trucks—pickup trucks or mini-buses—vans).

The increase in the popularity of street food and food trucks refers indirectly to the concept of a place, as it is formulated by Doreen Massey (1999). As it has already been written (see subsection 17.1.2.2.4.), while defining a 'place', it is important to take into account human mobility and commodities. But the usefulness of the Massey's concept is due to another reason. Empirical studies conducted in Warsaw and Kraków showed that dishes typical for other cuisines constitute a significant portion of the food offered by food truck owners. This means that our favourite food truck—which we treat as a 'place'—is linked to the culinary traditions typical for other cultures. And one more reason why Massey's views should be considered as particularly useful for explaining the food trucks phenomenon. In general, they are not attached to a specific geographical location. They are moving and can be found (usually depending on a day) at different points of the city (sometimes they even travel to other cities).

The second intellectual concept that explains the phenomenon of food trucks in the space of contemporary cities is the creative class theory (Florida 2005). Many food truck owners are people who are not professional chefs. Very often, they are people with higher education, who speak languages, travel a lot around the world, etc. Their decision to prepare Tex-Mex burgers, Vietnamese *pho* soup or sushi is the result of their creative approach to life.

Finally, it should be mentioned that the increase in the popularity of street food and food trucks in the urban space can be a challenge for 'traditional' gastronomy. The costs of its operation are generally much lower and it may turn out that stationary restaurants and bars lose competition with them and go bankrupt. We can imagine that traditional eateries will be replaced with street food stalls and food trucks moving around the city.

References

About carnes secas Heong Kei Iok Kon jerky, Heong Kei Jerky Macau since 1969. http://www.heongkei.com/en/content/about-carnes-secas-heong-kei-iok-kon-jerky. Accessed 02 July 2014

Ananberg E, Kung E (2015) Information technology and product variety in the city: the case of food trucks. J Urban Econ 90(November 2015):60–78

Bak kwa. http://eresources.nlb.gov.sg/infopedia/articles/SIP_1746_2010-12-30.html. Accessed 02 July 2014

Bhide S (2013) World's best cities for the most delicious street food, May 30, 2013. http://www.buzzle.com/articles/worlds-best-cities-for-the-most-delicious-street-food.html. Accessed 02 July 2014

Canini ND, Bala JJO, Maraginot EN, Mediana BCB (2013) Evaluation of street food vending in Ozamiz City. J Multidiscip Stud 1, 1(August):104–124. http://multidisciplinaryjournal.com/pdf/Evaluation.pdf. Accessed 20 April 2018

Draper A (1996) Street foods in developing countries: the potential for micronutrient fortification, April 1996. London School of Hygiene and Tropical Medicine, London

Etzold B (2014) Towards fair street food governance in Dhaka—moving from exploitation and eviction to social recognition and support. In: de Cassia Vieira Cardoso R, Companion M, Marras SR (eds) Street food. Culture, economy, health and governance, Routledge, New York, pp 61–82, Final draft. https://www.researchgate.net/publication/260227251_Towards_Fair_Street_Food_Governance_in_Dhaka_-_Moving_from_Exploitation_and_Eviction_to_Social_Recognition_and_Support. Accessed 20 March 2018

Fellows P, Hilmi M (2011) Selling street and snack foods, Diversification booklet number 18, Rural Infrastructure and Agro-Industries Division Food and Agriculture Organization of the United Nations, Rome

Fisher C (2014) The world's best cities for street food. http://www.foodandwine.com/slideshows/worlds-bestcities-for-street-food. Accessed 02 July 2014

Florida R (2005) Cities and the creative class, Routledge, New York–London

Hakam AN, Leong SM (1989) Foreign vs local tourist attitudes toward food vending places in Singapore. Proceedings of the 1989 Academy of Marketing Science (AMS) Annual Conference, pp 158–162

Hawk ZA (2013) Gourmet food trucks: an ethnographic examination of Orlando's food truck scene, Department of Anthropology, College of Science, University of Central Florida, Orlando, FL, MA thesie. http://anthropology.cos.ucf.edu/main/wpcontent/uploads/2013/09/Hawk_Zachart.pdf. Accessed 24 June 2015

Heelan Ch. A (2014) The world's best street food: 12 top cities. http://www.frommers.com/slideshows/818551-the-world-s-best-street-food-12-top-cities#slide837005. Accessed 02 July 2014

Henderson J. C., 2019, *Street food and gastronomic tourism*. In: Dixit, S.K. [Ed.], 2019, *The Routledge Handbook of Gastronomic Tourism*, Routledge, Oxon–New York, pp. 441–450

Henderson J. C., Yun O. S., Poon P., Biwei X., 2012, Hawker centres as tourist attractions: the case of Singapore, "International Journal of Hospitality Management", 31, 3 (September), pp. 849–855

Hou WC, Swinyard WR (1986) Comparative study of fast-food chains in Singapore. Singap Manag Rev 8, 1(January):23–36

Kowalczyk A (2014) From street food to food districts—gastronomy services and culinary tourism in an urban space. Turystyka Kulturowa 9(wrzesień):136–160. http://turystykakulturowa.org/ojs/index.php/tk/article/view/493/525. Accessed 15 May 2017

Massey D (1999) Power-geometries and the politics of space-time, Hettner-Lecture 1998, Hettner-Lectures 2, Department of Geography, University of Heidelberg, Heidelberg

Naidoo N, van Dam RM, Ng S, Tan Ch. S, Chen S, Lim JY, Chan MF, Chew L, Rebello SA (2017) Determinants of eating at local and western fast-food venues in an urban Asian population: a mixed methods approach. Int J Behav Nutr Phys Act 14(69). https://ijbnpa.biomedcentral.com/track/pdf/10.1186/s12966-017-0515-x. Accessed 12 Feb 2018

Polskie Food Trucki w liczbach (2014) Food Truck Portal, 30.07.2014. http://www.foodtruckportal.pl/polskie-food-trucki-w-liczbach. Accessed 20 May 2017

Sirpal R, Peng O (1995) Impact of food courts and other factors on tenants' businesses for a major shopping centre in Singapore. Prop Manag 13(4):13–20

Staniak I (2017) Zjawisko food trucks jako nowy trend w gastronomii, Wydział Geografii i Studiów Regionalnych, Uniwersytet Warszawski, Waraszawa, MA thesis

Street food. http://pl.wikipedia.org/wiki/Street_food. Accessed 02 July 2014

Taylor DS, Fishell VK, Derstine JL, Hargrove RL, Patterson NR, Moriarty KW, Battista BA, Ratcliffe HE, Binkoski AE, Kris-Etherton PM (2000) Street food in America—a true melting pot. In: Simopoulos AP, Bhat RV (eds) Street foods, World Review of Nutrition and Dietetics, 86, Karger AG, Basel, pp. 25–44

Chapter 18
Home Delivery Services

Andrzej Kowalczyk and Piotr Kociszewski

Abstract The concept of bringing food to the consumer's home is not new. Long time ago in cities peddlers supplied products in the streets or directly to houses. Initially, products were mainly raw or pre-processed goods. More recently, large grocery companies have started offering customers home delivery of virtually all food products. This trend has not been missed by gastronomy. New technologies related to the preparation of meals, and organisational changes in the restaurant sector have enabled consumers to order ready-made meals for their home or office. The growth in computer literacy and access to smartphones are the other factors affecting the development of home delivery business. This is important in the context of gastronomy in urban space, because it means that the whole city effectively becomes the gastronomic space, rather than just restaurants, bars and cafés.

Keywords Gastronomy · Technologies IT · Meal delivery services · Warsaw

18.1 The Origins and Growth in Popularity of Home Delivery

The concept of bringing food to the consumer's home is not new. In antiquity and mediaeval times, clients mainly came from the upper social classes but, over time, lower classes began to be included. Peddlers began to supply products in the streets or directly to houses. Initially, products were mainly raw or pre-processed goods.

A. Kowalczyk (✉)
Department of Tourism Geography and Recreation, Faculty of Geography and Regional Studies, University of Warsaw, Krakowskie Przedmieście 30, 00-927 Warsaw, Poland
e-mail: akowalczyk@uw.edu.pl

P. Kociszewski
Department of Tourism Geography and Recreation, University of Warsaw, Krakowskie Przedmieście 30, 00-927 Warsaw, Poland
e-mail: p.kociszewski@uw.edu.pl

© Springer Nature Switzerland AG 2020
A. Kowalczyk and M. Derek (eds.), *Gastronomy and Urban Space*,
The Urban Book Series, https://doi.org/10.1007/978-3-030-34492-4_18

For example, in Polish cities, potatoes, meat, eggs and dairy products. More recently, large grocery companies have started offering customers home delivery of virtually all food products. Experts argue that this is connected to cultural and technological change. Cultural changes include the growing professional aspirations of women, whose retreat from the kitchen is noticeable; technological changes such as the spread of the Internet and the popularity of mobile devices have also had a major impact (*Zamiast gotować—zamawiamy gotowe. Rynek dostaw jedzenia rośnie w dwucyfrowym tempie*, 2018). As Shukla et al. (2014, p. 9) noted, 'With the growth in computer literacy and access to smartphones and the Internet, the home delivery business is all set to grow'. The introduction of new technologies related to the preparation of meals, and organisational changes in the restaurant sector have enabled consumers to order ready-made meals for their home or office. Initially limited to individual restaurants and bars, there are now specialised companies that have become a link in the chain that starts at a restaurant or bar and ends at the client who eats the dish in their home or at their workplace. For many years, people ordered meals using a traditional (landline) telephone, but the expansion of information technology and the transition to the information society means that most people now order online using their computer or a mobile app. In practice, this means that it is possible to order a meal from your favourite restaurant via a mobile app, and have your lunch delivered to you in the park, on the embankment along the river or on the tennis court, for example. This is important in the context of gastronomy in urban space, because it means that the whole city effectively becomes the gastronomic space, rather than just restaurants, bars and cafés.

These services have become known as *meal delivery services*, which Bell (2016) defines as '…a service that sends customers fresh, prepared meals delivered to their homes'. They are most highly developed in the United States, with the most popular companies being Plated, Blue Apron, PeachDish, Chef'D, Purple Carrot, HelloFresh, Terra's Kitchen and Home Chef. Some services deliver individually packaged, pre-portioned meals to assist with a healthy diet, while others deliver meal kits, which include ingredients and recipes for customers to prepare food themselves (Bell 2016). A recent survey by UBS showed that in the United States, '…sandwiches are behind only pizza and Chinese food in the list of cuisines that people like to order for delivery' (Sion 2017).

In comparison, the situation in Poland is somewhat different. Majdan (2015, p. 18) noted that 55% of people visiting meal delivery websites ordered Italian cuisine (mainly pizza), 13% ordered Asian dishes and 12% American (e.g. burgers). It should also be noted that 20% of customers ordered traditional Polish dishes, most often pork chops and tomato soup. According to Hirschberg et al. (based on research by McKinsey in 16 countries) the business of delivering restaurant meals to the home appears to be undergoing rapid change. Online food delivery platforms are expanding choice and convenience, allowing customers to order from a wide array of restaurants with a single tap of their mobile phone. It has become a global phenomenon, with the market for food delivery now standing at €83 billion, representing 1% of the total food market and 4% of food sold through restaurants and fast-food chains. The most popular form of delivery (90% of the market) is still the

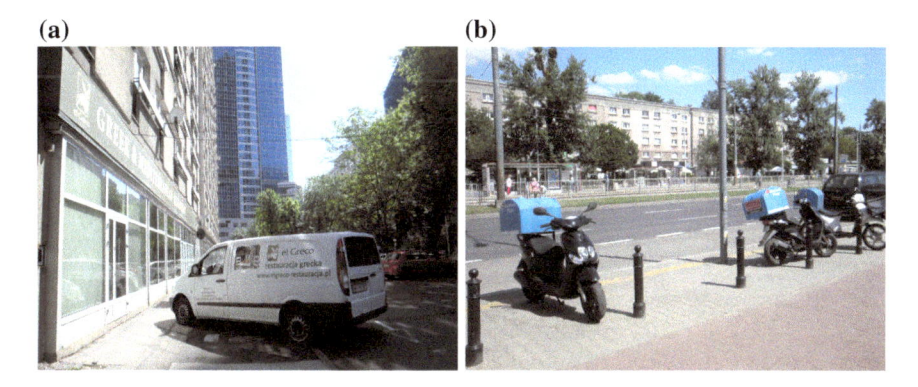

Fig. 18.1 Means of transport used to deliver food from restaurants to clients in Warsaw (Poland). A van belonging to an independent Greek restaurant (**a**). Scooters belonging to Dominos Pizza chain (**b**)

traditional model in which the consumer places an order with the local pizzeria or bar and waits for the restaurant to bring the food to the door. Most of those orders (almost 75%) are placed by phone, but recently there has been a rise in the number of customers (who are accustomed to shopping online) using mobile apps or websites to place their orders (Hirschberg et al. 2016).

There are two types of online platforms in the home delivery business. The first, which emerged around 2000, is known as aggregators (e.g., Delivery Hero, FoodPanda, GrubHub, Just Eat). The second consists of new players (e.g. Deliveroo and Foodora), who appeared in 2013. The difference between them is that aggregators take orders from customers and routes them to restaurants who then handle the delivery themselves, while new companies have built their own logistics networks, and provide delivery for restaurants that do not have their own drivers. These new players extend food delivery to new restaurants and new customers and offer a service that effectively substitutes for fine dining in a restaurant: consumers can dine at home with no loss of quality (Hirschberg et al. 2016). A study of how aggregators operate in India compared the growth and operating strategies of four such companies: Swiggy, Zomato, FoodPanda and TinyOwl. Zomato is the oldest; it was founded as a restaurant delivery platform in 2008 but expanded into the delivery space with Zomato Order in 2015. FoodPanda, which operates globally, began to expand in India in 2012. The others, TinyOwl and Swiggy, entered the market in 2014 (Bhotvawala et al. 2016, p. 141).

The delivery of meals from restaurants to homes mainly occurs in larger cities. In Poland, however, the biggest market is in medium-sized cities (50,000–80,000 inhabitants). In Siedlce, Biała Podlaska, Stalowa Wola, Bełchatów and Inowrocław around 30% of restaurants and bars offer this service, whereas it is only 18% in the ten largest urban cities (*Co piąta restauracja w Polsce dowozi jedzenie. Przodują średnie miasta* 2018) (Fig. 18.1).

18.2 Home Delivery in Poland—The Case of Pyszne.pl

Home delivery began in Poland in the early 1990s, but for a long time was largely restricted to the delivery of pizza. The companies involved usually had pizzerias, and generally belonged to global chains (e.g. Pizza Hut, and Dominos Pizza), although some were of Polish origin (e.g. Pizza Dominium S.A., Da Grasso Sp. z o. o.). Other restaurants very rarely delivered lunch or dinner to customers, a situation that has significantly changed in the past 10–15 years. Initially, individual restaurants provided services themselves, but later specialised companies appeared that acted as an intermediary between restaurants and clients. These companies took orders (initially by phone, but now mainly via the Internet) for lunch or dinner, and delivered to the customer's home. Examples include Szama.pl (established in 2011, the first in the business) and the Polish branch of the FoodPanda chain (which has German origins). The main player in this market is now Pyszne.pl.

Pyszne.pl was founded in Poland in February 2010 and later taken over by the Dutch Takeway.com group.[1] At the beginning of 2018, it delivered meals from nearly 6,000 restaurants and bars in 550 cities. While in 2015 clients placed 1.76 million orders, by 2016 this had risen to 4.33 million. The company charges a commission of 10–12% on each order. The Pyszne.pl portal is not as popular as similar operations in other countries. In Poland, only 3% of the population aged over 16 uses it; in Belgium, 6% use their local version of Takeaway.com; in Austria it is 7%; and in The Netherlands, 21%. In Poland, customers order Italian cuisine most often, followed by Turkish and Polish, although the picture is different across cities. In Warsaw and Gdańsk, for example, the most ordered item is kebabs (like Belgium), in Poznań and Wrocław it is pizza (like Austria), but in Kraków all options are ordered (more or less) equally frequently (Pallus 2018).

> In Poland, PizzaPortal.pl provides services that are similar to Pyszne.pl. The former company's origins are Swedish; in 2005, OnlinePizza Norden AB was founded. After a few years, it merged with Pizza.nu, and in 2009 with Mjam. at. In the same year, a Polish immigrant returned from Sweden to Poland to set up the Polish portal under the name Restaurant Partner Polska Sp. z o.o. In 2010, the Finnish portal Pizzaovelle.fi was created, which, in the same year, was consolidated with online-pizza.fi. The Online Pizza Norden Group consequently provided services in Sweden, Finland, Austria and Poland. PizzaPortal.pl has been very active in acquiring other companies offering similar services, including NetKelner.pl, Szama.pl and FoodPanda. In April 2017, however, 51% of the shares in PizzaPortal.pl were purchased by AmRest Holdings SE, the Polish company that runs KFC chains and PizzaHut (*Jak zmienia się rynek zamawiania jedzenia* 2017).

---

[1] In 2012, 100% of the shares in Pyszne.pl were taken over by yd.yourdelivery GmbH, from whom Pyszne.pl had previously bought Takeway.com in 2012 (Sędek 2017).

Table 18.1 The most popular cuisines offered by eating establishments on the Pyszne.pl website in the biggest seven cities in Poland (June 2017)

Warsaw (n = 861)	Kraków (n = 366)	Wrocław (n = 326)	Poznań (n = 284)	Łódź (n = 193)	Gdańsk (n = 158)	Katowice (n = 119)
Italian (32.8%)	Italian (45.6%)	Italian (43.9%)	Italian (49.3%)	Italian (49.7%)	Italian (52.5%)	Italian (36.1%)
Japanese (17.1%)	Polish (23.8%)	Polish (17.8%)	Polish (16.2%)	Polish (17.1%)	Polish (16.5%)	Polish (26.9%)
Polish (13.8%)	Japanese (8.7%)	Japanese (10.1%)	American (12.3%)	Japanese (10.9%)	American (4.4%) Turkish (4.4%)	Turkish (10.9%)

Source: Adapted from https://pyszne.pl (Accessed 01 June 2017)

Among Polish cities, the Pyszne.pl portal is most popular in Warsaw. In May 2017, 450 bars and restaurants used it. This rose to 861 a month later, and by May 2018 it had risen again to 1043, a rise of 131.7%. The numbers for other large urban areas are much more mixed: Kraków has 442 (up 16.0% from 381 in 2017); Wrocław 382 (up 11.2% from 324); Poznań 301 (down 1.2% from 307); in Łódź 251 (up 23.6% from 203), in Gdańsk 185 (down 9.8% from 205) and in Katowice 140 (down 31.0% from 203). In 2017, research was carried out across seven cities in Poland to find the largest number of restaurants and bars that use Pyszne.pl to provide delivery to the customer. Some of these results are summarised in Tables 18.1 and 18.2 and discussed below.

Table 18.1 shows that in all seven cities Italian cuisine is the most popular. In June 2017 in Gdańsk it was offered by 52.5% of the restaurants and bars included in the Pyszne.pl system, while in other cities it ranged from 32.8% (Warsaw) to 49.7% (Łódź). Polish cuisine was the second most popular, except in Warsaw, although its share was much smaller, ranging from 16.2% (Poznań) to 26.9% (Katowice). The relatively high share of restaurants with Polish cuisine in Katowice and Łódź could be because they are less of a destination for overseas tourists than other cities. It is not immediately apparent why Japanese cuisine came second in Warsaw. There are two possible explanations. The first is that Warsaw is the capital of Poland; it

Table 18.2 The most popular meals offered by eating establishments on the Pyszne.pl website in the biggest seven cities in Poland (June 2017)

Meals	Warsaw (n = 861)		Kraków (n = 366)		Wrocław (n = 326)		Poznań (n = 284)		Łódź (n = 193)		Gdańsk (n = 158)		Katowice (n = 119)	
	No.	%	No.	%	No.	%	No.	%	No.	%	No.	%	No.	%
Burger	87	10.1	44	12.0	34	10.4	25	8.8	11	5.7	14	8.9	13	10.9
Kebab	60	7.0	20	5.5	23	7.1	24	8.5	15	7.8	16	10.1	15	12.6
Pizza	203	23.6	153	41.8	137	42.0	114	40.1	63	32.6	77	48.7	38	31.9
Pasta	145	16.8	64	17.5	66	20.2	64	22.5	35	18.1	38	24.2	16	13.4
Sushi	117	13.6	24	6.6	25	7.7	23	8.1	16	8.3	9	5.7	9	7.6

Source: Adapted from https://pyszne.pl (Accessed 01 June 2017)

Table 18.3 Top national cuisines represented on the Pyszne.pl website in Warsaw (Poland) (May 2018)

Cuisine	Number	Per cent (n = 1043)
Italian cuisine	275	26.4
Japanese cuisine	139	13.3
Polish cuisine	131	12.6
American cuisine	87	8.3
Chinese cuisine	87	8.3
Indian cuisine	83	8.0
Thai cuisine	77	7.4
Turkish cuisine	70	6.7

Source: Adapted from https://www.pyszne.pl/jedzenie-na-telefon-warszawa (Accessed 01 May 2018)

attracts a lot of foreign tourists, and accommodates many expatriates. Japanese cuisine is globally considered fashionable among professionals. The second is that Warsaw residents are wealthier than residents of other cities and are more able to afford relatively expensive Japanese cuisine. The third most popular cuisine was less clear cut. In Warsaw it was Polish, in Kraków, Wrocław and Łódź it was Japanese, in Poznań, American, in Gdańsk a tie between American and Turkish, and in Katowice Turkish cuisine (Table 18.3).

The above was confirmed by research in 86 Polish cities in 2017. A survey carried out by one of the companies in the food delivery market (not only via Pyszne.pl) showed that there are very large differences in the popularity of dishes between cities in Poland. For example, the so-called Pizza index (population per pizzeria) varies from 1.643 (Jelenia Góra) to 17.947 (Koszalin). The average Pizza index for Poland was 4.000, the Burger index 23.477, the Sushi index 25.756 (but this fell to 12.096 in Warsaw), and the Vege index 49.235 (*Raport Stava o rynku dowozów jedzenia* 2018, pp. 9–10).

Table 18.2 shows which dishes are most popular among Pyszne.pl customers in seven major Polish cities. Although there is considerable variation across cities, it clear that sushi restaurants and bars are very popular in Warsaw, where pizza and pasta are relatively unpopular compared to other cities. It is not obvious, however, why Katowice has so many outlets that offer kebabs and so few that offer pasta. While this could be attributed to price (kebabs in Poland are generally cheap), it could also be down to differences in regional culinary preferences for meat dishes.

In spring 2018, a similar survey was conducted in Warsaw, to examine differences and similarities in the distribution of restaurants and bars covered by the Pyszne.pl system in the urban space. At that time, there were 1,043 eating establishments promoting their services via Pyszne.pl. They specialised in a range of cuisines, with most offering Italian food, although their share compared to the previous year was smaller (26.4% compared to 32.8%). The second most popular

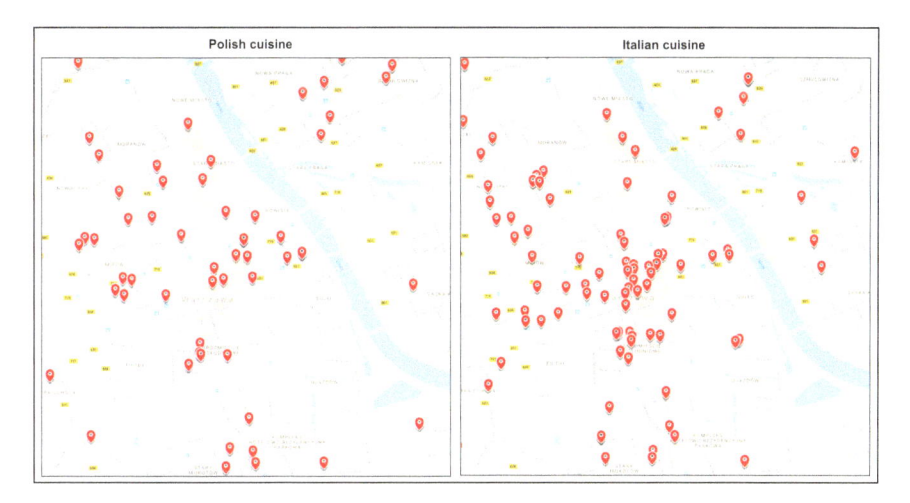

Fig. 18.2 Location of eating establishments offering Polish and Italian cuisine in downtown Warsaw (Poland) supplying customers via Pyszne.pl (May 2018). Source: Adapted from https://www.pyszne.pl/jedzenie-na-telefon-warszawa (Accessed 01 May 2018)

was Japanese cuisine, as in 2017, although its share had also decreased (from 17.1% to 13.3%). The third most popular cuisine (Polish) was also unchanged from 2017, which again had fallen (from 13.8% to 12.6%).[2]

The main goal of the study was to analyse the spatial distribution of restaurants and bars promoted by the Pyszne.pl website within Warsaw. The hypothesis was that their arrangement would reflect the overall distribution of gastronomy in the city. In other words, there would be more dining establishments in downtown and inner city areas. Furthermore, another hypothesis was that the distribution of eating establishments offering specific cuisines would vary depending on the cuisine. A partial summary of the findings is presented below; Figs. 18.2 and 18.3 show the location of restaurants and bars in the central part of Warsaw that offer Polish, Italian, Japanese or Chinese cuisine. The maps in Fig. 18.2 show the location of restaurants and bars offering Polish and Italian cuisine. It can be seen that they are scattered throughout the inner city. However, on closer inspection, both maps highlight a concentration of restaurants and bars offering Italian cuisine in the city centre and on the eastern bank of the Vistula River in Saska Kępa (middle-right on the map). In Praga (upper-right) there are more restaurants offering Polish cuisine. Some have opened recently, as Praga is being revitalised and is slowly becoming popular with tourists.

---

[2] A full analysis of the cusisines offered by Pyszne.pl in 2017 and 2018 is impossible, because in June 2017 the company changed its classifications. For example, in 2017, Vietnamese and Thai cuisines were in one category, as were Arabic and Turkish. Similarly, in 2017 only a few Warsaw restaurants offered American cuisine, but a year later there were 87 (fourth place), probably because the new classification included Mexican restaurants and bars.

Fig. 18.3 Location of eating establishments offering Japanese and Chinese cuisine in downtown Warsaw (Poland) supplying customers via Pyszne.pl (May 2018). Source: Adapted from https://www.pyszne.pl/jedzenie-na-telefon-warszawa (Accessed 01 May 2018)

Figure 18.3 shows the distribution of Japanese and Chinese restaurants in the same part of Warsaw. From a Polish perspective, these cuisines are exotic, but similar (in that they both use rice, tea and noodles), and it would make sense for them to coexist in the same urban space. This is not the case, however, as the maps show that Japanese restaurants and bars are concentrated in the western part of downtown Warsaw. In contrast, Chinese restaurants and bars are more widely scattered. The cluster of Japanese restaurants in the western part of the inner city can be explained by its rapid development and transformation into a modern business and financial centre, which includes the headquarters of leading banks and insurance companies as well as overseas corporations. New skyscrapers house the most expensive apartments and international hotel chains (e.g. InterContinental Hotels Group, Hilton Hotels & Resorts, Marriott International, Accor S.A., Westin Hotels & Resorts and the Radisson Hotel Group). This part of the city has become a prestige district, and Japanese restaurants are considered part of this prestige (Fig. 18.4).

Pyszne.pl is not the only intermediary providing restaurant food to the customer. Apart from PizzaPortal.pl, another major player has been very active on the Warsaw market for the past two years. Uber Technologies, Inc's Uber Eats (former UberEATS) is a smartphone app and website that allows customers to order meals from nearby restaurants. It has been operating elsewhere since December 2015, and

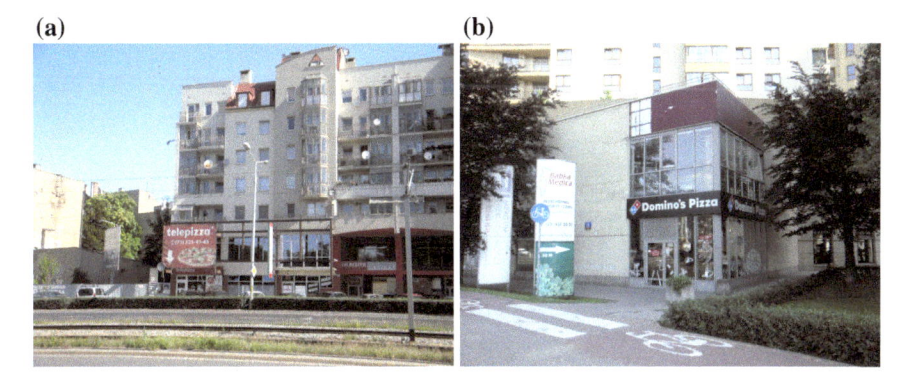

Fig. 18.4 Outlets of the leading pizza chains in Poland. A pizzeria belonging to Telepizza on Drobnera Street in Wrocław (**a**). A pizzeria belonging to Dominos Pizza on Słomińskiego Street in Warsaw (**b**). Source: Anna Kowalczyk 2017

in Poland since February 2017, first in Warsaw,[3] followed by Poznań (September 2017) and Kraków (February 2018). It runs almost a thousand restaurants in Poland —over 700 in Warsaw (up from 150 a year earlier), around 200 in Poznań and almost 100 in Kraków. Results from the first months of operation in Poland showed that customers preferred cheeseburgers, pad Thai (especially in the business district of Służewiec) and burritos. Immediately after the introduction of the service, it covered an area of 80 square kilometres, but after a few months this had expanded to cover 240 square kilometres (46% of the city's area). The system is based on the principle of territoriality: food is only transported to recipients from local restaurants and bars, which helps to avoid traffic jams. At the beginning of 2018, the average delivery time for ordered meals was 33 min in Warsaw and 29 min in Poznań (Fedoruk 2018). Meals are usually delivered by car, motorcycle, scooter or bicycle,[4] occasionally they can be delivered by public transport or on foot. Many delivery personnel are immigrants either from Ukraine (because they know Polish) or South Asia (India, Sri Lanka and Pakistan). Some of the Indian delivery staff have previously worked as *dabbawalas*, and can struggle because they are unfamiliar with the layout of the city, cannot speak Polish and, in some cases, are reliant on public transport, making deliveries with large bags on their back (Rakosza 2018) (Fig. 18.5).

---

[3]Warsaw was the tenth European city to benefit from the UberEATS service, after Amsterdam, Brussels, London, Lyon, Milan, Paris, Stockholm and Vienna (Wąsowski 2017).

[4]The main difference between Uber Eats and Pyszne.pl is that Uber Eats does not have its own transport.

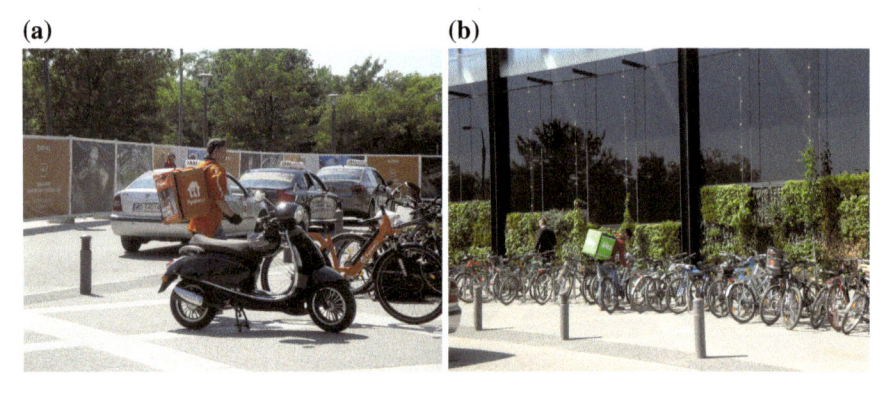

Fig. 18.5 Delivering food to homes in Warsaw. Biker working for Pyszne.pl with his orange bicycle (**a**). Cyclist working for Uber Eats (**b**)

The Indian concept of the *dabbawala* is somewhat similar to the home delivery service described above. The difference is that meals are not prepared in restaurants and delivered to the client's home or workplace, but are prepared at home and delivered to people at work. This system is most popular in Mumbai, with lunchboxes being picked up by couriers in the late morning (using bicycles or trains) and returned empty in the afternoon. The system started in 1890, when Mahadeo Havaji Bachche set up a lunch delivery service with about 100 staff. In 2006 his 5,000 dabbawalas were making about 200,000 lunch deliveries in Mumbai every day. Most dabbawalas come from Pune (the second-largest city in the state of Maharashtra, with over three million inhabitants) to the south-east of Mumbai. The name derives from the *dabba* (or *tiffin*) dish in which the meal is delivered. A dabba is a circular metal tin a bit like a small milk pail. It normally has two, three or four tiers. The bottom is the largest (for rice), while the others include a curry, vegetables, dal and flatbreads and a dessert. Dabbawalas collect around 30 dabbas from homes at around 10 o'clock and ride their bike to the nearest train station. These dabbas are collected at the destination station by other dabbawalas for lunchtime delivery. After lunch the whole process is reversed (Henderson 2017). Most dabbawalas are members of The Bombay Tiffin Box Suppliers Association, which in 2006 had 5,000 members (*Bombay dabbawalas go high-tech* 2006).

Both PizzaPortal.pl and Pyszne.pl posted losses in 2015, of PLN 18.5 million and PLN 16 million, respectively (*Raport Stava o rynku dowozów jedzenia* 2017, p. 4), although Pyszne.pl subsequently expanded significantly in 2016–2017.[5]

---

[5]Of the 16 administrative regions in Poland Pyszne.pl had a dominant market position in 13, compared to three for PizzaPortal.pl in 2017 (*Raport Stava o rynku dowozów jedzenia* 2018, p. 11).

Table 18.4 The dishes ordered most frequently from restaurants and bars for home delivery in Poland (in 2017)

Cuisine	Per cent
Pizza	40.6
Burgers	12.1
Oriental cuisine	11.1
Kebab	9.9
Thai cuisine	8.4
Polish cuisine	6.0
Sushi	5.3
Sandwiches	3.3
Vegetarian and vegan cuisine	3.3

Source: *Raport Stava o rynku dowozów jedzenia* (2017, p. 13)

Average receipts are primarily determined by the minimum order. The vast majority of restaurants do not accept orders below 20–25 PLN (around 4–6 euros) as below this threshold the cost of delivery in relation to the value of the order is financially non-viable. The average order was 40–60 PLN, and the most popular order was sushi (131 PLN on average) (*Raport Stava o rynku dowozów jedzenia* 2017, p. 14). The vast majority (80%) of orders were delivered to homes within four kilometres of the selected catering facility: 27% lived 1–2 km away, and 24% lived 2–3 kilometers away (*Raport Stava o rynku dowozów jedzenia* 2017, p. 16) (Table 18.4).

The recent expansion of Pyszne.pl, PizzaPortal.pl and Uber Eats in the Polish gastronomy market could be undermined if the leading restaurant chains set up their own delivery services. In the United States '…most restaurant chains just now adding delivery services, many of them prefer to rely on third parties in order to avoid the headache and cost of managing their delivery logistics' (Sion 2017). At the same time, again in the United States, 76% of consumers said that they preferred to order directly from the restaurant rather than through a third party (Sion 2017). A similar situation could happen in Poland. At this point, it is worth reiterating the factors that are deemed to be most important in determining the success of home food delivery (Shukla et al. 2014, p. 9):

- low startup costs;
- low rents;
- low expenditure on interior fittings, furniture, etc.;
- low labour costs;
- little dependence on quality real estate.

18.3 Conclusions

Shukla et al. (2014, p. 9) noted that '*Home delivery is an important lifeline (…). It is not surprising therefore, that even fine and casual dining formats are fighting to get a larger share of the pie. Deliveries help them realize higher revenues per square meters, better economies of scale, overcome the "seat constraint" during peak times, and enhance existing consumer loyalty'*. The evidence presented here suggests that home food delivery is not only set to continue but will expand in the near future. This means that it will significantly affect the current gastronomic landscape of cities.

Referring to the described phenomenon in the context of the aforementioned theories, it can be assumed that the idea of delivering ready dishes from restaurants to a home or an office refers to the smart city concept (see Sect. 1.2.5). Without new technologies as the Internet, smartphones and relevant applications, the phenomenon of bringing meals to the customer would remain at the level of the so-called takeaway services. These services consisted of ordering a dish by phone and picking food up by the client, or (less often) by delivering food by courier. This method naturally limited the possibility of receiving a selected dish by the customer.

Thus, it can be assumed that in the coming years, the phenomenon of delivering ready dishes to the customer's home will growth, which—unfortunately—may mean unfavourable changes in the functioning of restaurant business in its current form.

References

Bell KK (2016) The ultimate guide to the best meal kit delivery services. Forbes. Travel/ #WineAndDine", March 21, 2016. https://www.forbes.com/sites/katiebell/2016/03/21/the-ultimate-guide-to-the-best-meal-kit-delivery-services/#22145a8c7375. Accessed 10 March 2018

Bhotvawala MA, Bidichandani N, Balihallimath H, Khond MP (2016) Growth of food tech: a comparative study of aggregator food delivery services in India. Proceedings of the 2016 International Conference on Industrial Engineering and Operations Management, Detroit, Michigan, USA, September 23–25, 2016, pp 140–149

Bombay dabbawalas go high-tech, Phys.org, June 27, 2006. https://phys.org/news/2006-06-bombay-dabbawalas-high-tech.html. Accessed 10 Oct 2017

Co piąta restauracja w Polsce dowozi jedzenie. Produują średnie miasta. Business Insider, 19.04.2018. https://businessinsider.com.pl/wiadomosci/dowoz-jedzenia-z-restauracji-w-polsce-badanie-stavy/eybbey2. Accessed 15 May 2018

Fedoruk A (2018) Uber Eats debiutuje w Krakowie. Będą kolejne miasta. Business Insider, 19.02.2018. https://businessinsider.com.pl/firmy/strategie/uber-eats-juz-w-trzech-miastach-w-polsce-teraz-krakow/1kkp370. Accessed 15 May 2018

Henderson E (2017) How dabbawalas became the world's best food delivery system. The Independent, Friday, 4 August 2017. https://www.independent.co.uk/life-style/food-and-drink/dabbawalas-food-delivery-system-mumbai-india-lunchbox-work-lunch-tiffin-dabbas-a7859701.html. Accessed 10 May 2018

Hirschberg C, Rajko A, Schumacher T, Wrulich M (2016) The changing market for food delivery. McKinsey&Company High Tech, November 2016. https://www.mckinsey.com/industries/high-tech/our-insights/the-changing-market-for-food-delivery. Accessed 25 Feb 2018

Jak zmienia się rynek zamawiania jedzenia on-line? horecanet.pl, 28.06.2017. http://www.horecanet.pl/Jak-zmienia-sie-rynek-zamawiania-jedzenia-on-line-,wiadomosc,28,czerwiec,2017.aspx. Accessed 15 May 2018

Majdan K (2015) *Kliknij, zamów online, zjedz*. Gazeta Wyborcza 94(8427), 23.04.2015 (czwartek):18–19

Pallus P (2018) Polacy zamawiają coraz więcej jedzenia w sieci. Ale Pyszne.pl cały czas czeka na zysk. Business Insider, 10.01.2018. https://businessinsider.com.pl/wiadomosci/sytuacja-na-rynku-dowozu-zywnosci-i-plany-pysznepl-na-2018-rok/pyx6ghq. Accessed 15 May 2018

Pyszne.pl VS PizzaPortal.pl. Digital Mobile, 19.06.2017. http://digital.zecer.wi.zut.edu.pl/?p=157. Accessed 15 May 2018

Rakosza K (2018) *Gdzie jest moje jedzenie? Wkurzeni klienci UberEats narzekają na dostawców z Indii*, naTemat.pl, 18.04.2018. http://natemat.pl/235853,ubereats-wkurzeni-klienci-narzekaja-na-kurierow-z-indii. Accessed 15 May 2018

Raport Stava o rynku dowozów jedzenia 2017, Stava. Pierwsza w Europie franczyza gastro-kurierska, Stava Sp. z o. o., Opole

Raport Stava o rynku dowozów jedzenia 2018, Stava. Pierwsza w Europie franczyza gastro-kurierska, Stava Sp. z o. o., Opole

Sędek M (2017) *Sukces z dostawą do domu—historia Pyszne.pl*. Marketing Biznes, 25.04.2017. https://marketingibiznes.pl/start-up-zone/sukces-dostawa-domu-historia-pyszne-pl. Accessed 15 May 2018

Shukla R, Yadav R, Sharma V (2014) India food services trends. Trends Technopak 2014. Technopak Advisors Pvt. Ltd., Gurgaon

Sion L (2017) Restaurant chains must create in-house fleets before food delivery companies eat their lunch, "Bringg" Monday, April 24th, 2017. https://www.bringg.com/blog/industry-trends/restaurant-chains-must-create-house-fleets-food-delivery-companies-eat-lunch. Accessed 15 May 2018

Wąsowski M (2017) UberEATS jest już dostępny w Polsce. Na początek w Warszawie, "Business Insider", 07.02.2017. https://businessinsider.com.pl/lifestyle/ubereats-w-polsce-ruszyl-w-warszawie-w-lutym/08x1w1e. Accessed 15 May 2018

Zamiast gotować—zamawiamy gotowe. Rynek dostaw jedzenia rośnie w dwucyfrowym tempie, "Forbes", 12.01.2018. https://www.forbes.pl/opinie/rynek-dostaw-jedzenia-obecja-sytuacja-i-prognozy/6mc2ghf. Accessed 15 May 2018

Conclusion

Marta Derek

Abstract This short section concludes that the most important factors which shape the gastronomy–urban space nexus that emerged from the research presented in this book are patterns of urban (re)development, migration, tourism, and the ongoing technology shift. It also explains why Central and Eastern Europe is an interesting laboratory for exploring the development of gastronomy in urban space.

Keywords Gastronomy · Urban space

This volume focuses on the gastronomy–urban space nexus. Among the many different possible approaches, the book adopted a holistic, geographical perspective to highlight the changes that shape this relationship and the challenges it faces. The main focus was restaurants, bars, street food stalls, food trucks, cafés, etc., and how, where, why and when these eating establishments are organised in the urban space.

To a large extent, these issues were examined in relation to Central and Eastern Europe, notably the major Polish cities of Warsaw and Kraków, but also Prague, St Petersburg, and a few other regional cities. The geopolitical shift that took place in the region at the turn of the 1980s and 1990s was a jumping-off point for huge changes in the cultural, social, political and economic environment. These transformations are clearly reflected in the urban gastronomy landscape, which has seen unpreceded development in Central and Eastern Europe in the past thirty years. Under communism, the phenomenon of eating out was unevenly developed across the region (in Poland, for example, there were few restaurants and bars in urban

---

M. Derek
Department of Tourism Geography and Recreation,
Faculty of Geography and Regional Studies, University of Warsaw,
ul. Krakowskie Przedmieście 30, 00-927, Warsaw, Poland.
e-mail: m.derek@uw.edu.pl

© Springer Nature Switzerland AG 2020
A. Kowalczyk and M. Derek (eds.), *Gastronomy and Urban Space*,
The Urban Book Series, https://doi.org/10.1007/978-3-030-34492-4

areas until the late 1980s), and the shift from the centrally planned economy to the market economy resulted in huge (and relatively fast) changes in the gastronomic offer in cities. Consequently, as this volume demonstrates, it has become an interesting laboratory for exploring the development of gastronomy in urban space —especially as these aspects of geopolitical transformation have not been previously reported in broader studies.

Nevertheless, our hope is that the conceptual frameworks, concepts, problems, analyses and arguments presented in this book are relevant beyond the borders of Central and Eastern Europe. To this end, we also draw upon a diversity of other examples, ranging from Hong Kong, through Isfahan, Windhoek and Marrakesh, to European cities such as Barcelona, Berlin, Lisbon, Madrid, Vienna and many others. In each of these (and not only these) cities, globalisation—which has become a keyword of the end of the 20th century—meets local neighbourhoods, traditions, habits and people. Many of these 'meetings', together with their results and consequences, are identified and analysed in this volume.

As Andrzej Kowalczyk highlighted in the Preface, urban gastronomy is shaped by a wide range of factors, both geographical and non-geographical. But what emerges from the research presented here is that some of them are of particular importance.

Several chapters refer to **patterns of urban (re)development**. The example of St. Petersburg shows that gastronomy reflects the major phases of urban growth. Over its one hundred year history—from its status as the capital of the Russian empire, through its demotion to a Soviet city, to its current role as an urban agglomeration with more than five million inhabitants—the number of eating establishments has fluctuated. Interestingly, although the number of eating establishments per capita is currently increasing, it remains lower than at the very beginning of the 20th century, i.e. the end of the Imperial Era (Chap. 7). More broadly, two other contradictory, but intertwined, patterns were identified: the centralisation of eating establishments in central parts of cities, and their decentralisation towards residential and peripheral areas (Chaps. 7 and 8). Another pattern concerns the development of gastronomy in shopping malls. The evolution of malls—from places to shop to places to be and to eat—is becoming a significant challenge to traditional restaurants and bars situated in the urban space (Chap. 15). Finally, Chap. 16 highlighted that the development of bars and other catering outlets in waterfront areas has increased the attractiveness of these public spaces and encouraged users to stay longer. Gastronomy has become an important factor in making urban public spaces more liveable and vibrant, and contributes to their appeal. Evidence of this 'cultural' role of gastronomy can also be found on the outskirts of big cities, and in smaller cities and towns (Chaps. 9 and 10).

A very important driver of the gastronomy–urban space nexus is **migration**. Today's cosmopolitan metropoles welcome people from all around the world, which means that they are home to a huge range of culinary traditions, cuisines, tastes and flavours. In Amsterdam, for example, eating establishments offer around 40 different ethnic cuisines, which reflects the city's long history as a gathering place for immigrants from all over the world. Dominant cuisines are evidence of the

Netherlands' colonial past on the one hand, and the arrival and adoption of new, global trends on the other (Chap. 11). Similarly, the wide range of regional cuisines served in restaurants in Madrid show how the city has long been a destination for migrants from many different parts of Spain (Chap. 6). The blossoming of food trucks in cities with little experience of street food, such as Warsaw or Kraków in Poland, have helped to propagate new cuisines and new flavours among the population, as it is much easier to offer such niche cuisine from a food truck than to try to set up a restaurant. Food from other countries, prepared by immigrants, constitutes an important part of this business (Chap. 17).

Although **tourism** is a popular topic in many other books and articles focused on gastronomy, including those with a geographic background, it is not the main subject of any particular chapter in this volume. It is clear, however, that its impact on urban gastronomy cannot be overestimated. First, cities that offer specific kinds of food attract tourists. Several cosmopolitan metropoles, which are often hotspots for worldwide cuisines, have become renowned for their diverse gastronomic offer (rather than specific products or dishes), and are particularly attractive destinations for foodies (Chap. 5). Second, when these billions of travellers return home, they seek to extend their holiday by visiting a restaurant that serves the cuisine of the country or region they have just visited. Tourism changes their consumption patterns and creates demand for food tasted abroad in their place of residence (Chaps. 11 and 12).

Finally, many of the chapters in this volume revealed the role of the ongoing **technology shift** in the relationship between gastronomy and urban space. Changes include the development of transport links, and the global distribution of products, ideas and services (Chap. 13). New technological devices and virtual communication based on personal computers, tablets and smartphones have enabled people to order meals that are delivered to their home or workplace, find a niche restaurant, check where their favourite food truck is on a particular day, find out about culinary festivals or their favourite restaurant's daily special, or check the rating of a specific restaurant by the 'virtual community'. New social media and social network sites, which facilitate the exchange of information and communication, are a part of this phenomenon (see Chaps. 17 and 18 in particular).

Although the gastronomy–urban space nexus is not a new research problem, this volume sought to shed greater light on its various facets and dimensions. We argue that studying gastronomy adds to our knowledge of the urban landscape, while research on the urban space helps us to understand urban gastronomy. We hope that this book offers a jumping-off point for further exploration of the phenomenon.